POLYMER SCIENCE AND TECHNOLOGY
Volume 17

POLYMER APPLICATIONS OF RENEWABLE-RESOURCE MATERIALS

POLYMER SCIENCE AND TECHNOLOGY

Recent volumes in the series:

POLYMER SCIENCE AND TECHNOLOGY
Volume 17

POLYMER APPLICATIONS OF RENEWABLE-RESOURCE MATERIALS

Edited by

Charles E. Carraher, Jr.

Wright State University
Dayton, Ohio

and

L. H. Sperling

Lehigh University
Bethlehem, Pennsylvania

SPRINGER SCIENCE+BUSINESS MEDIA, LLC

Library of Congress Cataloging in Publication Data

Symposium on the Use of Renewable Resource Materials for Coatings and Plastics
(1981: New York, N.Y.).
  Polymer applications of renewable-resource materials.

  (Polymer science and technology; v. 17)
  "Proceedings of a Symposium on the Use of Renewable Resource Materials for
Coatings and Plastics, held at the American Chemical Society meeting, August 1981,
in New York"—T.p. verso.
  Includes bibliographical references and index.
  1. Polymers and polymerization—Congresses. 2. Renewable natural resources—
Congresses. I. Carraher, Charles E.   II. Sperling, L. H. (Leslie Howard), 1932 –
III. Title. IV. Series.
TP1081.S95 1981   668.9   82-10127
                                                                        AACR2
ISBN 978-1-4613-3505-4   ISBN 978-1-4613-3503-0 (eBook)
DOI 10.1007/978-1-4613-3503-0

Proceedings of a Symposium on the Use of Renewable Resource Materials
for Coatings and Plastics, held at the American Chemical Society
meeting, August 1981, in New York

© 1983 Springer Science+Business Media New York
Originally published by Plenum Press, New York in 1983
Softcover reprint of the hardcover 1st edition 1983

This book is dedicated to our wives
LOYALEA CARRAHER and BONNIE SPERLING

and to our respective natural products
CHARLES, SHAWN, MICHELLE, ERIN, HEATHER,
COLLEEN, and SHANNON CARRAHER
and
REISA and SHERI SPERLING

PREFACE

> For there is hope of a tree,
> If it be cut down,
> That it will sprout again
> And that the tender branch
> Thereof will not cease.

> Job XIV (7)

Mankind has been blessed with a multitude of resources. In the beginning he utilized almost soley replenishable items such as vegetation and animal protein, for both nourishment and shelter. Gradually, such metals as copper and iron were developed and replaced wood as a material of construction. Cement and glass, although more plentiful than other minerals, also replaced the use of growing substances. Coal and oil became the primary sources of heat and power.

Closer to the focus of this book, petroleum products began to replace the vegetable oils, tannin, wool, cotton, leather, silk, rubber, etc. in a host of applications. Surely, it was argued, the new materials did the job better and cheaper. What they didn't say is that soon we would run out of oil. In any case, research on growing natural products, now called renewable resources, slowed, and these industries sought only to maintain their status quo.

The 20th Century saw an unprecedented emphasis and dependence on nonrenewable resources as energy sources (petroleum, coal, uranium) and the fabric of technology (drugs, clothing, shelter, tires, computer parts). The predawn of the 21st Century brings a realization that a cyclic shift back towards the use of renewable resources for technological application is in order.

As with most ancient areas, technology has preceded science in the application of natural products. The area is fertile for the development of a science allowing us not only to understand why past applications worked, but to spark new ventures. The fulfillment of these new opportunities will require both extension from synthetic polymer and small molecule science as well as known science particularly developed for natural products

The purpose of this book is to present a balance between the technology and the science of renewable resources, emphasizing polymer applications. The book illustrates both the breadth of applications available, and the depth of science possible. Naturally, those areas where research activity is the greatest are emphasized.

Many of the chapters included in the book are based on material presented at the August, 1981 American Chemical Society meeting in New York. However, several chapters were contributed by people around the world who could not participate in the ACS meeting. Indeed, every continent except Antarctica is represented between these covers!

A number of authors sent us reviews of their particular sub-fields, rather than original research papers. Thus, the scope of the book was significantly broadened.

The book is divided into five divisions. The first section contains two reviews of the general area of renewable resource materials considering today's setting, historical treatment of renewable resources including specific governmental actions, problems and breadth of natural resource materials along with potential availability and a description of future research required.

The second section contains six chapters emphasizing saccharides and polysaccharides with natural feed stocks derived from sucrose, dextran, cellulose, cotton, xylan, chitin, starch, and bagasse for potential applications as commercial insulation, topical medical applications, fire retardant articles, durable roofing panels, and other building materials.

Section three contains six chapters dealing with the noncellulosic components of plant life including tall oil, wood and gum rosins, lignan, bark extracts, tannin, wood flour, and rice hull flour for use in the preparation of composites, resins, adhesives, and fillers. Filler properties are described in some detail. Substitutes for phenol-formaldehyde resins are described as well as the generation of the industrially important trimellitic anhydride from natural sources. A problem in the rubber elasticity of gutta percha networks is discussed.

Five chapters covering oils and lacquers comprise section four. The structural changes accounting for the extraordinary durability of Japanese lacquer are described along with expanded applications. A series of four chapters emphasize the past, present and future aspects of natural oils. Polymerized vernonia, linseed and tung oils are described as replacements for synthetic polymers. Coatings, adhesive and bulk applications are discussed.

The final section contains six chapters emphasizing the use of proteinaceous materials as leather, wool, polypeptides, and collagen. The chemical modification of leather through the use of radiation, and poly(gamma-methyl-L-glutamate) membranes is described as is the modification of collagen, wool, and leather by chemical means.

Throughout the book, attention is given to structure-property relationships particularly with regard to glass transition temperatures, modulus, tensile strength, impact resistance, and toughness. The majority of the data are given with ASTM cited for ready comparison of results with other materials.

The editors wish to take this opportunity to thank each of the authors for their splendid contributions. We also wish to thank Dr. R. Kleiman from the North Central Region USDA for sending to us a sample of tung oil fruits. Lastly, we want to thank the secretaries from around the world and our own laboratories who labored to translate many scentific and engineering ideas into well-typed book chapters.

Charles E. Carraher, Jr.
L.H. Sperling

January, 1982

CONTENTS

RENEWABLE RESOURCE MONOMERS AND POLYMERS

L. H. Sperling* and Charles E. Carraher, Jr.**

*Lehigh University, Materials Research Center #32
 Bethlehem, PA 18015
**Wright State University, Department of Chemistry
 Dayton, OH 45435

SETTING

Until the beginning of the ninteenth century, it was generally accepted that organic compounds could only be produced in living organisms. A special "vital force" was believed essential for the synthesis of complex organic compounds. Friedrich Wohler, in 1828, heated the inorganic substance ammonium cyanate to form the well known organic compound urea. This discovery is considered by many to mark the beginning of modern synthetic organic chemistry. The development of large quantities of inexpensive "black gold", petroleum technology (including cosmics, drugs, and plastics), allowed the emergence of a society based on synthetic materials. These two "happenings" coupled to minimize the impact of natural, renewable resources on our current society. With the exceptions of the traditional materials leather, cellulose, and wool, and to a small extent silk and natural rubber, modern society is devoid of large scale uses of renewable resource polymers.

The recognition that the supply of petroleum and coal is limited, and increasing costs, has pointed towards the need to emphasize natural renewable materials as replacements or substitutes for product materials now derived from petroleum and coal. However, going beyond the set idea of replacement or substitute materials, renewable resources have valuable, sometimes different properties of their own. So the natural materials, while "substituting" for the synthetic materials (which themselves substituted for the natural materials in the 1940's and 1950's), will have corresponding, rather than identical behavior profiles. Further, it must be also emphasized that the natural product research developments of the 1920-1940 era bears repeating according to the standards of the

1

1980's. Already, scientists and engineers are finding that materials impossible even to consider before World War II, may now be synthesized by modern techniques.

In this short review we wish to draw attention to the following: a) availability of renewable resources, b) (potential and real) problems associated with the use of renewable resources, c) variability of actual structure of most natural macromolecules with source and d) complexity of structure for many natural products. Regarding the latter, while the structure of many natural materials is complex it is knowable and in many cases already known.

PROBLEMS

Natural feedstocks must serve many human purposes. Select carbohydrates as raw materials are valuable due to their actual or potential nutritional value. For example, already operating are protein plants utilizing rapidly reproducible bacteria which metabolize cellulose wastes. Thus, bacteria are added to a nutrient broth emphasizing cellulose; the bacteria feed on the mixture, converting it to more protein rich bacteria; the bacteria are harvested and used as a protein feed meal. However, there is potentially available enough renewable carbohydrates to serve both the food and the polymer needs, and research into the modification of carbohydrates must continue at a heightened pitch.

However, we must meet the problem concerning the ability to produce industrial quantities of a number of interesting renewable materials and provide needed foodstocks. These difficulties are valid for some natural materials, but for many others a judicious choice of growing location and conditions will permit industrial quantities of the natural material to be raised without competition. The sea, for example, is a large cauldron of nutrients thus far singularly untouched; hydroponics can be considered based on the availability of necessary additives; multitiered growing areas may eventually be utilized to make fuller use of both growing area and sunlight.

Most of the potential natural feedstocks are actually complex products of varying composition, changing with source, pretreatment, etc. Further, the chemistry of these complex materials is not well known relative to petrochemical materials. While variation in processing will be necessarily dependent on the exact properties of a given material from a specified source, such problems are solvable. Indeed, the petroleum industry has faced and overcome this problem since the raw petroleum material derived from one source is unlike that derived from another source, differing, for example, in aromatic content. While defined structure-property relationships are desirable, they are not necessary in developing a useful product. Thus the polymer industry grew up, becoming a giant by the 1950's, before some scientists even accepted the concept of the macromolecule. For others, there exists sufficient chemical and physical data on analagous compounds, which may be transferred, allowing the use of such materials as feedstocks.

Finally, many potential renewable feedstocks are currently summarily destroyed or utilized in a noneconomical manner. Thus leaves and other plantstocks are "ritualistically" burned each fall. A number of these seemingly useless natural materials have already been utilized as feedstock sources and more must be included.

Furfural, obtained by the steam-acid digestion of corn cobs, bagasse, rice hulls, and oat hulls, acts as the precurser to hexamethylene diamine and adipic acid, the latter two utilized to form nylon-66.

$$\text{(furfural)} + CO + H_2O \longrightarrow HO_2C(CH_2)_4CO_2H \xrightarrow{NH_3} H_2NC(CH_2)_4CNH_2 \qquad (1)$$

$$H_2N(CH_2)_6NH_2 \longleftarrow$$

Sebacic acid is produced from the dry distillation of castor oil (triglyceride of ricimoleic acid) with sodium hydroxide, etc. Several natural oils, mostly triglycerides, have long been in use commercially. Thus, linseed oil and tung oil have formed the basis for paints; drying is actually a polymerization through the double bonds to make a crosslinked network.

A partly synthetic analog of the oil-based paints are the alkyd resins. The alkyds are esters formed from alcohols and acids through transesterification reactions. A direct source of unsaturated fatty acids is frequently used, for example, tall oil from the sulfate pulping process.

A word about the pricing problem must be said. Before the oil crisis of 1973-1974, petroleum products were frankly cheaper than their renewable resource counterparts. While all prices have inflated by a factor of two since then, petroleum oil and oil product prices have increased faster. Thus, while the renewable resource product is still uneconomical in many cases, modern industry must prepare now for a price inversion which may take place during the 1980's.

Potential Renewable Materials

When plant or animal tissues are extracted with a nonpolar solvent a portion of the material typically dissolves. The components of this soluble fraction are called lipids and includes fatty

acids, triacylglycerols, waxes, terpenes, prostaglandins and
steroids. The insoluble portion contains the more polar plant and
animal components and crosslinked materials including carbohydrates,
lignin, proteins and nucleic acids.

There are numerous natural materials and almost as many lists
and ways to partition such materials. Table 1 contains one such
listing along with suitable general references. Table 2 is a listing
for a number of natural products as a function of general availability. Brief descriptions follow of the more abundant natural
product groupings. Each has already shown itself viable in today's
industrial market place but with a potential for exceeding present
use.

LIGNIN

Lignin is a major polymeric component of woody tissue. During
synthesis of plant cell-walls, polysaccharides (principally cellulose)
are first deposited, with lignin filling the spaces between the polysaccharides fibers, cementing them together. Except for water transport, the lignified fiber appears to be relatively inactive in the
metabolic role of the plant, acting largely as a support. Lignin
constitutes about 25% of wood and is therefore an abundant renewable
resource material currently underused.

Lignin's structure is complex, varying with source and within
a given source. Phenol, aliphatic alcohol, ether, ketone and aldehyde functional groups are contained within each lignin and are
available for classical synthetic organic chemistry reactions. The
structure of lignin as a three dimensional network is shown in
Figure 1. Note its general relationship to phenol-formaldehyde-
type resins.

The vast majority of lignin, along with associated carbohydrates, is permitted to rot, returning to the biosphere. The
majority of lignin removed during wood processing is burned as a
fuel which serves to drive the pulping process. Some is converted
to lignin sulphonates and used as additives in oil-well drilling-
muds, adhesives, industrial cleaners, leather tanning, road binders,
cement products and in the production of vanillin. However, better
processing knowledge is needed, particularly a know-how of recovering linear high molecular weight lignins before large-scale polymer
uses can be instituted.

NATURAL POLYISOPRENES

While the best known polymer in this category is natural rubber,
a cis-polyisoprene originating from the tree Hevea brasiliensis,

Table 1. Renewable, Natural Material Groupings

Group	General Reference(s)
Illustrative Subgroups	
Alkaloids	1,2
Pyrrolidine, Pyridine, Pyrrolizidine, Tropane	
Quinolizidine, Isoquinoline, Piperidine, Indole	
Quinoline, Quinazoline, Acridone, Steroidal,	
Terpenoid	
Amino Acids	2,3
Carbohydrates	4-9
Simple (glucose, sucrose, fructose, lactose,	
galactose)	
Complex (starch, cellulose, glycogen)	
Drying Oils and Alkyd Resins	5,10-12
Linseed, Cottonseed, Castor, Tung, Soybean,	
Oiticica, Perilla, Menhaden, Sardine, Corn,	
Safflower, Vernonia	
Fossil resins-Amber, Kauri, Congo	
Oleoresins-Damar, Ester Gum	
Fungus, Bacteria and Their Metabolites	2
Heme, Bile and Chlorophylls	2
Lignins	5,9,13,14
Lipids	2,15
Simple (Glycerol Esters, Cholesterol Esters)	
Phosphoglycerides	
Sphingolipids (Mucolipids, Sulfatide,	
Sphingomyelin, cerebroside)	
Complex (Lipoproteins, Proteolipids,	
Phosphatidopeptides)	
Phenolic Plant Products	2
Phenols, Resorcinols, Anthraquinones,	
Naphthoquinones, Hydrangenol, Stilbenes,	
Coumarins	
Polyisoprenes	5,9,16-20

(continued)

Table 1.   Renewable, Natural Material Groupings
(continued)

Group	General Reference(s)

Proteins
  Enzymes (Lysozyme, Trypsin, Chymotrypsin)
  Transport and Storage (Haemoglobin, Myoglobin)
  Antibodies
  Structural (Elastin, Actin, Keratin, Myosin,
    Collagen, Fibroin)
  Hormones (Insulin)
	5,9,21-26

Purines, Pyrimidines, Nucleotides, Nucleic Acids	2,5,9,26-29

Steroids
  Cholesterol, Adrenocortical, Bile Acids
  Ergosterol, Agnosterol, Desmosterol
	2,30

Tannins	2

Table 2.   Relative Availability of Assorted Natural Products

Small scale - Biomedical, Catalysis
     Alkaloids
     Heme, Bile and Chlorophylls
     Phenolic Plant Products
     Steroids
     Tannins

Medium (many with potential for large) scale
     Amino Acids
     Fungus, Bacteria
     Lipids
     Proteins (specific)
     Purines, Pyrimidines, Nucleotides,
        Nucleic Acids

Large scale
     Carbohydrates
     Drying Oils, Alkyd Resins
     Lignins
     Polyisoprenes
     Proteins (general)
     Terpenes and Terpenoids

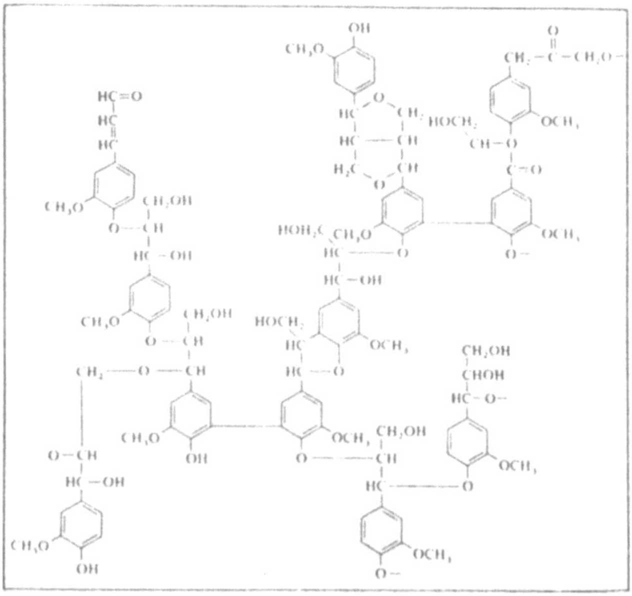

Figure 1. Summary of lignin structure. This natural thermoset plastic has a very irregular structure with various ether groups acting as crosslinks in many cases. (F. F. P. Kollman and W. A. Cote, Jr., Principles of Wood Science and Technology, Springer-Verlag, 1968; E. Adler, Papier, 15, 604 (1961).)

many other cis-polyisoprene sources are important, as well as sources for trans-polyisoprene. In the Hevea tree, cis-polyisoprene occurs as as a natural latex, or submicroscopic dispersion of the rubber in a sap-like material. All such latexes appear milk-white. The rubber can be recovered by coagulation processes. Polyisoprene has the structure

$$\{CH_2-C=CH-CH_2\}_n \overset{CH_3}{\underset{}{|}} \qquad (2)$$

with a molecular weight (for natural rubber) of $2x10^6$ gms/mole and a glass transition temperature of -72°C. Of course, the designation of cis- or trans- relates to the configuration about the double bond.

Vulcanization

The term vulcanization refers to the phenomenon now known as
crosslinking.* Substantially, vulcanization is the reaction between
rubber and sulfur, although many other compounds are added in commercial processes (32,33). A typical reaction is:

$$
\begin{array}{c}
CH_3 \\
|
\end{array}
$$

$$2\text{\scriptsize wwww}CH_2-C=CH-CH_2\text{\scriptsize wwww} + \text{sulfur}$$

$$\text{\scriptsize wwww}CH_2-\underset{\displaystyle\underset{\displaystyle S}{|}}{\overset{\displaystyle\overset{\displaystyle CH_3}{|}}{C}}-\overset{S-S-R}{CH}-CH_2\text{\scriptsize wwww}$$

(3)

$$\rightarrow \quad \underset{|}{\overset{|}{S}}$$

$$\text{\scriptsize wwww}CH_2-\underset{\displaystyle\underset{\displaystyle CH_3}{|}}{C}-CH-CH_2\text{\scriptsize wwww}$$

$$S-S-R$$

where R represents other rubber chains.

Vulcanization converts the material from a high viscosity liquid
to an amorphous solid, suppressing creep and flow, and enhancing
long-range rubber elasticity. Because cis-polyisoprenes crystallize
on extension, they form extremely tough, self-reinforcing materials.
Before the advent of synthetic rubber such as SBR, natural rubber was
the only material for automobile tires. Nowadays, it remains the
material of preference for heavy-duty tires such as are used by air-
planes and trucks.

Grafting Reactions

The use of natural rubber latex to form graft copolymers was one
of the earlier methods of preparing rubber/plastic compositons. The

---

*There are several older terms, still in wide use, which refer to
the crosslinking of natural polymer materials:

Crosslinking Term	Natural Polymer
vulcanization	rubber
tanning	leather
drying	oils
curing	inks

best known of these materials is Hevea plus (34,35). Hevea plus MG,
for example, is prepared by polymerizing methyl methacrylate in the
rubber latex, forming the graft copolymer. Depending on the amount
of MMA added, the product is harder or softer, but usually very tough.
The grafts of natural rubber with methyl methacrylate or styrene
result in materials which are similar to the ABS resins.

Other Natural cis-Polyisoprenes

Guayule (pronounced wy-oo-lee) rubber originates from Parthenium
argentatum Gray, which is a member of the sunflower family, growing
as a shrub in northern Mexico and the southern parts of California,
Arizona, New Mexico, and Texas (36,37). During the early 1900's,
about half of the rubber used in the United States was guayule
rubber, originating from Mexico. With the rise of Hevea plantations
in Asia guayule production decreased. While projects to grow guayule
began again during World War II, they were stopped in 1945 in favor
of SBR rubber.

Beginning with the oil crisis of the mid-1970's however, interest
in guayule revived again (38,39). While guayule rubber has substantially the same structure and molecular weight as Hevea, it contains
a resin content of 15-20%, which complicates processing. After
processing, the resin is reduced to approximately 2%, of which some
45% is fatty acids (which undoubtedly reduces crosslinking efficiency,
see below).

The properties of guayule rubber are compared with those of
Hevea rubber in Table 3. While the cure curves of guayule and Hevea
are similar, the degree of crosslinking is lower in the former (with
the same recipe), a major cause of the somewhat lower modulus and
poorer tensile and tear strengths (37).

Other sources of cis-polyisoprene are also interesting, and
number in the hundreds. The white sap of milk weed and dandelions
are actually a rubber latex. During World War II the Russians were
much interested in extracting the rubber from dandelions, especially
the roots, which contains 1.25-5% rubber.

In the 1920's Thomas Edison grew large fields of goldenrod.
His friend Henry Ford made tires out of the rubber, and drove around
on them.

Trans-Polyisoprenes

Gutta-percha and balata are two types of natural trans-
polyisoprenes, which comes from such tropical trees as the Palaquium
Gutta of Malaysia and the Bolle tree of Brazil, respectively. Because
the trans-polyisoprenes are crystalline at room temperature, they
form tough thermoplastics. Applications have included golf ball
covers and dental materials.

Table 3. Properties of Raw and Vulcanized Guayule Rubber Compared
with Hevea Rubber (37).

	Guayule	Hevea (SMR-5)
Raw guayule rubber[a]		
Mooney viscosity (ML-1 + 4 at 100°C)	105	85
Antioxidant (percent BHT)	0.6	
Acetone solubles (percent)	2	2.8
Wallace Rapid Plasticity (P_o)	47.5	
Plasticity Retention Index (percent)	41	60
Green Strength (kPa at 100 percent elongation)	13.8	13.8
Vulcanized guayule rubber[b]		
Initial Viscosity (newton-meter)	0.57	0.62
Minimum Viscosity (newton-meter)	0.42	0.45
Maximum Viscosity (newton-meter)	2.8	4.0
$T_s s(2)$, min.	10.5	7.0
$T_c,(90)$, min.	25	19.0
Cure Rate (newton-meter/min)	0.28	0.60
Cure Time at 140°C min.	25	19
Modulus at 300 percent (MPa)	7.24	12.2
Modulus at 500 percent (MPa)	16.9	--
Tensile strength (MPa)	25.1	27.9
Elongation (percent	635	490
Set at break (percent)	14	13
Bashore rebound (percent)	40	48
Shore A hardness	54	60
Swelling index (g. benzene imbibed/g. rubber)	3.44	2.94
M_c	13,000	9500
Tear strength (N/m)	3.12×10^4	7.63×10^4

[a]Guayule data based on early samples form the pilot plant at
Saltillo, Mexico.
[b]Vulcanized using recipe 2A given in ASTM D 3184-71. The vulcanization characteristics of each stock were determined on a Monsanto
Rheometer at 140°C using ASTM D 2084-71T.

CARBOHYDRATES

Carbohydrates are the most abundant (weight-wise) class of
organic compounds, constituting three-fourths of the dry weight
of the plant world. They represent a great storehouse of energy
as a food for man and animals. About 400 billion tons of sugars

are produced annually through natural photosynthesis, dwarfing the production of the other natural products with the exception of lignin. The vast majority of this is produced in the oceans, pointing out the importance of harnessing this untapped source of food, energy and renewable feedstocks.

The potential complexity of even the simple aldohexose monosaccharides is indicated by the presence of five different chiral centers giving rise to 2^5 or 32 possible stereoisomeric forms of the basic structure, only two of which are glucose and mannose. While these sugars differ in specific biological activity, their gross chemical reactivities are almost identical permitting one often to employ mixtures within chemical reactions without regard to actual structure with respect to most physical properties of resulting product.

Carbohydrates are diverse with respect to both occurrence and size. Familiar mono- and disaccharides include glucose, frutose, sucrose (table sugar), cellobiose and mannose. Familiar polysaccharides are listed in Table 4 along with natural sources. The purity, molecular weight, amount and location of branching and crosslinking, etc. vary as to source and even location of source. For instance cotton is a good source of cellulose yet the amount of cellulose varies from 85 to 97% depending on the variety of cotton plant, age of plant and location of growth. Again, the gross chemical reactivity and resulting physical properties are largely independent of the source of cellulose.

Major variables include the molecular weight and molecular weight distribution, residual hemicelluloses and lignins, and protein content. Uses include food (starch, frutose, mannose, sucrose), building material and clothing.

Table 4. Naturally Occurring Polysaccharides

Polysaccharide	Source	Monomeric Sugar Unit(s)
Amylopectin, Amylose	Corn, Potatoes	D-Glucose
Cellulose	Plants	D-Glucose
Chitin	Animals	2-Acetamido glucose
Glycogen	Animals (muscles)	D-Glucose
Gum tragacanth	Plant Resin	L-Arabinose and D-glactose
Inulin	Artichokes	D-Fructose
Laminaran	Seaweeds	D-Glucose
Yeast Mannan	Yeast	D-Mannose
Xylan	Plants	D-Xylose

Mercerized Cellulose and Rayon

The reactions of cellulose need to be briefly reviewed. The crystal structure of native cellulose originating from trees and cotton is known as cellulose I (40–42). While cellulose I is widely used in cotton fabrics, it is frequently mercerized by steeping the cellulose in concentrated alkali solutions, and then washing out the alkali. The crystal structure is changed to cellulose II during the process, and luster is increased. Mercerized cotton is slightly lower in density, has increased water absorption, better dyeability and higher extensibility. Some other properties, such as tensile strength, decrease, in part because of the degradation during the alkali treatment.

Addition of carbon disulfide to alkali cellulose, followed by solution in a sodium hydroxide solution produces sodium cellulose xanthate, used in the viscose process. Extruded into a sulfuric acid bath as a film, the material known as cellophane is produced. Spun into a sulfuric acid bath, this material regenerates the cellulose as cellulose II. The product is known as rayon.

A word should be said about the properties of rayon fibers. The materials were originally developed during the early 1900's as the only "synthetic" fiber. After the introduction of nylon, emphasis was on price. Rayon fibers could be made more cheaply than nylon, but with a sacrifice in properties necessitated by the low molecular weight and high spinning speeds concomitant with low price. Thus, rayon got a bad name at one point, unnecessarily.

During the 1950's and 1960's, however, rayon tire cords were being made which had much higher tensile strengths, 5-6 grams/denier compared to low-quality rayon cloth fiber at 2-3 grams/denier and nylons at 6-7 grams/denier. The point is that if the cellulose industry were to research rayon in the 1980's with the intensity that nylon was investigated in the period between Carother's work and the present, rayon fibers of equal (but not identical!) quality might be made.

Cellulose Esters and Ethers

Each glucoside residue in cellulose has three hydroxyl groups. These can be reacted in the ordinary way to make a wide variety of esters and ethers which are in current commercial use.

Cellulose nitrate (nitrocellulose) is used in gun powder and explosives. At low nitrate levels, it is used as a lacquer.

Cellulose acetate, or mixed esters such as acetate-butyrate are widely used in photographic film, safety film, and lacquers. Most of these materials are known as cellulose diacetates, etc., but actually have an average degree of substitution near 2.5 out of the possible 3.0. Because these materials are substantially random copolymers, they are amorphous.

Cellulose triacetate, which is highly crystalline, makes an excellent fiber sometimes known as "acetate", or "rayon acetate".

The melting and glass transition temperatures of the cellulose triesters varies with the length of the ester chain, as shown in Table 5 (43-45). While the lower esters are important commercially, much more could be done with the higher esters.

Sodium carboxy methyl cellulose (CMC) is an important cellulose ether. It is frequently sold at an average degree of substitution of about 0.5, where it stands on the edge of true water solubility. Two uses must be mentioned:

(1) Because of its very high viscosity, it is used as a thickener, particularly in foods such as ice cream. Thus, grandma's ice cream melts (or softens) at a low temperature, while commercial ice creams retain their shape longer during consumption.

(2) As an antiredeposition agent, it is widely used in detergents to prevent oil droplets from being redeposited on the fabric, once removed by the surfactant. In this case, ionic repulsion is called into play at the colloidal level.

It should be noted that at a higher level of substitution, where true water solubility is attained under a wide range of conditions, CMC will be less active, having a lower viscosity (reduced aggregation) and not depositing properly on the fabric to be protected.

While cellulose and cellulose derivatives have remained in use throughout the development of the synthetics their volume has tended to remain flat (at an amazing 50% of the total polymer industry) and the industry generally has been on the defensive. Cellulose is one of the few natural polymers with enough world-wide tonnage to replace the synthetics, when and as the need arises. Again, the need is to treat these as polymers for the 1980's, not as left-overs from the old days. The amount of research and development required is enormous, but we should begin now, to be ready for the future.

(NATURAL) OILS AND FATS

A major component in the organic soluble, lipid fraction is a group of compounds called natural oils and fats. While most natural oils and fats are triglycerides, a few such as tall oil are free

Table 5. T_g and T_m for Cellulose Triesters[a] (43)

Cellulose ester	T_g, °K	T_m, °K	T_g/T_m
Triacetate[b]	445	580	0.77
Tripropionate	406	507	0.80
Tributyrate	365	456	0.80
Trivalerate	289	395	0.73
Tricaproate	223	367	0.61
Triheptanoate	184	361	0.51
Tridecanoate	204	361	0.56

[a]Date of Malm et al. for T_m (45).

[b]Data of Russell for T_g (44).

acids. The triglycerides can be classified into three general groups. First are the highly saturated animal fats such as tallow. The second class are the unsaturated oils such as linseed oil. The third class are those triglycerides which possess active groups other than double bonds. Examples are castor oil, which has three hydroxyl groups, and vernonia oil, which contains epoxy groups. These will be discussed below.

Other natural oils are not triglycerides or their free acid counterparts. Japanese Lac is one, and turpentine is another. Each has a distinctive chemistry, the latter, for example, being composed of α- and β-pinenes.

Table 6 contains a listing of some common sources of triglycerides illustrating the complexity of the fatty acid content with respect to source. Table 7 lists ASTM and federal specification citations for some oils again illustrating that while the fatty acid composition of oils is complex, present industrial definitions and subsequent uses permit such dispersity.

Drying Oils

The triglyceride oils constitute the major portion of such materials as animal fats, butter, vegetable oils and margarine, and botanical oils from a range of oil-bearing seeds. While some triglyceride oils such as tallow are essentially saturated and cannot easily be polymerized, most oils contain at least one double bond per acid residue. Thus, a typical structure of linseed oil might be (46):

Table 6. Average Fatty Acid Composition of Common Fats and Oils

(Average) Acid Composition (mole-%)	Fat or Oil							
	Butter	Lard	Olive	Peanut	Cottonseed	Soybean	Linseed	Cod Liver
Saturated								
Butyric, C_4	3							
Caprylic, C_8	1							
Capric, C_{10}	2							
Lauric, C_{12}	4							
Myristic, C_{14}	8-16	1	1		1	1		6
Palmitic, C_{16}	25-30	25-30	5-15	7-12	17-25	5-10	48	3-10
Stearic, C_{18}	9-12	12-17	2-4	2-6	2	3	3	1
Unsaturated								
Palmitoleic, C_{16}	5	5			2			18-22
Oleic, C_{18}	18-33	47-60	65-85	30-60	15-40	20-30	15-30	25-33
Linoleic, C_{18}	3	6-12	8-12	20-40	45-55	50-60	15-25	27-32
Linolenic, C_{18}		1				5-10	45-60	

Table 7.　Selected ASTM and Federal Specification Citations for Some Common Oils.

Oil (Description)	ASTM	Federal
Linseed (Boiled)	D260-61 (1969)	TT-L-190C (Dec. 1964)
Linseed (Raw)	D234-68T	TT-L-215C (Jan. 1968)
Linseed (Heat polymerized)	–	TT-L-201 (May 1958)
Soybean (Refined)	D1462-60 (1969)	–
Tung (Raw)	D12-64	–
Safflower	D1392-58 (1969)	–
Castor (Dehydrated)	D961-58 (1969)	–
Castor (Raw)	D960-67	JJJ-C-86(1) (Dec. 1953)

$$
\begin{array}{ll}
CH_2\text{-O-CO}(CH_2)_7CH=CH\text{-}CH_2\text{-}CH=CH\text{-}CH_2\text{-}CH=CH\text{-}CH_2CH_3 & \text{linolenic acid portion} \\
CH\text{-O-CO}(CH_2)_7CH=CH\text{-}CH_2\text{-}CH=CH(CH_2)_4CH_3 & \text{linoleic acid portion} \\
CH_2\text{-O-CO}(CH_2)_7CH=CH(CH_2)_7CH_3 & \text{oleic acid portion} \quad (4)
\end{array}
$$

Linseed oil, tung oil, and other highly unsaturated oils are used as the basis for the oil-based paints. They dry (polymerize) by an oxidative mechanism, forming ether bonds between the triglyceride molecules, and through a series of oxidatively initiated free radical reactions attacking the double bonds (47,48). Since multiple points of oxidation are present, a three-dimensional polymeric network results.

As used in paints, the processes are slightly more complex. For example, it is common to boil some of the linseed oil in the presence of air, called kettle-bodying. Kettle bodied oils have some crosslinking already introduced (but not to the gel point), and a quantity of hydroperoxides at the α-position with respect to the double bond. The increase in viscosity during bodying is illustrated in Figure 2. When mixed with lead, cobalt, or manganese salts (drying agents, or catalysts), raw linseed oil, and other components, see Table 8 (49), a relatively rapidly drying product is formed.

Three reactive stages can be distinguished in the formation of a film with a drying oil:

(1) Autoxidation, where the oil reacts with molecular oxygen to form peroxy compounds.

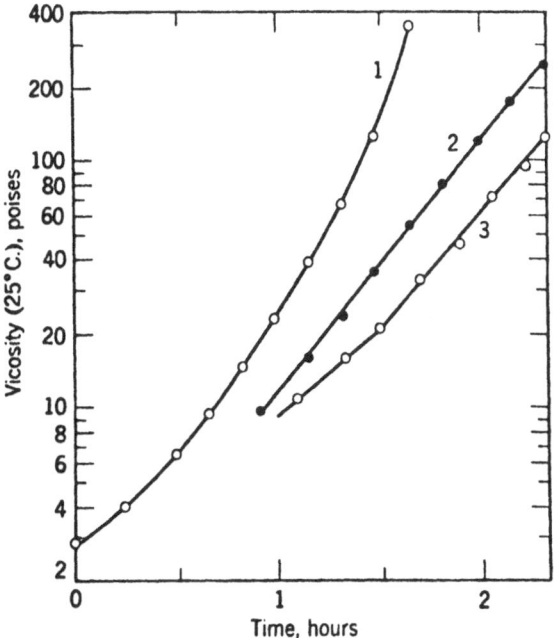

Figure 2. Viscosity versus bodying time in the heat bodying of
           conjugated acid oils. (1) Tung oil, in air, 450°F.;
           (2) oiticica oil, in air, 450°F.; (3) oiticica oil,
           in carbon dioxide, 450°F. (S. O. Sorensen, C. J.
           Schumann, J. H. Schumann, and J. Mattiello, Ind. Eng.
           Chem., 30, 211 (1938).

    (2) During network formation, peroxy compounds decompose to
create covalent bonds between the triglyceride molecules
    (3) The film continues to react by aging, forming additional
crosslinks, some volatile products, and eventually chalking.

Alkyd Resins

    Newer materials which contain acid residues plus synthetic struc-
tures are known as alkyd resins, from alcohol and acids. Reactions
can involve a range of transesterifications, or alternately free fatty
acids, especially tall oil, derived from the sulfate pulping of pine
chips.

Table 8. Brown House Paint (49).

Material	Pounds
Iron oxide brown	64.0
Acicular ZnO	254.0
Calcium carbonate	600.0
Soya lecithin	4.6
Raw linseed oil	280.0
Kettle-bodied linseed oil (X viscosity)	93.0
24% lead naphthenate	9.3
6% manganese naphthenate	1.1
Mineral spirits	108.9
Total	1414.9
Pigment volume 35.8%	
Consistency 83 KU	
Weight per gallon 14.1 lb	
Add puffing agent to increase viscosity	

Factices

The above discussion relates to the "drying" of oils by oxygen.
Materials like natural rubber (see Natural Polyisoprenes section)
are vulcanized with sulfur. The unsaturated oils can be made to
undergo a similar set of reactions (48). Brown factice is made by
first blowing a drying oil, then reacting it with 5-30% of sulfur
at 250-350°F for 1-2 hours.

White factice is made by reacting relatively saturated oils,
such as castor or rapeseed with liquid sulfur monochloride. The
reaction takes place at room temperature, forming a light-colored,
more or less crumbly material which sees wide use as lead pencil
erasers.

Functional Group Oils

Most of the triglyceride oils contain only double bonds as reactive sites. A few oils, however, contain other reactive groups,
usually in addition to unsaturation.

Castor oil is by far the most important oil in this category.
Each acid group contains an hydroxyl group:

$$
\begin{array}{l}
\underset{\text{2}}{\text{CH}}-\text{O}-\overset{\overset{\text{O}}{\|}}{\text{C}}-(\text{CH}_2)_7-\text{CH}=\text{CH}-\text{CH}_2-\overset{\overset{\text{OH}}{|}}{\text{CH}}-(\text{CH}_2)_5-\text{CH}_3 \\[2mm]
\text{CH}-\text{O}-\overset{\overset{\text{O}}{\|}}{\text{C}}-(\text{CH}_2)_7-\text{CH}=\text{CH}-\text{CH}_2-\overset{\overset{\text{OH}}{|}}{\text{CH}}-(\text{CH}_2)_5-\text{CH}_3 \\[2mm]
\underset{\text{2}}{\text{CH}}-\text{O}-\overset{\overset{\text{O}}{\|}}{\text{C}}-(\text{CH}_2)_7-\text{CH}=\text{CH}-\text{CH}_2-\overset{\overset{\text{OH}}{|}}{\text{CH}}-(\text{CH}_2)_5-\text{CH}_3
\end{array}
\tag{5}
$$

The hydroxyl group forms polyester and polyurethane polymers
in a variety of reactions (50). In the form of foamed polyurethane
rubber, polymerized castor oil enjoys wide use as seat cushions, for
example, as well as urethane paints, etc.

Incidently, castor oil derivatives include sebacic acid, used
as a monomer for nylon 610, and undecylenic acid, which forms 11-
aminoundecanoic acid, the monomer for nylon 11.

In a series of papers concerned with functional group oils (51),
natural castor oil was polymerized and formed into interpenetrating
polymer networks. (Using a 60/40 SIN of polystyrene and castor oil,
this research team made a pair of shoe heels which have seen wide
service on and off the lecture podium. The product is a very tough
elastomer.)

The USDA has examined a rather large number of triglyceride
oils, in a general search for interesting or unusual constituents.
While most of the oils (as expected) vary primarily in their chain
length and degree of unsaturation, a number of the oils contained
hydroxyl or keto groups (note castor oil above), epoxy groups, conjugated unsaturation, extra long chain fatty acids, etc., see Table
9 (52).

For purposes of this review, it should be emphasized that all
of these oils are polymerizable, perhaps by three different
mechanisms:

(1) The ordinary unsaturation, attacked through the oxygen-
free radical mechanism (discussed above);
(2) The special group reactivity, hydroxyl, keto, epoxy, etc.,
leading to esters, urethanes, etc.;
(3) After hydrolysis, the carboxylic group can be re-reacted,
as in the alkyd resins.

Steroids

The lipid fractions obtained from plants and animals contain

Table 9.　Unusual Triglyceride Oils (52).

Species	Oil in Seed, %	Component in Triglyceride Oil

Species with Hydroxy and Keto Fatty Acids

Species	Oil in Seed, %	Component in Triglyceride Oil
Lesquerella gracilis	32.7	$14\text{-OH-C}_{20}(70\%)$
Holarrhena antidysenterica	33.7	$9\text{-OH-C}_{18}(70\%)$
Cardamine impatiens	31-36	Dihydroxy C_{22} and C_{24} (23%)
Chamaepeuce afra	20-28	Trihydroxy $C_{18}(35\%)$
Lesquerella densipila	24-27	12-OH-C_{18} diene(50%)
Dimorphotheca sinuata	31-44	9-OH-C_{18} conj. diene (67%)
Coriaria myrtifolia	12.5	13-OH-C_{18} conj. diene (65%)
Cuspedaria pterocarpa	34.0	Keto acids (25%)

Potential Sources of Epoxy Fatty Acids

Species	Oil in Seed, %	Epoxy Acid Content, %
Vernonia anthelmintica	23-31	68-75
Vernonia pauciflora	40-42	73-80
Euphorbia lagascae	42-50	60-70
Stokesia laevis	36-49	75
Cephalocrton pueschellii	32-34	67
Erlangea tomentosa	21-31	50
Alchornea cordifolia	45.7	$50(C_{20})$
Schlectendalia luzulaefolia	31.9	45

Sources of Conjugated Unsaturation

Species	Oil in Seed, %	Type of Unsaturation
Valeriana officianalis	26-34	40% 9,11,13
Calendula officinalis	40-46	55% 8,10,12
Centranthus macrosiphon	28-32	65% 9,11,13
Impatiens edgeworthii	50-53	60% 9,11,13,15
Dimorphotheca sinuata	31-44	60% 10,12(+ hydroxy)
Coriaria myrtifolia	12.5	60% 9,11 (+ hydroxy)

Species with Long-Chain Fatty Acids in Seed Oil

Species	Oil in Seed, %	Component in Triglyceride Oil
Crambe abyssinica	30-40	60% C_{22}
Lunaria annua	30-40	40% C_{22}, 20% C_{24}
Limnanthes alba	20-30	95% $C_{22}+C_{20}$
Selenia grandis	18.5	58% C_{20}
Leavenworthia alabamica	18.0	50% C_{20}
Marshallia caespitosa	25.6	44% C_{20}

another important group of compounds called steroids. Steroids are important biological regulators and typically exhibit strong physiological effects when administered to living organisms indicating the potential use of such reagents in the synthesis for control release applications.

Cholesterol is one of the most widely occurring steroids, present in most animal tissues. While cholesterol was isolated as early as 1770, the last configurational details of its structure were not known until 1955. Much of this difficulty in assigning an absolute strucutre is that cholesterol contains eight different chiral centers, giving 2^8, or 256 possible stereoisomeric forms for the basic structural unit, only one of which is cholesterol.

The presence of functional groups is common in steroids; cholesterol contains an alcohol group, cortisone—two alcohol and three ketone groups, progesterone—two ketone groups, estradiol— two alcohol groups, norethynodrel—one alcohol, one ketone and one alkyne group. The presence of these functional groups allow ready modification of the steroid structure and incorporation into polymer structures. However, as yet, no important polymers have been made from this class of chemicals.

PROTEINS

Proteins are the essence of life to animals and plants alike. Wool, silk and leather, hair, feathers and ligaments are all common examples, besides muscle. In plant products, seeds are rich in protein (soy protein, for example).

While many proteins have been in long use, modern polymer science can be employed to make them better, or serve additional uses. Graft copolymers of wool and leather, now under investigation, yield softer and/or more water-resistant materials.

Proteins are mainly constructed of alpha-amino acids. These amino acids can be neutral as glycine, basic (containing an additional amine(s)) as lysine, acidic (containing an additional acid group(s) as aspartic acid or contain alcohol or thio functional groups- each representing a chemical "handle" for synthetic inorganic and organic chemists to ply their trade either on proteins containing these "available" functional groups or on the amino acids themselves.

The amino acid content and sequence varies widely from source to source and it is this variability that gives proteins unique structural configurations and conformations, and steric and electrical characteristics permitting them to literally be the "spice of life itself". For instance hemoglobin, with a minimum molecular weight of 68,000 contains 3.5 mole-% of arginine, 7.6-histidine, 8.0-lysine, etc. in a specific sequence and orientation yielding a complex fluted funnel structure with iron as the price for oxygen molecules successfully transgressing this geometric and electronic maze.

The messenger-RNA genetic code has been largely broken, allowing the synthesis of minute amounts of oligomeric proteins with specific acid sequencing. It should be possible to extend this utilizing a modified Merrifield automated protein synthesizer. In fact the templates presently utilized in the Merrifield synthesizer are themselves modified polymers permitting the analytical and polymer chemist to again practice their trade in constructing templates capable of generating pure homoproteins or co-, block, or alternating proteins each of which might exhibit desirable and somehow unique physical properties.

Fibrous Proteins

As mentioned above, wool, hair, leather, feathers, and silk are protein in content, all are crystalline. For example, the glass transition temperature of silk is 162°C., and its melting temperature is 250°C (53). Interestingly, these fibers benefit from water plasticization. When wool garments are creased and pressed, the material is exposed to steam, which at that temperature plasticizes T_g below 100°C. The material retains its shape on cooling and drying (54).

All of these fibers have a complex physical structure. For example, the outside of a wool fiber is covered with a layer of scales, with their free ends pointing toward the tip of the fiber. Inside is the cuticle, which surrounds an inner structure, called the cortex, which is made up of small spindles called cortical cells. Some of the coarser wool fibers have a hollow space in the center, called the medulla. The cortical cells are made up of keratin molecules, which are oriented into particular folds and crimps (54).

The structure of wool, a protein, interestingly enough, bears a significant resemblance to wood and cotton fibers, where fibrils of cellulose at the 50Å bundle diameter wind their way around the fiber in a complex pattern, several layers also being evident. The central hole in the cellulose fibers, corresponding roughly to the medulla, is called the lumen, and serves for water transport in trees.

Incidently, polymer scientists of all types, natural and synthetic, would do well to simulate the supermolecular structure of these fibers, which being hollow, are light in weight, and because of the particular angles of molecular orientation, are highly elastic and strong.

(NATURAL) ALKENES

By gentle heating or steam distillation of many plant materials
one obtains mixtures of odoriferous compounds called essential oils.
These compounds have a variety of uses including as the basic building blocks of adhesives, medication and the making of perfumes.
Terpenes and terpenoids are the most abundant constituent of essential oils. Most terpenes possess skeletons of 10 (monoterpenes),
15 (sesquiterpenes), 20 (diterpenes) and 30 (triterpenes) carbon
atoms. While plants do not synthesize terpenes from isoprenes, the
isoprene unit is a basic component of the structure of terpenes and
the recognition of this has greatly aided in the elucidating of their
structures. Most terpenes contain alkene units which can undergo
classical alkene reactions which can be taken advantage of by skilled
chemists in the synthesis of products from these natural resources.
Common terpenes include Beta-carotene (a tetraterpene from carrots),
squalene (a triterpene from shark liver oil), limonene (a monoter-
pene from lemon oil) and cembrene (a diterpene from pin oil).

Along with cellulose and lignin, natural alkenes are readily
prevelent in trees and are readily available in the thousands of
tons and upwards at the large scale lumber processing plants.

FUTURE RESEARCH REQUIRED

In modern industry, cost is often the overriding concern. It
can be argued, correctly, that as long as petrochemically derived
feedstocks are cheaper than "renewable resource" materials, the
former should be used. However, the price gap is narrowing, and we
should be prepared.

Many areas remain for future research projects. A few that can
be easily identified could be organized as follows:

(1) Isolation and purification of natural monomers and polymers.
Lignin is high on the list: a low-cost method of producing high
molecular weight, soluble lignins would revolutionize the forest
products industry.

(2) Modification of natural monomers and polymers: Research
on grafting and blending, perhaps with petrochemically derived products, will lead to new classes of materials. Some of the papers
in this book are concerned with grafted wool, leather, and cellulose.
Polymer impregnated wood is a related material. How about grafted
silks, or more research on blending natural fiber stocks with
plastic-forming polymers?

(3) New natural monomers. While nearly all natural materials
can theoretically be polymerized, perhaps after chemical modification,

there may be some pleasant surprises for researchers who go looking.

In conclusion, the authors see the growing need for research on natural products. However, this research should capitalize on modern ideas in polymer science.

REFERENCES

1. D. R. Dalton, "The Alkaloids," Dekker, NY, 1979.
2. P. Bernfeld (Ed.), "Biogenesis of Natural Compounds," Pergamon Press, NY, 1963.
3. E. Chargaff and J. N. Davidson, "The Nucleic Acids," Vols. I-III, Academic Press, NY, 1955-1960.
4. C. R. Wilke (Ed.), "Cellulose as a Chemical and Energy Resource," Interscience, NY, 1975.
5. R. Seymour and C. Carraher, "Polymer Chemistry," Dekker, NY, 1981.
6. W. Pigman (Ed.), "The Carbohydrates," Academic Press, NY, 1957.
7. R. L. Whistler and C. L. Smart, "Polysaccharide Chemistry," Academic Press, NY 1953.
8. C. Carraher and M. Tsuda (Eds.), "Modification of Polymers," American Chemical Society, Washington, DC, 1980.
9. E. A. MacGregor and C. T. Greenwood, "Polymers in Nature," John Wiley, NY, 1980.
10. J. K. Craver and R. W. Tess (Eds.), "Applied Polymer Science," American Chemical Society, Washington, DC, 1975.
11. C. L. Mantell, C. W. Kopf, J. L. Curtis and E. M. Rogers, "The Technology of Natural Resins," John Wiley, NY, 1942.
12. R. L. Whistler and J. N. BeMiller, "Industrial Gums," W. A. Benjamin, NY, 1966.
13. F. E. Brauns, "The Chemistry of Lignin," Academic Press, NY, 1952.
14. I. A. Pearl, "The Chemistry of Lignin," Dekker, NY, 1967.
15. R. J. Holman (Ed.), "The Chemistry of Fats and Other Lipids," Vol. 10, Pergamon, Oxford, 1970.
16. E. R. Yescumbe, "Plastic and Rubber," Applied Science Pubs., Essex, England, 1976.
17. M. Morton, "Rubber Technology," Van Nostrand-Reinhold, NY, 1973.
18. P. G. Cook, "Latex: Natural and Synthetic," Chapman & Hall, London, 1956.
19. H. L. Fisher, "Chemistry of Natural and Synthetic Rubbers," Reinhold, NY, 1957.
20. P. W. Allen, "Natural Rubber and Synthetics," Halsted, NY, 1972.
21. K. H. Gustovson, "The Chemistry and Reactivity of Collagen," Academic Press, NY, 1956.
22. G. E. Means and R. E. Feeney, "Chemical Modification of Proteins," Holden-Day, Ing., London, 1971.
23. R. S. Asquith (Ed.), "Chemistry of Natural Protein Fibers,"

Plenum, NY, 1977.

24. S. Fox and J. Foster, "Protein Chemistry," John Wiley, 1957.

25. P. S. Katsayannis, "The Chemistry of Polypeptides," Plenum, NY, 1974.

26. M. F. Perutz, "Proteins and Nucleic Acids," Elsevier, Amsterdam, 1962.

27. A. M. Michelson, "The Chemistry of Nucleoides and Nucleotides," Academic Press, NY, 1963.

28. E. E. Snell (Ed.), "Annual Reviews of Biochemistry," Annual Reviews Inc., California, 1968 onwards.

29. W. E. Cohn (Ed.), "Progress in Nucleic Acid Research and Molecular Biology, Vol. 1," Academic Press, NY, 1963 and subsequent volumes.

30. P. J. Scheuer, "Chemistry of Marine Natural Products," Academic Press, 1973.

31. J. L. Simonsen and others (Eds.), "The Terpenes," Cambridge University Press, Cambridge, England, V-1. 1, 1953; Vol. II, 1957; Vol. III, 1952; Vol. IV, 1957; Vol. V, 1975.

32. H. J. Stern, "Rubber: Natural and Synthetic," 2nd Ed., Maclaren & Sons, Ltd., London, 1967.

33. C. Goodyear, U.S. 3633 (1844).

34. Natural Rubber Producers Research Association, Technical Information Sheet No. 9, Revised (1977).

35. E. H. Andrews and D. T. Turner, J. Applied Polymer Science, $\underline{3}$, 366 (1960).

36. G. A. Schatz, News Report (National Academy of Sciences), $\underline{27(6)}$, 1 (1977).

37. F. A. Eagle, Rubber Chem. Tech., $\underline{54}$, 662 (1981).

38. "Guayule Reencuentra en el Desierto," Consejo Nacional de Ciencia y Technologia, Comision Nacionale de las Zonas Aridas, CIQA, Mexico, 1978.

39. Third International Guayule Conference, Pasadena, Cal., April 27–May 1, 1980. To be published.

40. E. Ott, Ed., "Cellulose and Cellulose Derivatives," Interscience, 1943.

41. E. Ott, H. M. Spurlin, and M. W. Graffin, Eds., "Cellulose and Cellulose Derivatives" 2nd Ed., Vol. 1-3, Interscience, 1954.

42. N. M. Bikales and L. Segal, Eds., "Cellulose and Cellulose Derivatives," Vols. 4 and 5, Wiley-Interscience, 1971.

43. A. F. Klarman, A. V. Galanti, and L. H. Sperling, J. Polym. Sci., $\underline{A-2}$, $\underline{7}$, 1513 (1969).

44. J. Russell and R. G. Van Kerpel, J. Polym. Sci., $\underline{25}$, 77 (1957).

45. C. J. Malm, J. W. Mench, D. L. Kendall, and G. D. Hiatt, Ind. Eng. Chem., $\underline{43}$, 688 (1951).

46. F. Rodriguez, "Principles of Polymer Systems," McGraw-Hill, 1970, p. 313.

47. D. H. Solomon, "The Chemistry of Organic Film Formers," Krieger, Huntington, NY, 1977.

48. D. Swern, Ed., "Bailey's Industrial Oil and Fat Products," 3rd Ed., Interscience, 1964.

49. D. H. Parker, "Principles of Surface Coating Technology,"
     Wiley, NY, 1965, p. 563.
50. F. C. Naughton, J. Am. Oil Chemists Soc., $\underline{51}$, 65 (1974).
51. L. H. Sperling, J. A. Manson, G. M. Yenwo, N. Devia-Manjarres,
     J. Pulido, and A. Conde, in "Polymer Alloys," D. Klempner
     and K. C. Frisch, Eds., Plenum, 1977.
52. L. H. Princen, J. Coatings Tech., $\underline{49}$, 88 (1977).
53. J. Magoshi, Y. Magoshi, and S. Nakamura, J. Appl. Polym. Sci.,
     $\underline{21}$, 2405 (1977).
54. G. E. Hopkins, "Wool as an Apparel Fiber," Rinchart & Co.,
     NY, 1953.

RENEWABLE RESOURCES FOR INDUSTRIAL APPLICATIONS

AN HISTORICAL PERSPECTIVE

Jett C. Arthur, Jr.

Technical Consultant
3013 Ridgeway Drive
Metairie, Louisiana 70002, U.S.A.

INTRODUCTION

Annually renewable natural products have always been major sources of polymers for industrial applications. Today, the world's most used non-food polymer is cellulose in fiber, film, paper, and housing applications. Starch and protein are useful both as food and feed products as well as industrial products. Natural products have been sources of chemicals by extraction of them from the products. Modification of these products through the actions of microorganisms and destructive distillation also yield industrial chemicals. With the development of internal combustion engines early in this century, natural products were proposed as sources of energy, particularly as a source of liquid motor fuels.

During the 1920's and 1930's, chemurgy developed as a division of applied chemistry concerned with the industrial use of annually renewable resources. Chemurgic research and development was conducted by industry and to a more limited extent by the state agricultural experiment stations and the U.S. Department of Agriculture. National chemurgic meetings were held to exchange information and to stimulate further work. Early research centers that were designed to

investigate the industrial utilization of annually renewable resources included: in 1914, The Citrus By-
products Laboratory in California; in 1934, The Cooperative Agricultural Byproducts Laboratory in Iowa; and
in 1936, The Cooperative U.S. Regional Soybean Industrial Products Laboratory in Illinois.

During the period 1910-1930, the production of agriculture in the United States was stimulated by the
demands of the export market and the increased market
value of the products. Mechanization of agriculture
and improvements in varieties of plants and in animals
led to increased productivity and to the output of agricultural products to record levels. However, with
decline in both export markets and market value of the
products in the 1930's, large surpluses of agricultural
products were generated.

Several national approaches to solve the surplus
problem of natural products were made. In one approach
the U.S. Congress, perhaps influenced by the promise of
chemurgy, in the Agricultural Adjustment Act approved
February 16, 1938, provided, as follows:
> The Secretary (of Agriculture) is hereby author-
> ized and directed to establish, equip, and main-
> tain four regional research laboratories, one
> in each major farm producing area, and at such
> laboratories to conduct researches into and to
> develop new scientific, chemical, and technical
> uses and new and extended markets and outlets
> for farm commodities and products and byproducts
> thereof. Such research and development shall
> be developed primarily to those farm commodities
> in which there are regular or seasonal surpluses
> and their products and byproducts.
> Public Law No. 430, 75th U.S. Congress.

This authorization provided the first large-scale
chemurgic effort that furthered research and development on renewable natural products from which new and
extended industrial uses could result. For example, in
the context of the times (1930's) before the large increases in the U.S. petroleum reserves, it was antici-
pated that the increasing numbers of internal

combustion engines could exhaust our national reserves
in perhaps the 1960's. Chemurgy seemed a reasonable
alternative to supply a liquid motor fuel derived from
natural products and also, if successful, to provide
the surplus natural products for additional industrial
markets. The regional research laboratories were opened
in 1941: eastern in Pennsylvania, northern in Illinois,
southern in Louisiana, and western in California. In
the long-range, the research programs of the regional
laboratories were flexible. As the surplus problems
were solved, and the national petroleum reserves were
increased, the scope and emphasis of the programs of
the laboratories were changed and broadened.

With the increase in the national petroleum reserves in the 1940's and 1950's, many new petrochemically based products were developed that competed with and
displaced many of the industrial products based on natural products. However, as the title of this monograph
suggests, perhaps there is again interest in renewable
resources for polymer applications and other industrial
uses. Petroleum availability and market value are also
major factors in this re-examination of natural products for industrial uses.

It is the purpose of this discussion to give a
historical perspective to renewable resources for industrial applications. In the limited space available,
the extensive survey reported in 1939 of all research
activities (present and suggested) relating to the industrial utilization of agricultural products will be
discussed. This survey is amazingly technically relevant to work on renewable resources for industrial applications in the 1980's and to work yet to be initiated. Additional information, both from published
works and from personal experiences, dating from the
opening of the regional laboratories for experimental
work in 1941 until to date, will be included.

COMPOSITION OF RENEWABLE RESOURCES

Annually renewable resources comprise literally
tens of thousands of species of plants. Plants are
products of stored solar energy. The light-catalyzed

reduction of carbon dioxide and water to more complex
chemical products by interaction of solar energy has
been the subject of numerous reports on photosynthesis.
Also, the initiation of additional chemical reactions
yields a wide variety of products. Consumption of these
plant products by animals and fishes also yields large
sources of annually renewable animal and aquatic products.

Obviously, there are a great number of annually
renewable resources. However, most of these resources
are composed principally of starches, sugars, proteins,
fats and oils, and/or celluloses. Numerous other minor
constituents have been isolated from renewable resources. The individualities in nature's packaging of
these resources are often equally as important as the
compositions of these resources. For example, cellulose in the form of straw or sugar cane bagasse has
value; however, cellulose in the form of cotton fibers
has much more valuable uses. Many natural sugars can
be fermented directly into alcohols and other chemicals.
However, when nature has packaged these sugars in the
form of polysaccharides, such as cellulose, lignocellulose, or starch, these resources must be depolymerized
to yield fermentable products, an often challenging
process step.

It can be concluded that both the composition and
nature's individual packaging of annually renewable resources must be considered in expanding their uses to
meet modern man's needs for food, fiber, shelter, and
energy.

STRUCTURE OF NATURAL PRODUCTS

Comprehensive examinations of the structures of
many natural products and their chemical compositions
to further industrial utilization of annually renewable
resources have been reported. Earlier research and development work can be extended to yield new industrial
products, particularly in modifying the properties of
natural products to give currently desired properties,
as reported in this symposium. A classification of important structures to be considered is outlined.

Morphological Structure

Renewable resources could be interpreted as composites in that natural products with high molecular weights are often interspersed principally with water, both free and bound, and constituents with lower molecular weights. The morphological structures of natural products are nature's individualized packagings of the same or very similar constituents in different products. As mentioned earlier, cellulose in the form of cotton fiber, straw, wood, or sugar cane bagasse has different morphological structures and uses. Similarly, proteins in plant products, wool, silk, or animal products have different morphological structures. Also, the uses of starches, particularly those uses that are related to the properties of their colloidal solutions, are dependent on the properties of the granules in the plants in which starches are deposited. The nature and properties of the granules vary with the plant source.

Macromolecular Structure

The macromolecular structures of celluloses, proteins, and starches in the same, and in different, natural products may have more than one degree of polymerization or molecular weight. Proteins in different natural products may also be composed of different types, amounts, and sequences of amino acids. Also, present in natural products, usually in relatively low concentrations, are specialized macromolecules containing proteins, such as enzymes, that catalyze specific reactions. For various industrial uses, it is important that the macromolecular structure of the desired constituent in a natural product be determined.

Molecular Orientation

The degree of molecular orientation, that is, degree of crystallinity, of natural products is also dependent on the source of the products. For example, cellulose in cotton fiber has a degree of crystallinity of about 80 percent. Celluloses from other sources are less molecularly oriented. The degree of molecular orientation in cellulose as well as crystal lattice type

can be altered when cellulose is immersed in solutions
that swell and interact strongly with cellulose. The
most commonly used commercial process to initiate these
changes in cotton cellulose is mercerization, first reported in 1842. For example, mercerized fibrous cotton
celluloses and regenerated cellulosic fibers have lower
degrees of molecular orientation and different crystal
lattice types from natural fibrous cotton cellulose.
The degrees of molecular orientation in non-cellulosic
renewable resources are generally low.

Organochemical Structure

The organochemical structures of annually renewable
resources range from simple chemical structures to complex structures that are not easily duplicated in the
test tube. The natural polymers useful for industrial
applications include cellulose, starch, and protein.
Cellulose is a polysaccharide with glucose linked as in
cellobiose. Cellulose usually occurs in a fibrous form.
Starch is also a carbohydrate polymeric compound that
consists of both linear polymer and branched polymer and
occurs in granules in plants. Proteins are found both
in plants and animals. Fats and oils and sugars are
usually monomers and have a range of compositions and
properties.

INDUSTRIAL APPLICATIONS OF NATURAL POLYMERS

Cellulose

The major constituent of almost every plant source
is cellulose. Cotton fibers are composed of cellulose
in almost a pure form. Wood fiber is the most harvested
plant cellulose and is often processed to wood pulp with
a high content of cellulose. Agricultural wastes that
contain cellulose are produced in enormous quantities
each year. Some major polymer applications are
indicated.

Cellulosic fibers have been and are still mankind's
greatest source of industrial and textile fibers, particularly since the industrial revolution. Cotton,
mainly from the United States, was the principal raw

material of the industrial revolution in the United
Kingdom. Regenerated cellulosic fibers, mainly from
wood fibers, have been and are still a large source of
industrial and textile fibers. Other important sources
for industrial cellulosic fibers include flax, hemp,
ramie, jute, and sisal.

Similar to reports in this monograph research and
development to modify cellulosic fibers to impart new
properties were programs proposed in the 1930's and
later implemented on a large scale in laboratories and
industry. Modifications of the properties of cellulosic
fibers by addition of finishes and by chemical reactions
to make them more resistant to fire, mildew, crushing,
creasing, soiling, abrasion, chemical action, and other
environmental and use conditions and to impart special
reactivities have continued to be of interest.

Wood fibers are converted into pulps that are useful both in the manufacture of paper and in preparation
of regenerated textile fibers. In the United States,
the annual per capita consumption of paper and paper
products is several times greater than that of other
non-food polymers. Modifications of the properties of
paper through chemical reactions are possibilities that
need further investigation.

Agricultural wastes usually contain high concentrations of cellulose. The use of processed agricultural
wastes, such as sugarcane bagasse, as fillers for wallboard, plastics, and insecticides and other biological
agents used in agriculture, has been reported. Reactions of wastes, for example, with vinyl monomers, can
impart second order transition temperatures to cellulosic products that are lower than their decomposition
temperatures, so that thermally moldable plastics are
obtained. This offers a possibility of using agricultural wastes as a raw material to make wallboard and
insulation products. Work to make such products resistant to fire and microorganisms could be done simultaneously.

The use of cellulose in the form of processed wood
for construction is, of course, a common practice.

However, today chemical treatment of wood can impart re-
sistances to fire, microorganisms, water, and abrasion
and yield wood products with increased plasticity that
facilitates forming of wood into desired shapes.

Protein

Proteins occur widely in both plant and animal
sources. Proteins in a fibrous form, such as silk,
wool, and hides and skins, are processed into textile
and other useful commercial products. Modifications of
the properties of wool and of leather derived from hides
and skins, some of which are discussed in this symposium, have been widely reported. As noted for cellulosic fibers, modifications of fibrous proteins to make
them resistant to detrimental effects of environmental
and use conditions have been reported.

Protein isolates from milk, other animal products,
and soybeans were widely used as adhesives and textile
sizes in industry. Competition with man-made adhesives
derived from petrochemicals has reduced the utilization
of these protein adhesives. After extraction of oils
from vegetable oilseeds, such as soybeans, peanuts, and
cottonseed, the residual protein-containing meals have
been useful in adhesives for the manufacture of plywood.

Protein isolates from milk, soybeans, peanuts, cottonseed, and corn have been extruded to form bristles
and textile fibers. For a number of years, these man-
made, wool-like fibers, principally derived from milk
casein, were on the commercial market in the U.S. and
Western Europe. Protein wastes, such as chicken feathers, have also been extruded to form fibers with textile
properties.

Starch

Starches occur in high concentrations in many
plant sources, such as corn, potato, rice, tapioca, and
wheat. Their compositions differ only slightly; however, their physical properties may differ. Starches
derived from corn, tapioca, and sweetpotatoes have been
used industrially as adhesives and textile sizes. These

uses for starches have continued for corn and some root starches.

NATURAL PRODUCTS AS ENERGY SOURCES

Liquid Fuels

With the invention and development of the internal combustion engine as a source of mobile and stationary power, the use of energy in the U.S. increased significantly in the twentieth century. Dependence on liquid fuels also increased. Early in the century, annually renewable products were proposed as a source of liquid fuels, particularly fuel-grade ethanol as a product of fermentation of carbohydrates and, to a lesser extent, methanol as a product of destructive distillation of wood.

Research and development programs were initiated in the 1920's and 1930's to make fuel-grade ethanol and to blend ethanol (10%) with gasoline (90%) to yield a motor fuel known as gasohol. In the late 1930's gasohol was test marketed as a motor fuel with no particular changes required in engine design. Initial research programs in the 1940's at the northern regional laboratory in Illinois included developing fuel-grade ethanol and testing in internal combustion engines. The Cooperative Agricultural Byproducts Laboratory in Iowa also had a leadership role in this area. A large industrial effort to convert renewable resources into liquid motor fuels was developed in Western Europe in the late 1930's and continued into the 1940's.

Some of the research and development areas emphasized in the 1930's which are relevant in the 1980's include: (1) saccharification of carbohydrates, particularly cellulose and starch, into fermentable forms using both biological agents and chemical agents; (2) design and operation of demonstration plants for production of liquid motor fuels; (3) utilization of fermentation plant byproducts and wastes; (4) energy balance economics; (5) food versus motor fuel directions and also feed versus fuel directions. In the 1930's corn and potato starches, as well as in the

1980's, were proposed as primary sources of fermentation
products. In the 1930's U.S. surplus agricultural products more than supplied the food and feed needs of that
market. In the 1980's there will be market competition
for agricultural products between processing for liquid
fuels and processing for food and feed needs. A conclusion reached in the 1930's -- that alcohols suitable for
use as motor fuels can be produced from renewable resources, however under present economic conditions and
present production methods such potential motor fuels
are relatively expensive -- has relevance in the 1980's.
In 1957 a commission on increased industrial use of agricultural products concluded -- in view of these facts,
the present state of scientific knowledge, and the costs
involved, the commission finds it impractical to recommend an alcohol motor-fuel program. With U.S. fuel
grade ethanol production from renewable resources in
1981 of about 125 million gallons per year, the impact
on production of agricultural food and feed products,
and also on petroleum utilization, is very small.

Solid Fuels

In 1981 the use of cellulose as a U.S. industrial
source of energy is very small. In wood processing
plants, some waste products are burned to produce steam.
Sugarcane bagasse may also be burned at the plant site
to produce steam. Several municipalities, in connec-
tion with waste disposal, have separated combustible
materials from garbage, compacted them, and burned them
with coal to produce steam.

An interesting research proposal in the 1930's was
to develop processes and motors for the direct use of
solid fuels that utilize either compacted or highly
pulverized materials from renewable resources. Although
commercial applications for mobile power have not been
reported, turbogenerator type motors based on the expansion and compression cycles in refrigeration have been
proposed to utilize solid fuels as the energy source.

Gaseous Fuels

Carbon monoxide, hydrogen, and methane, produced

from renewable resources may be used directly in engines or converted to liquid motor fuels.

NATURAL PRODUCTS AS CHEMICAL SOURCES

Renewable resources also yield an abundance of fats and oils, sugars, drugs, and other chemicals derived by fermentation processes. Some indications of the variety of these products are reported in this symposium. In addition to animal and fish derived fats and oils, the extraction and properties of fats and oils from babassu, castor, chia, china wood, colza, corn-germ, cottonseed, grapeseed, hempseed, jojoba, olive, peanut, safflower, sesame, soybean, sunflower, tea seed, tung, wheat-germ, and numerous other sources have been reported. Oils from grapefruit, lemon, orange, melon, pumpkin, tomato, wild radish, mustard, and locus seeds and from peach, apricot, cherry, walnut, and pecan kernels have been reported. These oils may be used for edible purposes and soap making and in some cases for drying oils.

Cane and beet sugars can be extracted in almost pure form and used as highly reactive chemical intermediates or for edible purposes. Numerous other sugars have been extracted from plant products or produced as hydrolysis products from hemicelluloses, celluloses, and starches.

Plants have been processed as a source for special chemicals. One of the most common drugs, quinine, is extracted from cinchona bark and used in medical treatment of malaria. Other examples of drugs from plants include: digitalis, cortisone, belladonna, henbane, stramonium, cascara, golden-seal, ginseng, licorice, ephedra, poppy, psyllium, nicotine. Insecticides include: pyrethrum, derris, cube, timbo, and devil's shoestring. Condiments and spices include: mustard, red pepper, sage, and thyme. Essential oils include: peppermint, spearmint, wormseed, wormwood, dill, lemongrass, citronella grass, vetivert, rose, geranium, lavender, coriander, caraway, fennel, and basil.

Some of the extractions of chemicals from

renewable resources are also reported in this monograph.
The numbers and types of useful chemicals from renewable
resources are obviously large.

PROCESSES AND PRODUCTS

    Processes and products developed to produce indus-
trial materials from renewable resources have been too
numerous to record here. For competitive reasons --
supply of raw materials and technical and economic considerations -- some of the products have varied widely
in industrial use. Major U.S. industrial consumption
of renewable resources have recently included: oils
and fats (animal and vegetable); industrial alcohol
(wheat, corn, grain sorghum); fibers (cotton lint,
flax, hides and skins); paper (forest products); isolated proteins (milk casein, animal glues, soybean,
corn); turpentine and rosin (naval stores); and other
chemicals (monosodium glutamate--wheat; starch and
dextrin--corn; lactose--milk; molasses and pulp resi-
dues--sugarcane and beet; tannin; lecithin; pectin;
furfural).

CHALLENGES IN THE 1930's AND THE 1980's

    Research and development on renewable resources
for industrial applications in the 1930's and in the
1980's had one stimulus in common, namely the potential
diminishing availability of petroleum. In the 1930's
the U.S. and world markets for agricultural products
for food and feed did not totally utilize U.S. production. Then there was little or no competition between
food and feed uses and industrial products from renewable resources. In the 1980's the U.S. and world markets for agricultural products for food and feed uses
are large and competitive with any diversions of these
products into industrial uses. Limitations on food and
feed uses are largely production, processing, distribution, and economics.

    However, no harvested agricultural product is
wholly edible. Processing of harvested products for
food and feed yields inedible residues. Also, large
residues, usually cellulosic in nature, are left in the

fields after harvesting. Forest products also yield
residues at the point of harvest and later at the point
of processing. Industrial utilization of these residues
would not decrease food and feed availability. In fact,
in some processes byproducts would be produced that
could have feed value. However, diversion of acreage
for food crops to acreage for other agricultural crops
that could be used with greater facility industrially
would require technical, social, and economic evaluations. Growing sugar or starch producing crops primarily for liquid motor fuel production versus growing
similar crops for food and feed would be an example.

Continuing challenges of renewable resources as
industrial raw materials include: geographic distribution of the resources; transportation and concentration
of the resources; storage of the resources to insure
supplies for year-round operation of industrial plants;
and coordinated research and development programs ranging from fundamental and applied research to economic
evaluation of the processes and products. An ever present challenge to industrial products developed from renewable resources is possible future competition from
similar products derived from nonagricultural materials
other than petroleum. For example, the development of
a greater number of industrial products from the large
U.S. supplies of coal is a possibility. Also, short-
term increases in availability and decreases in costs
of petroleum could cause long-term investment problems
in industrial plants based on renewable resources.

The human's consumption of non-food products, that
is, industrial products, has been limited generally by
availability and economic considerations. Research
work to develop and to increase the number of industrial
products derived from renewable resources, in the long
run, could give mankind more flexibility to meet these
needs. However, the rhetorical question in the 1980's,
as it was in the 1930's, is how far can chemurgy
advance?

SUGGESTED READING

A Report of a Survey Made by the Department of

Agriculture Relative to Four Regional Research
Laboratories, One in Each Major Farm Producing
Area, 76th U.S. Congress, 1st Session, Senate
Document No. 65, 1939, U.S. Government Printing
Office, Washington, D.C., 429 pp.

Report of Commission on Increased Industrial Use
of Agricultural Products, 85th U.S. Congress,
1st Session, Senate Document No. 45, 1957, U.S.
Government Printing Office, Washington, D.C.,
135 pp.

Arnold, L. K., and Kremer, L. A., Corn as a Raw
Material for Ethyl Alcohol, Iowa Eng. Expt. Sta.,
Ames, Iowa, Bull. No. 167, 1950, 103 pp.

Arnold, L. K., Plagge, H. J., and Anderson, D. E.,
Cornstalk Acoustical Board, Iowa Eng. Expt. Sta.,
Ames, Iowa, Bull. No. 137, 1937, 47 pp.

Leslie, E. H., in Motor Fuels, Chapt. XII. Alcohol -- A Motor Fuel of the Future, Chem. Cat. Co.,
New York, 1923.

Stefferud, A., Ed., Crops in Peace and War, The
Yearbook of Agriculture 1950-1951, U.S. Department
of Agriculture, 1951, U.S. Government Printing
Office, Washington, D.C., 942 pp.

Sweeney, O. R., and Arnold, L. K., Plastics from
Agricultural Materials, Iowa Eng. Expt. Sta., Ames,
Iowa, Bull. No. 154, 1942, 52 pp.

Sweeney, O. R., and Arnold, L. K., Furfural Utilization, Iowa Eng. Expt. Sta., Ames, Iowa, Bull.
No. 169, 1950, 47 pp.

Wiley, H. W., Industrial Alcohol, U.S.D.A.
Farmer's Bull. No. 429 (1911).

ENHANCING THE USE OF CELLULOSE BY

SOLUTIONS IN DMSO AND DMAc

Raymond B. Seymour

Department of Polymer Science
University of Southern Mississippi
Hattiesburg, MS   39401

Wood (xylem) is a vascular tissue which occurs universally in gymnosperms (soft wood trees) and dicotyledonous angiosperms (hard wood trees). The wood consists of lignin and holocellulose. The former is a polymer consisting of repeat phenolpropane units and the latter consists of cellulose and hemicellulose. Hemicellulose is a mixture of pentosans, hexosans and polyuronides.

Wood is a renewable resource, which is used for lumber and paper. The production of the latter, which is primarily cellulose, requires the removal of hemicellulose and some of the lignin as lignosulfonic acid and alkali lignin. Cellulose is also the major component of cotton, flax, hemp, jute, ramie, and sisal.

The ancient Chinese produced paper from rags and this source was used in America's first paper mill in 1690. However, Keller developed a mechanical process for making pulp from wood in 1841. This invention was followed by the Watt and Burgess' soda process in 1853, Tilghams's sulfite process in 1874, and Dahl's sulfate or kraft process in 1879[1]. Cotton, linen, and rags are still used for paper making, but the principal raw material is cellulose obtained from wood.

In 1664, Robert Hooke, predicted that silk could be produced by artificial means but man-made cellulosic fibers were not produced until Count Chardonnet used the cuprammonia process to produce rayon in 1891[2]. Viscose rayon was produced in the U.S. from cellulose xanthate in 1910 and is still in production[3].

Cellulose is soluble in aqueous mineral acids, in aqueous solutions of zinc chloride, calcium thiocyanate, aqueous alkalies,

quaternary ammonium hydroxide and in metal ammonia hydroxide complexes, such as copper ammonia hydroxide (Schweitzer's solution)[4].
Aqueous zinc chloride has been used for the production of vulcanized fiber, aqueous sodium hydroxide has been used for mercerization of cotton and sulfuric acid has been used to produce parch-
ment paper. However, most modern solutions of cellulose are based
on the formation of soluble derivatives, such as sodium cellulose
xanthate, which is used to produce regenerated cellulose (rayon
and cellophane) in the viscose process. Attempts have been made
to dissolve cellulose in dimethylformamide (DMF) in the presence
of nitrogen tetroxide (N_2O_4), sulfur dioxide-amine or nitrosyl
chloride and other systems containing polar solvents[5]. Rayon
has been produced experimentally from solutions of cellulose in
amine oxides.

Cellulose is also soluble in dimethyl sulfoxide (DMSO) in the
presence of formaldehyde (usually as paraformaldehyde)[6,7]. This
system would be competitive with the viscose system if a suitable
recovery system was developed[8]. The solute in this system is
methylolcellulose.

Chitin, which is soluble in the DMSO-H_2CO system, is also
soluble in dimethylacetamide (DMAc) in the presence of lithium
chloride[9,10,11]. Cellulose is soluble at room temperature in
DMF-chloral and pyridine[12]. The solute in this system is cellulose
trichlorate.

Cellulose may also be dissolved in DMAc-LiCl solutions.[13]
In a typical illustration, 10g of cellulose is suspended in 1
DMAc containing 50g LiCl and the cellulose is dissolved by stirring
for 1 hr. at 150°C. In addition to being useful for the production
of regenerated cellulose these solutions may also be used for the
homogeneous production of cellulose derivatives.

Cellulose acetate was produced by the room temperature homogeneous acetylation of cellulose with acetic anhydride in the presence of amines in DMSO-H_2CO systems[14,15,16]. Cellusose acetate
was also prepared by adding 7ml acetic anhydride to 100ml of a
solution of 5 percent cellulose in DMAc-LiCl to which had been
added 0.5ml of 71 percent perchloric acid solution. The solution
was allowed to stand for one day at room temperature and was then
heated at 40°C for 2 hrs. before the ester was precipitated from
the solvent by the addition of methanol. Cellulose and chitin have
also been esterified and carbanylated at room temperature by the
addition of phenyl isocyanate and pyridine to a solution in DMAc-
LiCl systems.[17]

Cellulose ethers, such as carboxymethylcellusose (CMC) and
benzylcellulose, have also been produced in DMSO-H_2CO[18] and
DMAc-LiCl systems[19]. Carboxymethylcellusose was prepared by

adding 5g sodium chloroacetate to 100ml of a solution of 5 percent cellulose in DMAc-LiCl to which 3.5ml of 80 percent tetraethyl ammonium hydroxide in methanol had been added. The solution was stirred for 8 hrs. at room temperature before the CMC was precipitated from its solvent by the addition of methanol.

In contrast to the classic method for the production of cellulose acetate with a D.S. of 2 by the partial saponification of cellulose triacetate, the acetylation of cellulose, in homogeneous solutions, may be stopped at any desired D.S. value.

These systems which permit the production of cellulose derivatives homogeneously, should enhance the use of this renewable resource. Topochemical reactions may be run on pulp or cellulose fibers. The solutions of cellulose or the cellulose derivatives may be used for coatings or for producing blends with other soluble polymers such as polyacrylonitrile.

The 1981 Marcus Wallenberg prize of $10,000.00 was awarded to H.H. Holton for his discovery that the rate of delignification in alkaline pulping could be increased by the addition of small amounts of anthraquinone. [20] Paper making is an old art which has been improved by Holton. The solutions of cellulose and its derivatives are relatively new. Hence, one may anticipate many new discoveries as the result of additional investigations of these systems.

Cellulose is the world's most abundant renewable resource. The availability of new solvent systems should increase the importance of this naturally-occurring polymer.

REFERENCES

1. M. Heath, Chemical Ind 62 405 (1948).
2. H. de Chardonnet, French Pat 165,349 (1884).
3. C.F. Cross, E.T. Bevan, Brit Pat 8700 (1892).
4. A.F. Turbak, R.B. Hammer, R.E. Davies, N.A Portnoy, Chapter in ACS Symposium Series No. 58, A.F. Turbak, Ed., American Chemical Society, Washington, D.C., 1977.
5. N.A. Portnoy, D.F. Anderson, Chapter 5 in ACS Symposium Series No. 58, A.F. Turbak, Ed., American Chemical Society, Washington, D.C., 1977.
6. D.C. Johnson, M.D. Nicholson, F.C. Haigh, Appl Polym Symp 28, 931 (1976).
7. R.B. Seymour, E.L. Johnson, J. Appl. Polym. Sci. 20, 3425 (1976)
8. R.B. Hammer, A.F. Turbak, R.B. Davies, N.A. Portnoy, Chapter 6 in ACS Symposium Series No. 58, A.F. Turbak, Ed., American Chemical Society, Washington, D.C., 1977.

9.   R.A. Rutherford, P.R. Austin, P 182 in Proceedings of First
     International Conference on Chitin at MIT, Cambridge, MA,
     1978.

10. P.R. Austin, C.J. Brine, J.E. Castle, J. Zikakis, Science 212,
     749 (1981).

11. P.R. Austin, U.S. Patents 4,059,457,4,062,921 (1979).

12. K. Kamide, K. Okajima, T. Matsui, Manabe, S-I., Polymer J. 12,
     (8) 54 (1980).

13. C.L. McCormick, D.K. Lichatowich, J.A. Pelezo, K.W. Anderson,
     Chapter 24 in ACS Symposium Series No. 124, C.E. Carranher
     Jr. and M. Tsuda, Eds., American Chemical Society, Washing-
     ton, D.C., 1980.

14. R.B. Seymour, E.L. Johnson, J. Polym. Sci, 14, 670 (1976).

15. T.J. Baker, L.R. Schroeder, D.C. Johnson, Carbohydrates Rec, 67
     C4 (1978).

16. R.B. Seymour, E.L. Johnson, J. Poly Sci--Chem. Ed., 16, 1
     (1978).

17. C.L. McCormick, D.K. Lichatowich, J. Polym Sci., Letters Ed.,
     17, 479 (1979).

18. R.B. Seymour, J. Coat. Technol, 49 (626), 36 (1977).

19. C.L. McCormick, T. Shen, ORPL Preprints, 1981.

20. M. Heylin, Chem. Eng. News, 59 (27), 38 (1981).

MODIFICATION OF DEXTRAN THROUGH REACTION WITH ORGANOSTANNANE HALIDES AS A FUNCTION OF REACTION SYSTEM - A MODEL FOR POLYSACCHARIDE MODIFICATION

Charles E. Carraher, Jr. and Timothy J. Gehrke

Department of Chemistry
Wright State University
Dayton, OH 45435

INTRODUCTION

Introduction and Historical

Today, when sources of feedstocks, such as petroleum based compounds, are being consumed at a faster rate than they are being discovered, there is a need for conservation, renewing of feedstocks and development of readily available, regenerable natural feedstocks. Polysaccharides are one such readily available, natural feedstock.

Carraher and co-workers have previously formed a number of metal containing polymers through reaction with a wide variety of diols including ethylene glycol and hydroquinone (such as 1-6). Furthermore, polyvinyl alcohol has been successfully modified

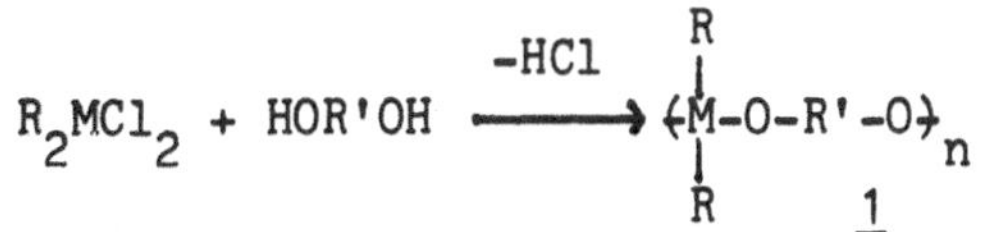

$$R_2MCl_2 + HOR'OH \xrightarrow{-HCl} \{M\text{-}O\text{-}R'\text{-}O\}_n \quad \underline{1}$$

through condensation with a wide variety of organometallic halides (for instance 7-9). In general there is a direct correlation between

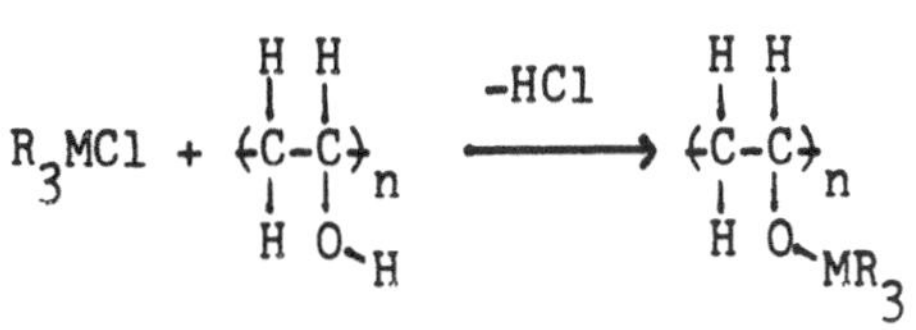

$$R_3MCl + \{C\text{-}C\}_n \xrightarrow{-HCl} \{C\text{-}C\}_n \quad \underline{2}$$

45

reactions with the hydroxyl contained on a polymer and the reactions employing diols (3,10). An extension is the modification of polysaccharides through condensation of the hydroxyl groups of the polysaccharides giving products of varying exact structure containing a variety of varying metal content including unreacted units. For polysaccharides derived from 1 → 6 hexoses this can be depicted as follows for M=Sn.

Carbohydrates are the most abundant class of naturally occurring organic compounds. They constitute three-fourths of the dry weight of the plant world and are widely distributed in both plant and animal forms of life. Thus, the vegetation of the earth contains large quantities of chemically combined carbon, mostly in the form of carbohydrates, with a major portion being polysaccharides. Polysaccharides are typically high molecular weight (ca 25,000 to 15,000,000) polymers composed of monosaccharide units. These polysaccharides are exceedingly complex with the general composition varying from chain to chain even when derived from the same source. An exact structure for these polymers cannot be written but the general structure can be described in a reasonable degree of completion.

The chemical and physical modification of polysaccharides is one of man's oldest technologies typically focusing on cotton since the 1850's. Most of these modifications are topochemical in nature, occurring through reactions involving cellulosic reactive groups which are available in the amorphous regions and on the surfaces of crystalline areas. Our group has chosen to attempt more thorough, homogeneous modifications of polysaccharides in the belief that such modification will yield material more homogeneous with respect to subsequent physical and chemical properties.

Carraher, et al. recently effected modification of cellulosic material utilizing bisethylenediamine copper II hydroxide solutions to effect solution of cellulose derived from cotton with subsequent reaction with solutions containing organotin halides. Through use of this modified interfacial condensation system, cellulose derived from cotton was successfully modified with yields and tin content varying from low to high (5 to 95% yield; 6 to 23% tin) dependent on the particular reaction conditions. Copper-amine solutions were used for a number of reasons including a) prior knowledge concerning the behavior of "cotton" in such solutions, b) such liquids permit good solution of the "cotton", c) the system is easily handled and can be utilized on a milligram to ton scale, and d) effectiveness of removal of unreacted copper-diamine is easily followed through analysis of the blue coloration of the modified cellulose.

There are avilable commercial quantities of a number of poly-saccharides from a wide divergence of sources. Further, the technology exists to permit the commercial usage of chemically modified polysaccharides which can replace currently used poly-meric materials.

Purpose

The major purpose of this research involves the evaluation of a number of potential condensation systems for the purpose of determining which systems might be useful in the modification of various polysaccharides through condensation with organostannane halides.

Rationale

Dextran was the polysaccharide chosen for preliminary study. It is water soluble, permitting the evaluation of aqueous reaction systems. Also, it is readily available on a large scale in a wide

variety of molecular weights, the latter permitting modification
of dextran to be studied as a function of chain length. Dextrans
are primarily found in yeast and bacteria consisting of branched
storage polysaccharides of D-glucose. They differ from glycogen,
starch and cotton in having a variety of backbone linkages which
may be 1→2, 1→3, 1→4, or 1→6 depending on the particular source.

An underlying assumption is that dextran is a representative
polysaccharide source and that results derived from studying its
modification can be directly applied to other sources of polysaccharides. If dextran can be successfully modified utilizing a
specific reaction system, other polysaccharides can likewise be
modified utilizing the same reaction system if that particular
polysaccharide is soluble in the given reaction phase utilized.
Thus, a key to the transfer of the results from this study to other
polysaccharides, and to the successful "intimate" modification of
dextran, is to achieve solution of the particular polysaccharide.
The literature has been scanned for suitable potential solvents
for polysaccharides.

Organostannanes of the form R_2SnCl_2 (where R is an alkyl or
aryl group) were employed as the metal containing reactant.
Reasons for utilizing such stannanes in the present study include
the following:

1. The modification products obtained through modification
   of cellulose derived from cotton resulted from condensation with R_2SnX_2 compounds and all showed some biological activity with respect to inhibition of common
   fungi and bacteria (9,11,12). For instance, compounds
   derived from dibutyltin dichloride and cellulose showed
   good activity against A. flavus, A. niger, A. fumagatus,
   T. reesei and C. Globosum. Potential application
   involves mildew and rot resistance including commercial
   insulation, bandages and in topical medical formulations.

2. The use of difunctional reactants typically ensures
   formation of insoluble products if reaction occurs since
   some degree of cross linking should occur. This allows
   a more ready preliminary evaluation of the success or
   failure of effecting modification of the dextran. The
   use of R_3SnCl monofunctional reactants was included
   permitting the possible formation of soluble modified
   polysaccharides.

3. The presence of tin permits the ready identification of
   the extent of tin-moiety inclusion through elemental
   analysis for tin content.

4. There are commercially available a wide number of R_2SnX_2 compounds including R = methyl, propyl, butyl, hexyl, cyclohexyl, octyl, benzyl, phenyl, ethyl, and X = F,Cl,Br permitting trends to be established with regard to the steric and electronic nature of the employed organostannane. The use of dibutyltin dichloride was emphasized since it has been successfully condensed with diols and with polyvinyl alcohol and it is the least expensive organostannane halide.

EXPERIMENTAL

Chemicals

The following chemicals were used without further purification: dibutyltin dichloride (Fisher Scientific Co., Fairlawn, N.J.), diphenyltin dichloride (Metallomer Laboratories, Maynard, Mass.), dibutyltin dibromide, dilauryltin dichloride, tri-n-propyltin chloride, diethyltin dichloride, dimethyltin dichloride, dioctyltin dichloride, tricyclohexyltin bromide, triphenyltin dichloride and tribenzyltin chloride (all from Ventron-Alfa Inorganics, Danvers, Mass.), tri-n-butyltin chloride (Aldrich Chemical Co., Milwaukee, Wis.) and dextran (molecular weight - 200,000 - 300,000; United States Biochemical Corp., Cleveland, Ohio).

Reaction Procedures

Interfacial Condensation Procedures. Dextran modification reactions were carried out using both the aqueous and nonaqueous interfacial condensation techniques. In a typical aqueous procedure, dextran and base were dissolved in H_2O and stannane was dissolved in carbon tetrachloride. The latter solution was added to the open Kimex emulsifying jar. The metal lid was screwed on, the blender stirring begun, and the aqueous solution was added to the stirred organic phase through a powder funnel. Stirring was stopped after 30 seconds, and the reaction mixture, typically a white gel, was filtered using a Buchner filter with suction. The filtered product was washed with water and carbon tetrachloride to remove unreacted monomers. The gelatinous product was left to dry at room temperature on the filter paper. It was removed from the filter paper by scraping gently with a spatula. The nonaqueous technique was conducted in an analogous manner except that two immiscible organic solvents were employed.

Solution Condensation Procedures. The solution condensations were carried out using a single organic solvent employing either rapid stirring in an emulsifying jar or slow stirring in a beaker. The typical procedure involved dissolving dextran and base in a solvent and then pouring this solution into a stirred solution containing stannane. In some cases salts were added to effect the solubilization of dextran in the solvent.

Instrumentation

A one pint Kimex Emulsifying Jar fitted onto a Waring Blender (700 Model 31BL46) was utilized in conducting the polymerizations. The stirring rate was standardized using a Strobatac type 1537-A (General Radio Co., Concord, Mass.) with the blender connected to a Powerstat type 116 (Superior Electric Co., Bristol, Conn.) and the "no load" speed measured as a function of powerstat reading. The apparatus is fully described in reference 11.

Elemental Analysis

Elemental analysis for tin was conducted as described in reference 12 except that one drop of water and concentrated nitric acid (10 drops) are added subsequent to the sodium fusion and this mixture is heated until brown fumes are evolved. Also, the stannic sulfide is heated in air converting it to stannic oxide which is weighed and this weight utilized in determining the percentage tin. This process is not suitable for extremely volatile organostannanes such as dibutyltin dichloride, but is suitable for tin polyesters, polyethers and polyamines. Tin polyesters were utilized as test compounds. Thus, the polyester derived from disodium adipate and dibutyltin dichloride was found to contain 31 percent tin and the calculated value is 31 percent.

Solubility

Solubility tests were conducted by placing about 1 mg of sample in 3 ml of liquid at room temperature. The mixtures were periodically shaken and observed for one to three days.

Summary

The structural proof is given in a separate paper and is based on results of analogous studies, elemental analysis, control reactions, infrared spectroscopy, solubility and coupled thermogravimetric analysis-mass spectroscopy, TG-MS (13).

Briefly, then, physical characterization of the products resulting from combining dextran and cellulose with organotin halides under various reaction conditions is consistent with a tin-modified product composed of units such as those depicted in 3 through 6. Results from the elemental analysis shows tin present consistent with the presence of the organotin moiety. Results from studying control reactions are consistent with the product

being derived from both the organotin halide and polysaccharide.
The presence of the organotin moiety, polysaccharide and the tin-
ether linkage is indicated by infrared spectroscopy. TG-MS
results are also consistent with the presence of both reactants.

Values obtained from tin analysis vary from sample to sample
depending on the nature of the organotin halide and reaction conditions, even so the values are often indicative of a fairly high
proportion of cellulose hydroxyl's being condensed with the
organotin halides. Furthermore, products derived from dihalostannanes typically show no Sn-OH groups and only small to
moderate amounts of Sn-Cl groups indicating that complete reaction
of both tin-halide sites is common. Thus, the calculation of
theoretical yields (under the heading of Yield (%) and maximum tin
percentages listed in Tables 1-4 are based on structures $\underline{3}$ for
products derived from monohalostannanes and $\underline{5}$ for products derived
from dihalostannanes. Structures $\underline{3}$ and $\underline{5}$ represent chains where
all potential reactive sites are reacted. Because of the wide
variety of potential structural assumptions (i.e., monosubstitu-
tion, disubstituted, chloride end-groups, etc.) the actual yield
weight is noted allowing ready calculations based on alternative
structures.

RESULTS AND DISCUSSION

Operating Parameters

As previously noted, the major purpose of this project is the
preliminary evaluation of a number of potential condensation
systems for the purpose of determining which systems can be used in
the modification of polysaccharide derived from other sources.

There are three general condensation techniques: melt, solution and interfacial. The melt method will be omitted from this
study for at least three reasons. First, true solution of polysaccharide is not effected in the "neat" state through heating
because degradation occurs before melting. The goal of this project is the homogeneous modification of a polysaccharide which
requires the polysaccharide to be in solution so modification can
occur throughout the polysaccharide and not just at the surface of
exposed hydroxyl moieties. Thus, "neat" melt condensation systems
are unsuitable for this particular study. Second, polysaccharides
undergo degradation at temperatures of $250^{\circ}C$ and above while
glucose units experience ring opening at about $200^{\circ}C$. Moreover,
many organotin halides begin to degrade at about $120^{\circ}C$. Thus,
reaction temperatures of above approximately $100^{\circ}C$ should be
avoided to minimize undesirable, thermally induced side reactions.
Third, from an energy conservation standpoint, systems utilizing

Table 1. Yield and Tin-Content as a Function of Organotin Halide

Organotin halide	Organotin halide (mmoles)	Dextran (mmoles)	Yield (%)	Yield (g)	Sn (%)	Sn-% at 100 substitution
$\emptyset_2SnCl_2$	6	4	77	1.7	12	30
Me_2SnCl_2	6	4	65	1.0	33	47
$(C_8H_{17})_2SnCl_2$	6	4	92	2.6	23	26
$(C_9H_{17})_2SnCl_2$ a.	6	4	67	1.9	18	26
$(C_{12}H_{25})_2SnCl_2$	6	4	76	2.6	15	21
$(C_4H_9)_2SnCl_2$	6	4	81	1.7	23	35
$\emptyset_3SnCl$	12	4	95	4.7	14	29
$\emptyset_3SnCl$	8	4	100	3.7	--	29
$(C_4H_9)_3SnCl$	12	4	2	0.05	--	35
$(C_6H_5CH_2)_3SnCl$	12	4	13	0.7	14	16
$(C_6H_5XH_2)_3SnCl$	6	2	11	0.3	14	26
$(C_6H_{11})_3SnBr$	6	2	11	0.3	16	29

Reaction Conditions: Dextran and sodium hydroxide (held at a 3:1 NaOH:Dextran ratio) in 40 ml of aqueous solution is added to stirred (18,000 rpm) chloroform (40 ml) containing organotin halide at $25^{\circ}C$ for a stirring time of 30 seconds

a. In CCl_4 instead of chloroform

less energy are favored for industrial acceptance if other factors are the same.

The two remaining reaction technique categories are the solution and interfacial techniques, collectively referred to as the low temperature condensation procedures. Both techniques encompass a wide variety of reaction systems. The following are brief descriptions of a number of these various reaction systems as pertaining to the present study. The systems were developed utilizing both knowledge derived from previous studies by Carraher and co-workers and from literature references related to reactions and solution of polysaccharides and organotin halides.

Throughout the preliminary evaluation of these reactions systems, the use of dihalo reactants was emphasized as the metal-containing reactant. Since reaction with the polysaccharide will lead to crosslinked products which are insoluble in the reaction systems, the products can be easily recovered for further evaluation.

Interfacial Condensation Systems

Interfacial polymerization systems generally employ an aqueous phase containing the Lewis base, and a water immiscible organic phase containing the Lewis acid. Reaction occurs at or near the interface of the two immiscible phases. The interfacial method provides a way to partially control the rate of reaction by varying the stirring rate thereby controlling the interfacial contact area. When the stirring is stopped, the interfacial contact area is greatly reduced thus effectively stopping the reaction (10,14,16).

For the present study, the now classical aqueous interfacial system utilizes the dextran dissolved in water along with base, typically sodium hydroxide, with the acid chloride, organotin halide, dissolved in a suitable organic solvent (which is largely immiscible with water) such as chloroform or hexane.

Modification was carried out utilizing the classical interfacial system over a wide variety of reactant conditions with yields being from low to high, dependent on the stannane utilized (Table 1).

A modified aqueous interfacial system, which was utilized by Carraher, Schroeder, and McNeely (9,17) to effect the modification of cellulose derived from cotton, a nonaqueous soluble source of cellulose was studied. Bis-ethylenediamine copper (II) hydroxide solutions containing dextran were added to stirred solutions of an organic solvent containing the organometallic halide. The copper-diamine system was capable of giving poor to good (5% to 95%) yields of modified dextran (9,17).

It is of interest to compare results derived from utilizing the classical aqueous interfacial (CAIS) with those derived from utilizing the bisethylenediamine copper (BEDC) system reported in references 9 and 17. The yield of modified product is higher for the CAIS systems but the general overall yield trend as the stannane is varied is similar, indicating that the overall reaction pathways are similar for the two systems (yield trends: CAIS - triphenyl >dioctyl >dibutyl > diphenyl > dimethyl; BEDC-triphenyl > dioctyl > dimethyl > dibutyl > diphenyl). These trends have been established over only a narrow set of reaction conditions and should be more fully developed before acceptance as overall yield trends.

An inverse interfacial system utilized by Carraher and co-workers in the reaction of aqueous soluble Group IV B metallocenes, Cp_2MCl_2, to form polyethers would contain the dextran in the organic phase and the metallocene in the aqueous phase (4,6). This system is not suitable for use with the stannanes due to the insolubility and lack of hydrolytic stability of stannanes in aqueous solution.

Two nonclassical interfacial systems have been developed by Carraher and co-workers (3,10). One utilizes a typical aqueous phase containing a Lewis base (here the dextran) and an added base brought together with a liquid Lewis acid, here a liquid stannane. Thus liquid stannanes such as tributyltin chloride and dibutyltin dibromide can be condensed with aqueous solutions containing the cellulose. Modification was successfully carried out for the majority of liquid stannanes (Table 2).

This system is severely limited in that it is applicable to only reactions employing liquid Lewis acids. Thus a second non-aqueous interfacial system was developed (2,10). Here the Lewis base (dextran) is dissolved in a suitable organic solvent, typically a dipolar aprotic liquid such as DMSO and DMF, and the Lewis acid is contained in a long chain hydrocarbon liquid or carbon tetrachloride (these form immiscible liquid pairs; reference 10 contains a listing of other suitable liquid pairs). This non-aqueous interfacial system was able to give good modification utilizing all attempted organostannanes. Results as a function of added base for dibutyltin dichloride appear in Table 3.

Solution Condensation Systems

A homogeneous reaction medium is necessary for the solution method. Either a single solvent or a mixture of solvents that are

Table 2. Yield and Tin Content for Products Synthesized Employing Layered Liquid Organostannane Interfacial Systems

Reactants	Yield (%)	Yield (g)	Sn (%)
Bu_3SnCl[a.]	2	0.1	-
Pr_3SnCl[a.]	0	0	-
Bu_2SnBr_2[b.]	62	1.3	10
Bu_2SnBr_2[c.]	43	0.9	10
Bu_2SnBr_2[d.]	71	1.5	9

Reaction Conditions: a. Dextran (4 mmoles) and sodium hydroxide (12 mmoles) in 40 ml of water are added to liquid stannanes (6 moles) at 25°C for a stirring time of 30 seconds and a stirring rate of 18,000 rpm. b. Ibid. a., except the stannane is added to the stirred aqueous phase. c. Ibid. b., except employing triethylamine as the added base.

Table 3. Yield and Tin-Constant for Products Synthesized Using "Modification Two" Interfacial Systems

Added Base	Solvents	Yield (%)	Yield (g)	Sn (%)
NaOH[a]	Decane/DMSO	48	1.0	16
Triethylene diamine	Decane/DMSO	43	0.9	21
Triethylamine	Decane/DMSO	0	0	-

Reaction Conditions: Dextran (4 mmoles) and added base (12 mmoles) in 40 ml of decane are added to stirred (18,000 rpm) DMSO (40 ml) solutions containing dibutyltin dichloride (6 mmoles) at 25°C for a stirring time of 30 seconds.

[a]Added as a solid.

immiscible with each other can be used (under the conditions employed). Solution systems employing simple single organic liquids as benzene, toluene, diethylether and hexane are excluded due to the lack of dextran solubility in these solvents.

Dextran is soluble in a number of dipolar aprotic, pure liquids and a wider variety of salt-dipolar aprotic liquid combinations. Suitable (solvent systems which dissolve both reactants) solvent systems were developed with this in mind.

A wide variety of solution condensation systems were devised and tested including one and two liquid combinations and salt-containing systems. Only the system using the stannane dissolved in DMSO and dextran with sodium hydroxide in water yielded the desired modified cellulosic product (Table 1). Other results indicate that the presence of a strong base is conducive to effecting the modification of the cellulose. The DMSO systems had the NaOH present only as a largely undissolved solid. Further studies should consider the use of DMSO, DMF, etc. soluble strong bases such as triethylenediamine.

Table 4. Yield and Tin Content for Products Derived Using Solution Condensation Systems

Brief Description of System	Yield (%)	Yield (g)	Sn (%)
DMSO/DMSO(Et_3N,Me_4NCl)	0	0	–
DMSO(Bu_2SnCl_2)/H_2O(Dextran,NaOH)	71	1.5	17
DMSO/DMSO(Me_4NCl)	0	0	–
DMF(LiCl)/DMF(LiCl)	0	0	–
DMF(LiCl)/DMF(LiCl,Et_3N)	0	0	–
DMSO/DMSO(NaOH,Me_4NCl)	0	0	–

Reaction conditions: Dextran (4 mmoles) and added base and/or salt (12 mmoles of each) in 40 ml of liquid are added to stirred (18,000 rpm) solutions containing dibutyltin dichloride (6 mmoles) at $25^\circ C$ for a stirring time of 30 seconds.

Summary

Modification of dextran through condensation with organostannane halides was studied as a function of a variety of low temperature condensation systems. Modification was general for the classical interfacial and bisethylenediamine copper-interfacial systems, less general for two nonaqueous interfacial systems and largely unsuccessful for solution systems (Table 5). The presence of a strong base enhances the chances for successful modification. The diversity of systems which lead to the modification of dextran should permit the modification of polysaccharide derived from many other sources. Finally, this study was cursory and mainly employed only dibutyltin dichloride as the representative stannane, thus further study should be done before given systems are discarded, particularly emphasizing the use of a soluble, strong base.

Finally, as a cursory observation, inclusion of the organostannane appears to occur within a 40 to 60 mole-% range (relative to available hydroxyl groups) regardless of product yield. This range may represent a sterically-limiting proportion of stannane that can be accomodated by the dextran. Further, this indicates that once modification begins for a select dextran chain, that it continues, giving only chains of high stannane content. The latter may be due to several factors including the presence of a dextran within the "reaction zone" having available suitable quantities of organostannane halides for further reaction, initial reaction with the dextran changing (lessening) the hydrophilic nature of dextran making it more "approachable" to the hydrophobic organostannane, etc.

Table 5. Summary of Systems Evaluated

System	Description	Evaluation
Classical Interfacial	Organic/H_2O	Good
Bisethylenediamine Copper Interfacial	Organic/H_2O,BEDC	Good
Layered Liquid Nonaqueous Interfacial	Liquid Stannane/H_2O	Satisfactory
Nonaqueous Interfacial	like Decane/DMSO	Satisfactory
Solution (One and Two Solvent Systems and Salt Solvent Systems)	like NaI-DMSO	Largely Unsatisfactory

REFERENCES

1. C. Carraher and G. Scherubel, J. Polymer Sci., A-1, 9, 983 (1971).
2. C. Carraher and G. Scherubel, Makromolekulare Chemie, 152, 61 (1972) and 160, 259 (1972).

3. C. Carraher, Inorganic Macromolecules Reviews, 1, 271 (1972).
4. C. Carraher and S. Bajah, Br. Polym. J., 7, 155 (1975).
5. C. Carraher and G. Burrish, J. Macromol. Sci. - Chem., A10(8), 1457 (1976).
6. C. Carraher and L. Jambaya, Angew, Makromol. Chemie, 52, 111 (1976).
7. C. Carraher and J. Piersma, Angew. Makromol. Chemie, 28, 153 (1973).
8. C. Carraher and J. Piersma, J. Macromol. Sci. - Chem., A7(4), 913 (1973).
9. C. Carraher, J. Schroeder, C. McNeely, D. Giron and J. Workman, Organic Coatings and Plastics Chemistry, 40, 560 (1978).
10. C. Carraher, Interfacial Synthesis, Vol. II: Polymer Applications and Technology, edited by F. Millich and C. Carraher, Chapter 21, Dekker, N.Y., 1977.
11. C. Carraher, J. Chem. Ed., 46, 314 (1969).
12. C. Carraher, "Chemistry in Everyday Life," Dakota, Vermillion, S.D., 1972.
13. C. Carraher, H.M. Molloy, T. Tiernan, M.L. Taylor and J. Schroeder, J.Macrom. Sci. - Chem., A16(1), 195 (1981).
14. P.W. Morgan, Condensation Polymers: By Interfacial and Solution Methods, Wiley, N.Y., 1965.
15. C. Carraher and F. Millich, Editors, "Interfacial Synthesis," Vol. I, Marcel Dekker, N.Y., 1977.
16. C. Carraher and J. Preston, Editors, "Interfacial Synthesis," Vol. III, Marcel Dekker, N.Y., 1981.
17. C. Carraher, J. Schroeder, C. McNeely, J. Workman and D. Giron, Modification of Polymers, edited by C. Carraher and M. Tsuda, Chapter 25, Am. Chem. Soc., Washington, D.C. 1980.

Acknowledgements

     The authors are pleased to acknowledge partial support from the American Chemical Society - Petroleum Research Foundation Grants 13084-B3-C and 9126-B3-C.

STARCH-g-POLY(METHYL ACRYLATE)--EFFECTS OF GRAFT LEVEL AND MOLECULAR WEIGHT ON TENSILE STRENGTH

Charles L. Swanson, George F. Fanta, Robert G. Fecht, and Robert C. Burr

Northern Regional Research Center, Agricultural Research Service, U.S. Department of Agriculture,* Peoria, Illinois 61604

INTRODUCTION

Starch, a hydrophilic polymeric material produced by plants as a food reserve, is produced in amounts in excess of that required for nutritionally balanced diets. In the United States about 4 billion pounds are used annually in industrial markets where a water-soluble polymer is needed, as in paper sizes and pastes. Interest has been shown in chemical modification of starch to make it hydrophobic to increase its potential as a replacement or extender for certain polymeric materials derived from nonrenewable resources.

Most starch is a mixture of linear and branched polymers of 1,4-α-D-glucopyranosyl (anhydroglucose) units. Amylopectin, the branched polymer, is the major component in most common starches and has a molecular weight on the order of several million.[1] Branching occurs at carbon atom six on about 4% of its anhydroglucose units (AGUs), resulting in about 25 AGUs between branches.[2] The linear starch polymer, amylose, has a molecular weight on the order of several hundred thousand.[3] Its linearity allows it to crystallize and gives it somewhat different properties from amylopectin. For example, amylose crystallizes from dilute aqueous solutions while amylopectin forms stable dispersions.

---

* The mention of firm names or trade products does not imply they are endorsed or recommended by the U.S. Department of Agriculture over other firms or similar products not mentioned.

Starch is normally recovered from plants as granules that
are about 5-40 microns in diameter. Native granular starch is
insoluble in water unless the granules are disrupted, e.g. by
heat or alkali. Starch granules can be disrupted and the
starch made soluble in many organic solvents by esterifying or
etherifying a portion of its hydroxyl groups with suitable
organophilic reagents.

At high degrees of substitution of ester or ether groups
on the starch backbone, these polymers become thermoplastic,
although starch itself does not melt or have a measurable T_g
below its decomposition temperature. The industrial potential
of various starch esters was examined by Mullen and Pacsu, who
concluded that whole starch esters would never produce molded
articles of great strength because of the branched structure
of amylopectin.[4] Unplasticized starch acetate triesters tend
to be brittle, whereas higher esters of starch are softer and
weaker. Linear amylose triacetate has properties similar to
those of cellulose triacetate but has never been produced
commercially, probably because it has no advantage over cellulose
triacetate and also because amylose is not produced commercially
in large quantities. Properties of starch triethers are very
similar to those of the triesters.

Recently, water-insoluble polymers have been produced by
grafting synthetic hydrophobic polymers onto the starch
backbone.[5] Grafting can be accomplished by a variety of
methods; however, most graft polymerizations involve addition
of unsaturated monomers to free radical sites generated on
starch by irradiation or chemical treatment. The reaction of
starch with ceric ion is perhaps the most commonly used method
of chemical initiation and one generally accepted reaction
path is shown in Figure 1.[6,7] Initially, ceric ion complexes
with the vicinal hydroxyls on carbons 2 and 3 of an AGU and is
reduced, whereas the bond between carbons 2 and 3 is broken
with formation of a free radical, an aldehyde, and a hydrogen
ion. The free radical may initiate polymerization of monomers
to form a grafted branch or it may be consumed by reaction
with additional ceric ion. For acrylonitrile (AN) and methyl
acrylate (MA), yields of polymer of 98% are frequent, with
more than 90% of the synthetic polymer being grafted to starch.
Graft copolymers containing more than 75%, by weight, of
grafted synthetic polymer (75% add-on) have been produced by
ceric initiation. Grafted side chains usually have molecular
weights on the order of 1×10^5-1×10^6.

A lower graft molecular weight and more frequent grafting
usually results from anionically initiated graft copolymerization.
Zilkha et al. initiated graft polymerization of ethylene and
propylene oxides with alkali metal oxides of starch.[8-10] Their

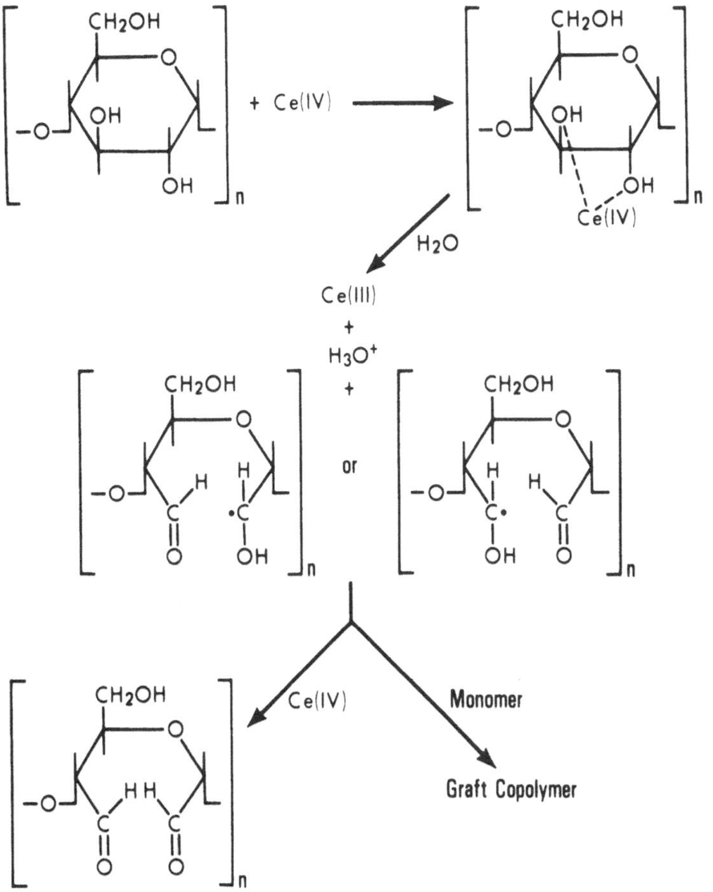

Fig. 1. Initiation of graft copolymerization by Ce(IV).

graft copolymers melt and behave more like simple ester and ether derivatives of starch than the graft copolymers resulting from radical initiation.

Graft polymerization may be carried out with either gelatinized (dispersed) or granular starch; however, most work has been done with granular starch, because one can work at higher solids levels without developing high viscosities. Granular starch graft copolymers can have complex structures, since grafting may be initiated either on the surface or inside of individual granules. Fanta et al. have shown that ceric-initiated grafting of AN takes place not only on the surface but within the starch granules as well.[11] If starch is removed from their graft copolymer particles, e.g. through acid hydrolysis, the residual particles of poly AN (PAN) still resemble the original granules in size and shape. At PAN levels near 50%, by weight, even the centers of the residual

granule-like particles contain PAN. Homopolymer may either be
in the external solution or within the granules.

Typical free radical-initiated graft copolymers with
thermoplastic grafts are recovered as powders that are insoluble
in water and most organic solvents. However, they can be
dispersed and/or dissolved in the few solvents that disperse
both starch and the synthetic polymer, e.g. starch-
g-poly(methyl acrylate) (S-g-PMA) is dispersed by DMSO.

Like starch, these graft copolymers do not melt, although
the grafted thermoplastic portions exhibit T_gs near those of
ungrafted synthetic polymers having the same composition as
the grafted side chains.[12,13] Ray-Chaudhuri made starch-g-
poly(ethyl acrylate) and starch-g-poly(butyl acrylate) materials
thermoplastic by esterifying the starch component in pyridine
solution, to give moldable acetate, propionate, and butyrate
esters.[14,15]

Bagley et al. grafted styrene, methyl methacrylate, MA,
and methyl acrylate-butyl acrylate (1:1) to starch at levels
of 39-47% synthetic polymer, and these polymers were extruded
into continuous, translucent leathery ribbons without further
derivatization.[16,17] Extrusion temperatures ranged from 95-
190°C, higher temperature and torque being required to extrude
graft copolymers containing the higher T_g synthetic polymers.
Continuous plastic ribbons are believed to result from high-
pressure fusion of the individual heat-softened particles
rather than by melting. The extrudates had tensile strengths
in the range 5.2-6.3 Kg/mm^2 for the brittle polystyrene graft
copolymers and 1.7-2.1 Kg/mm^2 for the more leathery S-g-PMA
materials. Perhaps the most outstanding characteristic of
these extrudates is their low die swell, which can be ascribed
to formation by sintering rather than by melt-flow. Die swell
values were typically less than 15% of the sample thickness.

Dennenberg et al. suggest the suitability of films of S-
g-PMA for use by vegetable farmers as biodegradable mulches.[18]
They find that extruded S-g-PMA (40% PMA) swells, whitens, and
loses strength on soaking in water but maintains its integrity.
The extrudates support growth and sporulation of Penicillium
funiculosum, Aspergillus niger, and Trichoderma viride.
Weight losses during growth of the microorganisms for 22 days
were in the range 12-38%, whereas decreases in tensile strength
were 25-75%.

Dennenberg et al. also present the ultimate tensile
strength (UTS), %PMA, and viscosity average molecular weight
($\bar{M}_v$) data given in Figure 2.[18] They ascribe the observed
changes in UTS to changes in $\bar{M}_v$ of the grafted side chains;

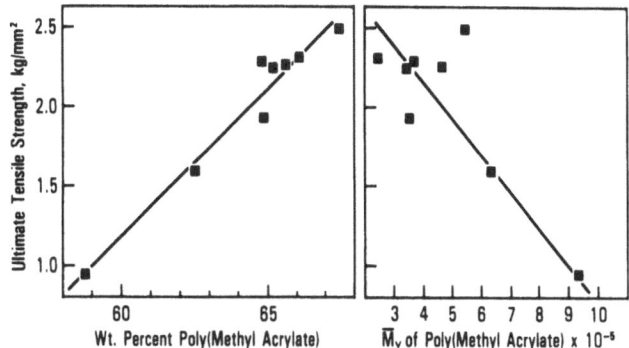

Fig. 2.   Ultimate tensile strength of starch-g-poly(methyl
          acrylate) from Dennenberg et al. as functions of
          poly-(methyl acrylate) content and molecular weight
          of the grafted side chains.[18] Homopolymer is
          retained in the samples.

however, as shown in the figure, the data fit more closely a
line relating UTS to %PMA. The degree of change in UTS with
either variable is not consistent with the usual behavior of
polymeric materials. Most of the increase in tensile strength
of polymers occurs at lower molecular weights, and little
increase in UTS results on increasing molecular weights above
1×10^5. Also, a 10% change in the concentration of filler
(starch in Fig. 2) typically changes the tensile strength of
conventional filled elastomers by only a few percent.[19,20]

    Since neither the UTS-$\bar{M}_v$ nor the UTS-%PMA relationship
could be predicted, these graft copolymers warranted further
study to learn the reasons for their behavior. One obvious
possibility is that starch may not function as a conventional
filler in these grafted systems, but rather as an integral
part of the plastic matrix. Thus, the thermoplastic component
might loosen the granular structure and permit it to deform
and sinter under the heat and pressure of extrusion. Further
study was also needed so products with optimum properties
could be produced. In this paper, we report the effects on
UTS from changes in %PMA (42-77), $\bar{M}_v$ of grafted PMA (5×10^5-
1×10^6), polymerization temperature (27 or 40°C), homopolymer
content (5-15%), post-polymerization oxidation by ceric ion
[1.32 or 3.96 Ce(IV)/100 AGU], and reaction with nitric acid
(2 or 6 times normal levels).

EXPERIMENTAL

Graft copolymerization was conducted as reported by
Dennenberg et al.[18] Granular corn starch (Globe 3005, CPI,
International) was dispersed in cooled, freshly boiled water
and stirred for 1 h while sparging with N_2. Distilled MA was
added all at once. The ceric ammonium nitrate initiator
solution (0.2M in N HNO_3) was added in four equal portions at
15-min intervals. The following reactant combinations were
used [MA/AGU, Ce(IV)/100 AGU]: 1.88, 2.0; 2.35, 2.0; 3.77,
2.0; 4.31, 0.92; 4.31, 1.77; 4.31, 4.9; and 7.70, 4.0. Reaction
temperature was maintained at either 27 or 40°C by use of a
water bath. Reaction was continued for 2 h after the reaction
exotherm subsided. Initially, the total product was washed
with acetone to remove homopolymer, but later products were
divided into two portions, and only one was freed of homopolymer
before air drying. Dennenberg et al. maintained the MA:starch
ratio at 4.31 MA/AGU but varied the initiator:starch ratio
(0.92, 2.4, 3.8, 4.9, 7.4, 9.95, 12.2, or 19.9 Ce(IV)/100 AGU)
to control both $\bar{M}_v$ and %PMA.[18] They began their reactions at
room temperature and the reaction exotherm took the reaction
temperature to about 40°C.[21] They retained a reported 5-10%
homopolymer in their test materials.[18]

Grafted side chains were recovered from the graft copolymer
by perchloric acid oxidation as outlined by Dennenberg and
Abbott.[22] A gel fraction of 0.8-15%, insoluble in acetone,
was removed by centrifugation. Intrinsic viscosities determined
in acetone at 20+0.1°C were used to estimate $\bar{M}_v$, using the
relationship, $[\eta]=7.4\times10^{-5}M^{0.76}$, derived by Staudinger.[23]

Effect of Ce(IV) reagent on the graft copolymer was
evaluated by treatment of portions of a single product
[2.37 MA/AGU, 1.32 Ce(IV)/100 AGU] with large amounts of ceric
reagent [1.32 or 3.96 Ce(IV)/100 AGU] for 4 h at 40°C. For
comparison, portions of graft copolymer were also treated with one
N nitric acid at the same levels used for the ceric treatments.

Air-dried S-g-PMA powder ($\sim$5% H_2O) was extruded from a
C. W. Brabender Plasticorder extruder equipped with a 3/4-in.
diameter 20:1 length-to-diameter barrel. The screw had a 3:1
compression ratio and was operated at 40 RPM. Temperature of
the slit die (1x0.020 in.) and the end of the barrel was 140°C
whereas the inlet zone was 90°C. Dumbbell-shaped tensile
specimens were tempered 4 days at 22.2°C and 50% relative
humidity before testing on an Instron testing machine at 5 or
50 cm/min cross head speed. Average UTS values were determined
from five specimens of each sample. Relative standard deviation
for the reported UTS values is 3.7%.

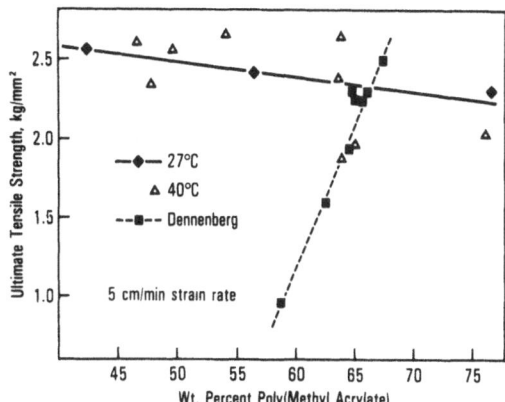

Fig. 3. Effects of graft content and temperature of polymerization
of starch-g-poly(methyl acrylate) on its ultimate tensile
strength at 5 cm/min strain rate. Dennenberg's data is
from samples containing homopolymer and polymerized over
the temperature range 27-40°C[18]; other data is for
homopolymer-free materials polymerized at the specified
temperatures.

RESULTS AND DISCUSSION

Temperature of Polymerization

Graft copolymerization at 27 and 40°C produced S-g-PMA
with about the same UTS values, as is evident for tensile
tests run at both 5 cm/min (Fig. 3) and 50 cm/min (Fig. 4)
crosshead speeds. Data were obtained from homopolymer free
samples, with the exception of points from Dennenberg et al.
in Figure 3.[18] Scatter is much greater for UTS values of
samples copolymerized at 40°C than for those prepared at 27°C.
Although the small scatter of UTS values for the lower
temperature products may be fortuitous, it seems likely that
the slower reactions at 27°C are more controllable and,
therefore, can produce more reproducible products.

Percentage of PMA

In Figure 3, the UTS values of Dennenberg et al. are
compared with the results of our study, which show an obvious

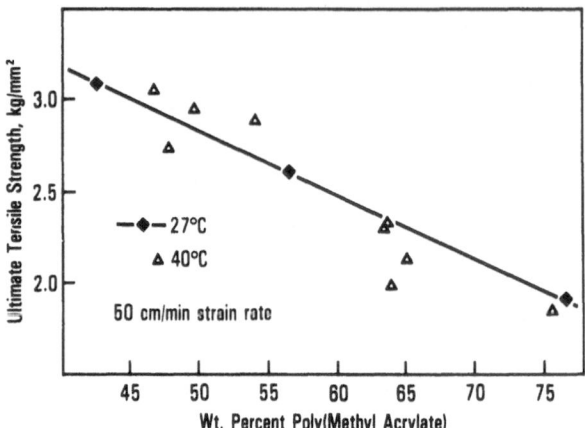

Fig. 4.   Effects of graft content and temperature of
          polymerization of starch-g-poly(methyl acrylate)
          on its ultimate tensile strength at 50 cm/min
          strain rate. The test specimens are homopolymer free.

decrease in tensile strength as %PMA level increases. However,
if Dennenberg's two samples with lowest UTS are discarded as
outliers (abnormal variants), disagreement between the test
results disappears. Attempted replication of the outliers in
reactions at 40°C produced materials having somewhat higher
%PMA levels (63.8 and 65.1 vs. 58.8 and 62.5 for the outliers).
UTS values for the replicates fell within the envelope
established by the majority of the samples.

    A remnant of Dennenberg's 6.35×10^5 $\bar{M}_v$ sample contained
19.1% extractable homopolymer, which may account for a portion
of the observed depression of UTS for this sample. However,
some factor other than homopolymer content or, as will be seen
later, molecular weight of the grafted copolymer must cause
the depressed UTS of the outliers. Differences in distribution
of the grafted side chains within and on the surface of the
granules are likely to influence UTS greatly and may be the
factor controlling UTS of these materials. Rate of stirring
during polymerization might be expected to influence this
distribution, and UTS was low (1.0 Kg/mm^2 at 5 cm/min strain
rate) for a copolymer that was allowed to react unstirred
after polymerization set in. Removal of 28.8% homopolymer
increased its UTS to only 1.4 Kg/mm^2, which is still outside
the envelope of expected UTS values. This unstirred product
had low add-on (52.7 %PMA) and the grafted PMA had unusually
high molecular weight (1.33×10^6).

Strain Rate

Higher UTS values are normally observed at higher test rates, and more brittle materials tend to have greater differences in UTS when testing at different strain rates. Thus, comparison of the UTS values at 5 and 50 cm/min (Figs. 3 and 4) confirm subjective conclusions that S-g-PMA graft copolymers having lower percentages of PMA are more brittle. As the ductile PMA replaces brittle starch in the copolymer, difference in UTS at the two test speeds tends to decrease.

Unexpectedly, there is an apparent reversal of the effect of test rate at 63% PMA so that the two specimens having higher %PMA levels give higher tensile values at 5- than at 50-cm/min. The reversal was also observed for a third S-g-PMA copolymer, from which starch was removed enzymatically to give a 85.1 %PMA level. This polymer exhibited a UTS of 1.26 Kg/mm^2 when tested at 50 cm/min but had a UTS of 1.42 Kg/mm^2 at 5 cm/min. Differences between the UTS values for the two test rates are on the order of 2-4 standard deviation intervals. Smith reports a similar reversal in the effect of strain rate on UTS at low temperatures for Adiprene L-100 cured with 4,4'methylene bis(2-chloroaniline).[24] This is a tough noncrosslinked elastomeric block copolymer. He attributes the reversal of rate effect to stress-induced formation of superstructures.

As %PMA in the graft copolymers increases, elongation of the test specimens before failure increases more rapidly than the tensile strength decreases. For example, as %PMA increased from 42-75%, elongation increased almost fivefold (65-320%) but UTS declined only about 1/3. Consequently, the area under the stress-strain curve increases with increasing levels of PMA in the graft copolymer, and total energy absorbed by the tensile specimens increases from about 41 Kg-m/m^2 at 45% PMA to 214 Kg-m/m^2 at 75% PMA. The greater stretch and flexibility of films having higher PMA levels gives them a rubber-like quality far different from the stiffer and more leathery films having less than 60% PMA.

$\bar{M}_v$ of PMA in Side Chains

Points on plots of UTS vs. $\bar{M}_v$ of the side chains scattered badly because there were many samples with similar $\bar{M}_v$ values but different %PMA levels. Consequently, a paired sample test was used to isolate the effects of changing molecular weight. UTS values were compared for pairs of samples having less than 2% difference in PMA contents, but more than 1×10^5 difference in molecular weight. Samples differing by less than 0.19 Kg/mm^2 (approximately two standard deviation units) were assumed to be the same. The results tabulated in Table 1 show no evidence

Table 1. Effect of %PMA and $\bar{M}_v$ on UTS-Paired Sample Test

| Variable | Selection Rules | Numbers[a] | | | | Test Rate |
		Pairs	-	↑	↓	(cm/min)
$\bar{M}_v$	$\Delta\%PMA<2$, $\Delta\bar{M}_v>1\times10^5$	9	4	2	3	5
"	"	11	6	1	4	50
%PMA	$\Delta\bar{M}_v<1\times10^5$, $\Delta\%PMA>10$	5	1	0	4	5
"	"	11	1	0	10	50

[a]Pairs were classified according to the direction of change in UTS as the test variable increased: -, unchanged (difference less than two standard deviation units); ↑, increased; ↓, decreased.

of effect of $\bar{M}_v$ on the UTS at either 5- or 50-cm/min strain rate. In contrast, the same test on effect of %PMA levels shows definite reduction in UTS as %PMA increases, in agreement with Figures 3 and 4.

Homopolymer Content

Homopolymer contents of crude S-g-PMA from stirred reactions fell in the range 3.4-14.8% of the weight of the corresponding homopolymer-free graft copolymers. Higher homopolymer content tended to correlate with higher grafting levels so that 7-20% of polymerized PMA was homopolymer. Effect on UTS of retained homopolymer (5-15%) in the S-g-PMA graft copolymers was small (Table 2). Typically, extraction of the homopolymer improved UTS less than 5% and would be unnecessary in a production situation.

Low-molecular-weight solvents added at a 10% level as plasticizers for one or both polymers in the graft copolymers had a detrimental effect on the UTS values. A S-g-PMA copolymer containing 65% PMA had an initial UTS of 2.12 Kg/mm^2, but lost strength as follows (agent, % loss): glycerol, 40; N-butanol, 52; DMSO, 54; dioctylphthalate (DOP), 79. DOP is incompatible with S-g-PMA; it lubricates the particles so they slide through the die without developing enough pressure to sinter well. The additives all required extrusions at lower temperatures than the pure graft copolymer to assure good interparticle bonding. UTS values for all these extrudates are far below those of pure

Table 2. Effect of Homopolymer on Tensile Strength of Starch-
g-Poly(Methyl Acrylate)

Grafted PMA (%)	Homopolymer (%)	UTS (Kg/mm^2)		UTS Increase (%)
		Crude	Extracted	
76.6	12.5	1.98	1.91	-3.5
75.6	14.8	1.80	1.85	2.7
64.0	11.0	1.75	1.98	13.1
63.9	10.7	2.23	2.32	4.0
46.7	5.2	2.95	3.04	3.1
42.5	7.0	3.00	3.08	2.6

graft copolymer and suggest that the strength of S-g-PMA extrudates would be seriously impaired by contact with organic solvents.

Attempts to measure UTS after soaking tensile test specimens in acetone, acetic acid, and benzene were discontinued when the specimens disintegrated into white powder after less than 1/2 h in these solvents. Failure of this magnitude suggests that bonding of the graft copolymer particles must be almost entirely through PMA interactions. Immersion in water, which is a dispersant for starch, causes only swelling, whitening, and some loss of strength. Possibly the surface of the S-g-PMA particles is primarily PMA. The starch is not entirely encased, however, since the particles stain on treatment with iodine solution. Staining by iodine is much slower than the staining of unreacted starch granules.

Ceric ion and HNO_3

Post-polymerization treatment of S-g-PMA with nitric acid and Ce(IV) ion was used to study the effects of initiator level on the properties of the graft copolymer. Nitric acid can hydrolyze the acetal linkages between AGUs while Ce(IV) cleaves C-C bonds oxidatively. Attack by the nitric acid increased the amount of extractable homopolymer by up to 55% but had little

effect on UTS (Table 3). The combination of nitric acid
and ceric ion doubled the amount of homopolymer that could be
extracted and reduced UTS by about 9% at the highest treatment
level. Unless partial depolymerization of the starch is
desirable for an end use, both contact time and concentration
of these reagents should be minimized. Yield of graft copolymer
and molecular weight of the grafted side chains seem not to be
greatly influenced by the amount of initiator used, so
minimizing the amount of initiator has no detrimental
consequences.

CONCLUSIONS

    UTS of S-g-PMA is adversely affected by high levels of
grafted PMA, very high homopolymer content, extended contact
with nitric acid-ceric ammonium nitrate initiator solution
and, probably, insufficient agitation during polymerization.
Of these, %PMA had the greatest effect on strength; however,
the energy consumed in breaking the tensile specimens and the
elongation at break were greatest for specimens having the
highest percentages of PMA. $\bar{M}_v$ of the grafted sidechains
had no apparent effect on UTS. There was drastic reduction in
tensile strength of S-g-PMA graft copolymers on incorporation
of 10% levels of polar organic solvents, such as glycerol,
butanol, or DMSO. The disintegration of tensile specimens
on immersion in solvents for PMA suggests that the surface of
these graft copolymers is primarily PMA, which is responsible
for interparticle bonding. Differences in distribution of the
grafted PMA within the starch granules may cause tensile strength
differences that are unexplained by the effects of the variables
we have studied.

Table 3. Effect of Ce(IV) and N HNO$_3$ on UTS[a] at S-g-PMA

		UTS		Homopolymer	
Treatment	Ce(IV)/100 AGU	(Kg/mm^2)	% Less	%	$\bar{M}_v \times 10^{-5}$
Control A	---	2.94	0.0	3.4	2.7
81 ml N HNO$_3$	---	2.92	0.7	4.4	2.9
243 ml N HNO$_3$	---	2.93	0.3	5.3	3.1
Control B	---	2.84	0.0	7.8	4.3
81 ml Ce Soln.	1.32	2.74	3.5	18.2	3.8
243 ml Ce Soln.	3.96	2.42	14.0	15.5	3.9

[a]50 cm/min strain rate.

REFERENCES

1. J. F. Foster, in "Starch: Chemistry and Technology,"
   Vol. 1 (R. L. Whistler and E. F. Pascall, Eds.), Academic
   Press, New York, 1965, p. 375.
2. R. L. Whistler and W. M. Corbett, in "The Carbohydrates,"
   (W. Pigman, Ed.), Academic Press, New York, 1957, p. 675.
3. J. F. Foster, in "Starch: Chemistry and Technology,"
   Vol. 1 (R. L. Whistler and E. F. Pascall, Eds.), Academic
   Press, New York, 1965, p. 356.
4. J. W. Mullen and E. Pacsu, Ind. Eng. Chem., $\underline{35}$, 381 (1943).
5. G. F. Fanta and E. B. Bagley, Starch Graft Copolymers, in
   "Encyclopedia of Polymer Science and Technology," Suppl.
   Vol. 2 (H. F. Mark, Chm. Ed. Bd.), Wiley-Interscience,
   New York, 1977.
6. H. L. Hintz and D. C. Johnson, J. Org. Chem., $\underline{29}$, 556 (1967).
7. C. R. Pottenger, Diss. Abstr., B., $\underline{29}$, 1988 (1968).
8. A. Zilkha, Ben-Ami Feit and A. Bar-Nun, U.S. Pat. 3, 341, 483
   (1967).
9. A. Zilkha, M. Tahan and G. Ezra, U.S. Pat. 3, 414, 530
   (1968).
10. M. Tahan and A. Zilkha, J. Polym. Sci., A-1, $\underline{7}$, 1815 (1969).
11. G. F. Fanta, F. L. Baker, R. C. Burr, W. M. Doane and
    C. R. Russell, Staerke, $\underline{25}$, 157 (1973).
12. V. A. Kargen, P. V. Kozlov, N. A. Plate and I. I. Konoreva,
    Vysokomol. Soedin., $\underline{1}$, 1547 (1959).
13. V. A. Kargin, J. Polym. Sci., C, $\underline{4}$, 1601 (1964).
14. D. K. Ray-Chaudhuri, U.S. Pat. 3, 332, 897 (1967).
15. D. K. Ray-Chaudhuri, Staerke, $\underline{21}$, 47 (1969).
16. E. B. Bagley, G. F. Fanta, R. C. Burr, W. M. Doane and
    C. R. Russell, Polym. Eng. Sci., $\underline{17}$, 311 (1977).
17. E. B. Bagley, G. F. Fanta, W. M. Doane, L. A. Gugliemelli
    and C. R. Russell, U.S. Pat. 4, 026, 849 (1977).
18. R. J. Dennenberg, R. J. Bothast and T. P. Abbott, J. Appl.
    Polym. Sci., $\underline{22}$, 459 (1978).
19. L. Bateman (Ed.), "Chemistry and Physics of Rubber-Like
    Substances," John Wiley & Sons, Inc., New York, 1964.
20. J. A. Brydson, "Plastics Materials," D. VanNostrand Co.,
    Inc., Princeton, New Jersey, 1966.
21. R. J. Dennenberg, personal communication.
22. R. J. Dennenberg and T. P. Abbott, J. Polym. Sci., B, $\underline{14}$,
    694 (1976).
23. H. Staudinger, Z. Prakt. Chem., $\underline{155}$, 216 (1940).
24. T. L. Smith, J. Polym. Sci., A-2, $\underline{12}$, 1825 (1974).

BAGASSE-RUBBER COMPOSITE TECHNOLOGY

A. M. Usmani and I. O. Salyer

University of Dayton Research Institute

Dayton, Ohio   45469

INTRODUCTION

Natural rubber chemically consists essentially of cis-1,4-polyisoprene that is produced as a latex, primarily by the tropical tree Hevea brasiliensis. Natural rubber is of special interest in most developing countries where the climate is suitable for its cultivation, since it constitutes a valuable indigenous, renewable, resource. This natural polymer, and the analogous synthetic materials, can be crosslinked (vulcanized) with varying percentages of sulfur to produce materials ranging in physical properties from true elastomers to high modulus rigid materials (hard rubber). The use of high concentrations of sulfur raises the glass transition temperature (T_g) from well below 0°C to as high as 100°C. In practice, the hard rubbers will contain a minimum of 20-23 parts per hundred of rubber (phr) of sulfur[1] and usually 30-50 phr. Flexible elastomers are made with lower 2-5 phr of sulfur.

The cis-1,4-polyisoprene has very low crystallinity, low T_m (28°C) and T_g (-73°C) values, and is an excellent elastomer over a temperature range including room temperature. More than two billion pounds per year of cis-1,4-polyisoprene are used in the U.S. for tires, coated fabrics, molded articles, adhesives, rubber bands, and other elastomeric applications. Trans-1,4-polyisoprene known in commerce as gutta percha is harder than natural rubber since it can crystallize to a greater degree due to symmetry and has relatively high T_m (74°C) and T_g (-58°C) values.

73

Bagasse is composed of the sheath and pith material from the sugar cane stalk, and is a byproduct of sugar cane processing. It is basically cellulose but contains hemicellulose, lignin, and extractives as well, with a high fiber content and a minor quantity of amorphous pith. Bagasse has a small percentage of inorganic silica that contributes to inherent fire retardant properties. Finally, bagasse also contains some residual sugars, pentosans, hexosans, and other reactive low molecular weight products.

In a separate report[2], we described materials made of bagasse (sugar cane processing residue) that were bonded with phenol/formaldehyde thermosetting resins to form strong low cost composites useful as building materials. This work describes similar rigid composite materials that also utilize bagasse as major component (filler), but are bonded with natural rubber and cured to the rigid hard rubber state (using relatively high concentrations of sulfur). The results of this development were successfully demonstrated in pilot plant manufacture of the material and installation on roofs of houses in the Philippines and Ghana. However, further optimization of the composition and formulation, process scale-up, and plant manufacturing studies are still needed before the material and process will be ready for commercial production.

BAGASSE-RUBBER COMPOSITE DEVELOPMENT

Composite Filler Selection

We examined a large number of agricultural and other low cost residues as fillers in composites with natural rubber as the binder. Typical fillers examined in raw form were bagasse, jute sticks, rice straw, rice hulls, coconut husks, palm fronds, water hyacinths, balsa wood, wood shavings, sawdust, and excelsior. The evaluation was done by determining the effects of the raw fillers, at high volume percent loading on the mechanical properties of the composite, initially and after 1,000 hours accelerated aging in a weatherometer.

The bagasse filled composites retained a higher percentage of their initial strength, after weatherometer accelerated aging tests than any other filler materials tested. Sawdust was second best in overall merit and therefore a good candidate filler in areas where bagasse is not available. Moreover, bagasse has other inherent features which make it especially attractive including:
● Renewable natural resource
● High fiber content
● Readily "powdered" and fibrillated

- High silica content and good fire retardant properties
- Compatibility with binder resins in high volume percentages
- Availability in quantity
- "Gathered" to a central place
- Low cost

Accordingly, we did most of our research and development on natural rubber bonded composites using bagasse filler. As a binder resin, natural rubber has some of the same advantages as bagasse in that it is a naturally renewable resource that is widely produced in large quantities in tropical countries at a "reasonable" cost.

However, rubbers other than "natural" were also briefly tested and could be used wherever natural rubber is not available, although higher in cost. Alternate rubber binder resins tested included:

- Synthetic cis-polyisoprene
- Polybutadiene
- Styrene/butadiene copolymer
- Butadiene/acrylonitrile copolymer
- Ethylene/vinyl acetate copolymer
- EPDM terpolymer

Formulation and Processing Studies in Natural Rubber Bonded Bagasse Composites

One of the inherent advantages of natural and other diene rubbers is that products ranging from elastomers to rigid plastics can be obtained by reacting (curing) the rubber with various percentages of sulfur. For many building materials a strong, high modulus material is needed. Early studies were therefore concentrated on establishing the concentration of sulfur and the curing time/temperature conditions to produce a rigid product (T_g>R.T.).

Since it was known that relatively high concentrations of sulfur (e.g., ∿35 phr) would be required to obtain high modulus products, the effects of sulfur concentration was investigated over the range of 19-45 phr (as shown in Table 1) for composites cured at 325°F and 600 psi for 60 minutes. Highest flexural strength (7000 psi) was obtained at 35 phr of sulfur. The corresponding percentages by weight of all components in the composite is also tabulated in Table 1.

The effects of curing time and some other formulation variables are summarized in Table 2, along with initial flexural strength, and tensile strength and modulus before and after 1,000 hours weatherometer accelerated aging. It is readily apparent

Table 1. Flexural Strength and Modulus of Vulcanized
Rubber Bagasse Composite as a Function of
Sulfur Content

Sample Number	Composition (a) Part Per Hundred of Rubber					Flexural Strength (psi)	Deflection at Break (inch)
	Natural Rubber (b)	Sulfur	Sundex 790 Oil(c)	Bagasse (d)	Iron Oxide #477(e)		
1	100	19.2	22.0	394.7	13.2	1700	0.29
2	100	23.4	0.0	395.3	13.3	2900	0.14
3	100	23.2	21.5	394.2	13.2	2500	0.22
4	100	45.1	43.4	351.4	12.8	5000	0.10
5	100	35.4	30.4	324.9	9.2	7000	0.20

(a) Stearic acid 1 part per hundred (phr), and A-100 accelerator
(Monsanto) 1.5 phr used in all compositions. Processed in
Banbury. Panels pressed at 325°F for 60 minutes under
615 psi pressure.

(b) MR-5 of Malaysian origin (Source: Firestone Rubber and
Chemical Company).

(c) Aromatic petroleum oil from Sun Oil.

(d) Whole and dried.

(e) Source: Cities Service.

(samples 1-4) that curing times of up to 60 minutes are required
to obtain hard rubber products of highest strength and modulus.
The generally good retention of tensile strength after 1,000 hour
accelerated aging is especially pronounced in sample No. 3
(3050 psi initial versus 2940 psi aged) and is at least partially
attributable to the red pigment (iron oxide) that largely
restricts UV damage to the surface. The effects of ozone inhibi-
tor (Flectol H), and asphalt and rubber dust as additive fillers
are shown in samples 6-9. The use of sulfur as the only binder
(Sample No. 5) gave a weak brittle product with poor flow, at the
concentration tested.

The effects of cure time at slightly higher (350°F) temperature are shown in Table 3 where the Shore hardness and color of
comparable bagasse and sawdust filled composites are compared.
At the higher temperature, 45 minutes was sufficient to produce
high modulus products with both filler materials. It is also
noteworthy that flexible products, suitable for applications not
requiring rigidity, can be produced with shorter (more economical)
15-minute curing cycles.

Table 2. Preliminary Hard Rubber Compositions and Test Data to Determine Feasibility of Rubber as a Bonding Resin for Bagasse

Sample Number	Composition, parts by weight									Curing(h) Minutes	(°F)	Remarks
	Natural Rubber (a)	Sulfur	Whole Bagasse	Sundex 790 Oil(b)	Iron Oxide #477(c)	Zinc Oxide	Stearic Acid	A-100(d) Accelerator	Other Additives			
1	100.0	45.0	348	87.0	12.5	---	---	1.5	---	7	350	Partial vulcanization; panel is flexible.
2	100.0	45.0	301	45.0	12.0	---	---	1.5	---	35	290	Panel extremely flexible.
3	100.0	45.0	301	45.0	12.0	1.0	1.0	1.5	---	60	325	Rigid panel.
4	100.0	100.0	800	119.8	24.5	2.0	2.0	3.0	---	45	325	Good surface characteristics, but brittle and breaks easy.
5	0.0	408.0	1634	245.0	50.0	4.0	4.0	6.1	---	30	325	No strength and poor flow.
6	100.0	45.0	1000	120.0	25.0	2.0	2.0	3.0	100(e)	45	325	Good flow and look; slightly flexible.
7	100.0	50.0	726.4	---	22.0	1.0	1.0	1.5	2(f)	45	325	---
8	100.0	50.0	726.4	---	22.0	1.0	1.0	1.5	2(f)	15/60	225/325	---
9	100.0	45.0	301.2	---	11.3	1.0	1.0	1.53	75(g)	45	325	---

(a) Natural rubber, MR-5 (Malaysian origin) - Firestone
(b) Sundex 790 Oil - Sun Oil
(c) Mapico #477 iron oxide - Cities Service
(d) A-100 Accelerator - Monsanto
(e) Asphalt - Local Vendor
(f) Flectol H stabilizer - Monsanto
(g) Rubber dust - American Hard Rubber
(h) Pressure 600 psi

Table 2. (continued)

Sample Number	Flexural Strength (psi)	Tensile Strength			
		Initial		1000 hours weatherometer exposure	
		Strength (psi)	Modulus (10³psi)	Strength (psi)	Modulus (10³psi)
1	---	---	---	---	---
2	---	---	---	---	---
3	4350	3050	---	2940	---
4	2840	1520	---	1370	---
5	---	---	---	---	---
6	1400	1080	210	996	266
7	---	1430	220	1670	310
8	---	1570	235	1840	314
9	3960	1640	---	---	---

Table 3. Effect of Cure Time and Filler Type on Development
of Hardness in a Rubber Formulation

Curing Time (min)	Bagasse Composite[a,b]			Sawdust Composite[a,b]		
	Shore Hardness	Color of Sample[c]	Remarks	Shore Hardness	Color of Sample[c]	Remarks
15	40	3	flexible	52	3	flexible
30	52	5	flexible	63	5	flexible
45	53	7	hard	81	7	hard
60	109	9	hard	110	8	hard
75	106	10	hard	116	9	hard
90	106	10	hard	118	10	hard

(a) Hard Rubber Formulation:

Ingredient	Percent	Parts Hundred Rubber
MR-5 natural rubber	20.0	100.0
Sulfur	9.0	45.0
Sawdust or whole bagasse	58.5	292.5
Sundex Oil 790	9.0	45.0
Mapico 477 iron oxide	2.4	12.0
Zinc oxide	0.2	1.0
Stearic acid	0.2	1.0
A-100 Accelerator	0.3	1.5
Flectol H	0.4	2.0

(b) Curing Conditions: Prepress test samples cold and cure at
350°F in an oven under ambient pressure.

(c) Color Rating 1-10; 1=light red, 5=dirty red, 10=black.

Optimized Composition

Table 4 gives the formulation for the optimized bagasse-
rubber composite material on both a percent by weight and parts
by weight based on 100 parts of natural rubber.

The system is made up of three basic parts: (1) components
of a hard rubber matrix material; (2) constituents needed to introduce fire resistance; and (3) the fillers (both to reduce
costs and increase durability and toughness).

The hard rubber component includes natural rubber, sulfur,
stearic acid, zinc oxide, stabilizer, and accelerator. Sulfur
crosslinks the natural rubber into a rigid state. The sulfur
reaction is promoted by a unique accelerator (Monsanto Company's
A-100 accelerator was used). Stabilizers are used to prevent

Table 4. Bagasse-Rubber Roofing Formula

Ingredient	Weight	
	(Parts)	(Percent)
Natural rubber	100	20.5
Sulfur	46	9.4
Chlorinated paraffin	42	8.6
Stearic acid	1	0.2
Zinc oxide	1	0.2
Dry whole bagasse	265	54.2
Iron oxide pigment	15	3.0
Magnesium oxide	15	3.0
Stabilizer	0.25	0.05
Accelerator	1.5	0.31
Pentachlorophenol	0.1	0.02
Antimony oxide	2.5	0.51

Chlorinated paraffin - Chlorowax 80, Diamond Shamrock Company.

Iron oxide pigment - Mapico No. 477, Cities Service Company.

Stabilizer - Monsanto's Flectol H rubber stabilizer or equivalent.

Accelerator - Monsanto's A-100 rubber accelerator or exact
                equivalent.

Antimony oxide - fire retardant grade.

degradation during thermal curing of the rubber and also prevent
ozone oxidation during long-term outdoor exposure. The stabilizers used were Monsanto Company's Flectol H, zinc oxide, and
stearic acid.

Rubber systems normally include a processing oil which provides a variety of benefits. Among them is reduced cost. In the
fire-retarded formulation, the oil is a chlorinated paraffin
(40% chlorine). The oil aids in making the rubber resilient,
tough, and compatible with the filler.

Process

The process outline for making the dense, bagasse-rubber
material is shown in Figure 1.

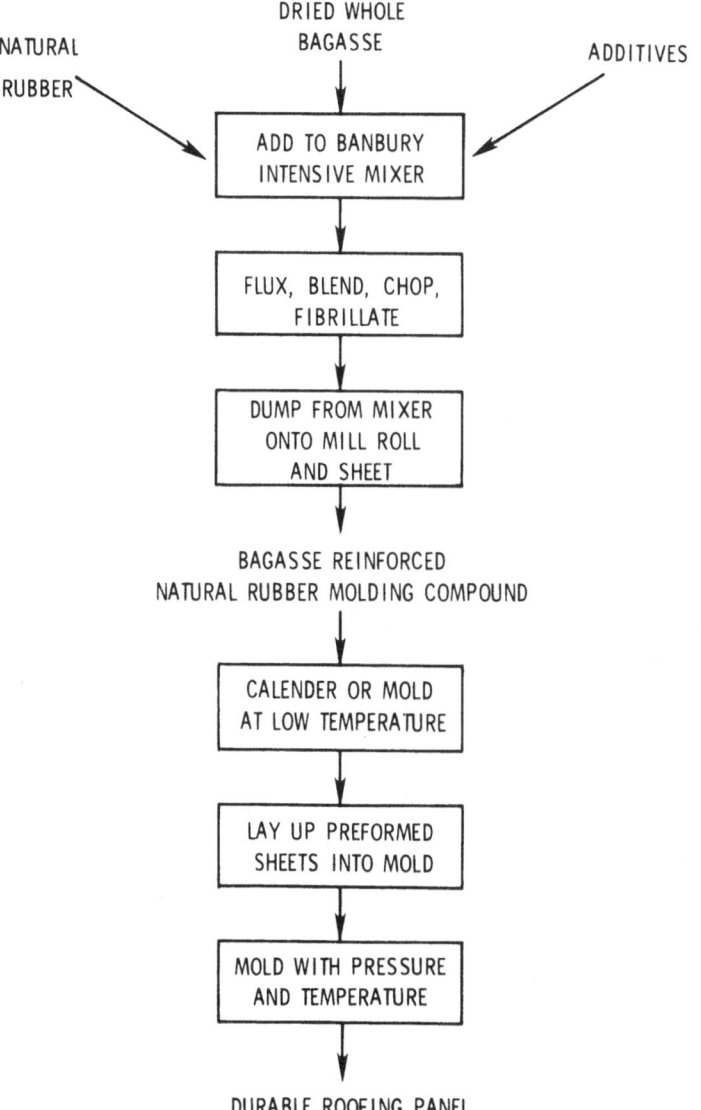

NATURAL RUBBER

DRIED WHOLE BAGASSE

ADDITIVES

ADD TO BANBURY INTENSIVE MIXER

FLUX, BLEND, CHOP, FIBRILLATE

DUMP FROM MIXER ONTO MILL ROLL AND SHEET

BAGASSE REINFORCED NATURAL RUBBER MOLDING COMPOUND

CALENDER OR MOLD AT LOW TEMPERATURE

LAY UP PREFORMED SHEETS INTO MOLD

MOLD WITH PRESSURE AND TEMPERATURE

DURABLE ROOFING PANEL

Figure 1.   Process Outline for Making Dense, Bagasse-Rubber Composite.

The first step is to oven-dry the whole bagasse to reduce moisture below 0.1%. The second step, processing in an intensive mixer (e.g., Banbury), accomplishes intimate blending of all of the materials into a homogeneous mass, reduction in size of the bagasse, and removal of water. Temperatures as high as 250°-300°F are generated purely by the mechanical action of the intensive mixing, even though cold water is run through jackets in the Banbury mixer. Temperatures as high as 325°F were sometimes achieved when large batches were mixed in manufacturing-scale Banbury mixers. In such cases it was advisable not to add the sulfur and/or the accelerator to the product in the Banbury mixer and preferably to delay their addition to the roll mill, calendering stage.

The hydrocarbon or chlorinated paraffin oil is well distributed throughout the bagasse as a result of the mixing. This assists both in making the bagasse compatible with the natural rubber, and in achieving the desired intimate mixing of these two materials.

The product from the Banbury-type intensive mixing is a homogeneous, red, rubbery compound with very little visible fiber. The compound is sheeted on a mill roll into manageable sheets for handling.

The third step, calendering, converts the roofing compound into a uniform continuous sheet that has almost the final desired thickness and width for a roofing panel. Calendering consists of placing the bulk compounded material onto mill rolls (set of from 3 to 4) which grab the material and convert it into a sheet configuration. Such sheets, when formed at moderate speeds, have few residual stresses. The product of this process step is a roll of thin (0.13 inch), flexible rubber that can be trimmed with scissors or knives to fit into any desired mold.

The final (fourth) process step is compression molding at elevated temperatures and pressures. In this step, the flexible calendered sheet is laid into a mold that can be flat, corrugated, or any other desired shape. Pressures in the range of 300 psi to 600 psi are applied, and the materials are heated to at least 325°F. This temperature must be maintained for at least 30 minutes to cure the rubber into a relatively hard state.

Composition Latitude

Lower cost general purpose bagasse-rubber composite formulations are identical to those for the fire-retarded formulation, except that a non-chlorinated rubber processing oil (e.g., Sundex 790) is substituted for the chlorinated paraffin oil, no antimony oxide is included, and magnesium oxide is not necessary. Additionally, a slightly lower amount of sulfur is required to cure the rubber.

Additional latitude exists in formulating the bagasse-rubber composite in the following respects. First, the natural rubber can be of lower cost brown grades usually available in commerce, and it need not be of the highest quality as normally required for automobile tires. Second, a variety of chlorinated paraffins can be used in the system, depending on the level of fire retardancy desired. Those having chlorine contents greater than 40%, however, caused severe difficulties in processing at 350°F. Finally, the stabilizer, Flectol H, could probably be reduced in quantity as time and experience in the processing and long-term outdoor durability are better quantified.

Product and Properties

The bagasse-rubber product is a rigid-tough, dark red panel that can be molded into flat, corrugated, or other roofing element shapes. It can be nailed, drilled, cut, sawed, etc. for installation of a roof.

The outdoor durability of this roofing product is very good, as determined both by accelerated weatherometer exposure and outdoor exposure of panels for three years in Jamaica, Ghana, and the Philippines. The dimensional stability of flat bagasse rubber roofing panels was good in actual use, although slight warping was found after three years of service. This undesirable defect can be eliminated by use of other roof panel geometries (e.g., corrugation).

The mechanical characteristics of the bagasse-rubber roofing products are shown in Table 5. Results are for the initial material and following 1,000 hours of weatherometer exposure with cyclic UV, water spray, temperature, and humidity. The tensile strength retention of 97% is considered excellent.

Applications

Roofing applications require very high performance from materials with respect to both strength and outdoor durability. But materials that are adequate for roofing may also be useful to construct other parts of a building. Bagasse-rubber composite appears to be functionally useful in this regard, but the economics should be studied.

Examples of other potential applications in home building are wall panels, ceilings, flooring, counter tops, fences, siding,

Table 5. Physical Properties of Bagasse-Rubber Composite[a,b]

	Initial	Following Exposure[c]
Flexural Properties		
Strength, psi	4,500	---
Modulus, psi	400,000	---
Tensile Properties		
Strength, psi	3,000	2,900
Durability		
Weatherometer - 5,000 hours		
Outdoor - Jamaica satisfactory		
Philippines performance		
Ghana for >3 years		
Density, lb/ft^3	60	

(a) Process: Banbury melt blending, sheeted on mill roll, compression molded at ∿500 psi and 325°F.

(b) Flexural - ASTM 790; tensile - 638; rate - 0.05 in./min.

(c) Tensile strength following 1,000 hours of weatherometer exposure with cyclic UV (carbon arc), water spray, temperature, and humidity.

acoustical panels, sinks, furniture, doors, and shutters. For some uses (e.g., floor tile, counter tops) high modulus and rigidity may be unnecessary, or even undesirable. In these cases, flexible formulations containing lower sulfur concentrations can be used. The flexible products can also be cured on shorter, more economic (e.g., 15 minute) curing cycles.

Mechanism of Vulcanization

Crosslinking of natural rubber is an essential requirement for producing desirable properties. Vulcanization of diene polymers can be achieved by using sulfur, peroxides, other reagents, or radiation. Our work involved crosslinking of natural rubber by heating with sulfur and therefore we will restrict our discussion to vulcanization by sulfur.

Sulfur vulcanization although studied since its discovery in 1839 by Goodyear is not yet fully understood. Free radical mechanisms were originally proposed[3-5] but recent evidence including studies on model compounds indicate that vulcanization more likely proceeds by an ionic mechanism[4-6]. Radical initiators and inhibitors do not influence sulfur vulcanization. Sulfur vulcanization is, however, accelerated by organic acids and bases as well as by solvents of high dielectric constant. Below we depict the ionic mechanism of sulfur vulcanization involving initial formation of sulfonium ion.

$$S_8 \xrightarrow{\text{heat}} \overset{\delta+}{S_m} \cdots \cdots \overset{\delta-}{S_n} \quad \text{or} \quad S_m^+ + S_n^-$$

Polarized Sulfur   Sulfur Ion Pair

$$\downarrow \sim\!\sim CH_2C(CH_3)=CHCH_2 \sim\!\sim$$

$$\sim\!\sim CH_2C(CH_3)CHCH_2 \sim\!\sim + S_n^-$$
$$\overset{+}{S_m}$$

The sulfonium ion abstracts hydride from cis-1-4-polyisoprene as follows.

$$\sim\!\sim CH_2C(CH_3)CHCH_2 \sim\!\sim$$
$$\overset{+}{S_m}$$

$$\downarrow \sim\!\sim CH_2C(CH_3)=CHCH_2 \sim\!\sim$$

$$\sim\!\sim CH_2CH(CH_3)CHCH_2 \sim\!\sim$$
$$\underset{S_m}{\mid}$$

$$+$$

$$\sim\!\sim \overset{+}{C}HC(CH_3)=CHCH_2 \sim\!\sim$$

The polymeric allylic ion undergoes crosslinking as follows.

$$\sim\!\!\!\overset{+}{C}HC(CH_3)=CHCH_2\sim$$

$$\downarrow S_8$$

$$\sim\!\!\!CHC(CH_3)=CHCH_2\sim$$
$$|$$
$$\overset{+}{S_m}$$

$$\sim\!\!\!CH_2CH=C(CH_3)CH_2\sim$$

$$\sim\!\!\!CHC(CH_3)=CHCH_2\sim$$
$$|$$
$$S_m$$
$$|$$
$$\sim\!\!\!CH_2\overset{+}{CHC}(CH_3)CH_2\sim$$

$$\sim\!\!\!CH_2C(CH_3)=CHCH_2\sim$$

$$\sim\!\!\!CHC(CH_3)=CHCH_2\sim$$
$$|$$
$$S_m$$
$$|$$
$$\sim\!\!\!CH_2CHCH(CH_3)CH_2\sim$$

$$+$$

$$\sim\!\!\!\overset{+}{C}HC(CH_3)=CHCH_2\sim$$

Vulcanization of natural rubber by heating with sulfur alone is very inefficient with about 40-50 sulfur atoms attached into the polymer per crosslink. Much sulfur can be wasted by formation of long polysulfide crosslinks, vicinal crosslinks, and intramolecular cyclic sulfide structures.

Thus sulfur vulcanization must be carried out in presence of additives that increase the rate and efficiency of the vulcanization process. The most widely used accelerators are derivatives of 2-mercaptobenzothiozole. Other frequently used accelerators are zinc dimethyldithiocarbamate and tetramethylthiuram disulfide.

CONCLUSIONS

We have developed a composite material and associated processes that utilize renewable agricultural raw materials; namely, natural rubber and bagasse. Other agricultural residues (e.g., sawdust) can also be used but compositions should be examined and optimized.

ACKNOWLEDGEMENT

The work reported herein was a part of a large 5 year research program sponsored by the U.S. State Department (USAID). Besides the authors, important team members whose major contributions should be recognized include Mr. George L. Ball (Monsanto), Mr. Dennis W. Werkmeister (Monsanto), Prof. J. P. R. Falconer (Washington University), and Prof. B. S. Bryant (University of Washington). Special thanks to Mrs. Jeanne Drake of the University of Dayton for organizing this chapter.

REFERENCES

1. S. Kimura and N. Namikawa, J. Soc. Chem. Ind. Japan, 32, 196 (1929).
2. I. O. Salyer and A. M. Usmani, Ind. & Eng. Chem. Products Res. & Dev., in press.
3. G. Alliger and I. J. Sjothun (Eds.), "Vulcanization of Elastomers," Van Nostrand Reinhold, New York, 1964 .
4. L. Bateman, "The Chemistry and Physics of Rubber-Like Substances," Wiley-Interscience, New York, 1963 .
5. E. H. Farmer, J. Chem. Soc., 1519 (1947).
6. L. Bateman, C. G. Moore, and M. Porter, J. Chem. Soc., 2866 (1958).

CHEMISTRY AND TECHNOLOGY OF IN-SITU GENERATED RESIN BONDED

BAGASSE COMPOSITE MATERIALS

A. M. Usmani and I. O. Salyer

University of Dayton Research Institute

Dayton, Ohio   45469

INTRODUCTION

The sugar cane residue bagasse is an underutilized, renewable agricultural material that consists of two distinct cellular
constituents. The first is a thick-walled, relatively long,
fibrous fraction derived from the rind and fibrovascular bundles
dispersed throughout the interior of the stalk. The second is a
pith fraction derived from the thin-walled cells of the ground
tissue.

Bagasse is generally gray-yellow, to pale green in color.
It is bulky and quite nonuniform in particle size. As it is produced at the sugar mills, bagasse is used as fuel to power the
plant. However, wherever there is large scale production of
sugar cane, there will be excess bagasse which must be disposed
by other methods such as burning, dumping, land burial, etc.
Thus, an economic use of the excess bagasse could both increase
the profits of the sugar mill and solve an environmental problem.
"Bagasse board" and bagasse pulp have not been too successful and
other applications need to be found.

It has been suggested that much of the U.S. polymer needs
and chemical needs could be supplied from wood and agricultural
residues[1,2].

We feel that the value of agricultural residues such as
bagasse can be upgraded by bonding with resins to produce composites suitable for building materials. In earlier work we
bonded bagasse with phenolic, natural rubber, and styrenic resins

to produce building material composites[3,4]. This work describes
a process for generating phenolic, melamine, urea, and urethane
resins from bagasse and the concurrent formation of bagasse reinforced composites with these resins as they are produced. The
in-situ generation of resins from a few other agricultural residues (pine needles, water hyacinth, rice hulls, corn cobs) was
also briefly studied.

CHEMISTRY AND COMPOSITION OF BAGASSE

    Bagasse contains ∿40% cellulose, ∿30% hemicellulose, and
∿15% lignin. Bagasse fiber is similar to cotton fiber in that it
has a spiral structure (length = 1-4 mm, width = 0.01 to 0.04 mm).
Cellulose is a linear polyglucose of DP about 7,000-10,000 and
highly hydrogen bonded. Therefore, it is a highly crystalline,
stable polymer and extremely resistant to solvent and chemical
attack. Hemicelluloses are also polymers but may contain several
different sugar units and their degree of polymerization is low
(∿200). They are not crystalline and are not resistant to attack.
Lignins are complex, crosslinked polymers of phenylpropanoid units
joined by benzylic and phenolic ether linkages and C-C linkages.
DP of lignin is only several hundred. Both lignin and hemicellulose sugar can be used to make resin intermediates. Hemicellulose hydrolysis can be done either with 0.1-1% H_2SO_4 at
95-120°C for short reaction times, or with 2-4% acid at 90°C for
about 4 hours to yield primarily pentoses[5].

    Cellulose which can be thought of as polyglucose forms a
crystalline structure which protects the internal bonds from
hydrolysis[6]. Also, cellulose in bagasse is protected by lignin.
Lignin forms a seal around the cellulose for further protection
against hydrolysis. The structure of cellulose in bagasse is
shown in Figure 1. Thermal treatment of bagasse above 200°C for
long periods will result in loss of fiber structure. Minor
degradation or crosslinking can even be observed at temperatures
of 150°C. Reactions of cellulose with various monomers are
shown in Figure 2 to show full potentials of bagasse.

    An analysis of whole bagasse from several different geo-
graphic sources (Table 1) shows the material to be of comparable
composition regardless of origin[7].

IN-SITU RESIN GENERATING CONCEPT AND PROCESS

    Phenol/formaldehyde, melamine/formaldehyde, and urea/formal-
dehyde prepolymer resins (B-stage) are widely used as thermosetting adhesives for plywood and other wood products. In prior

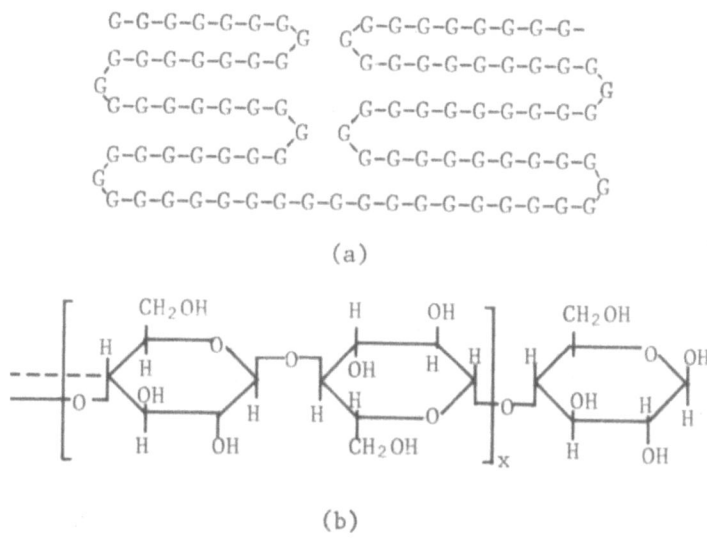

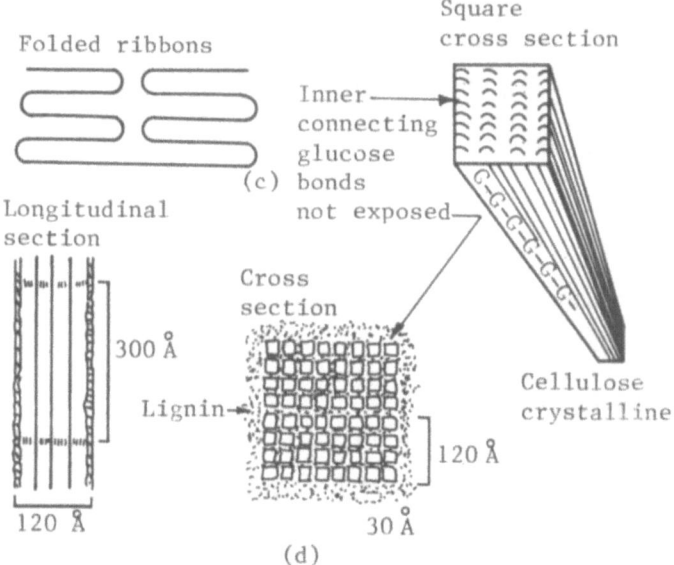

Figure 1. Structure of Cellulose in Bagasse: (a) Cellulose as
Polyglucose, (b) Chemical Structure of Cellulose,
(c) Cellulose Crystallite, and (d) Cellulose Crystallites in Microfibril.

(a) Vinyl Monomers

- Butadiene
- Divinyl benzene

$$2R_{cell}OH + CH_2=CH-CH=CH_2 \rightarrow$$
$$R_{cell}OCH_2CH_2 \cdot CH_2CH_2OR_{cell}$$

(b) Isocyanate Monomers

- Toluene diisocyanate

$$2R_{cell}OH + OCN-\phi-NCO \rightarrow R_{cell} \cdot OCONH\phi \cdot$$
$$NHCO \cdot OR_{cell}'$$

(c) Urea/Formaldehyde

$$H_2NCONH_2 + 2CH_2O \rightarrow HOCH_2NH \cdot CONHCH_2OH$$
$$2 R_{cell}OH + HOCH_2NHCONHCH_2OH$$
$$\downarrow$$
$$R_{cell} \cdot OCH_2NHCONHCH_2OR_{cell}$$

(d) Dicarboxylic acid, e.g., - oxalic acid

$$2R_{cell}OH + HOOC(CH_2)_nCOOH \rightarrow$$
$$R_{cell} \cdot OOC(CH_2)_n \cdot COOR_{cell} + H_2O$$

(e) Formaldehyde

$$CH_2O + H_2O \rightarrow HOCH_2OH$$
$$2R_{cell}OH + HOCH_2OH \rightarrow R_{cell}OCH_2OR_{cell} + 2H_2O$$

(f) Epoxy compounds

- diepoxides
- epichlorohydrin
- ethylene oxide
- propylene oxide

$$2R_{cell}OH + CH_2 - CH-(CH_2)_n-CH-CH_2$$
$$\underset{O}{\diagdown}\qquad\qquad\underset{O}{\diagdown}$$
$$\downarrow$$
$$R_{cell}OCH_2CHOH(CH_2)_n \cdot CHOH \cdot CH_2OR_{cell}$$

Figure 2.   Reactions of Cellulose.

research[3,4] we have shown that these thermosetting resins, as well as natural rubber and styrene copolymers, could be used in low concentrations as binder adhesives for whole and fibrillated bagasse to make strong, fire and water resistant composites suitable for building materials. Obviously, it would be economically advantageous if one of the two monomer components of the resin (the aldehyde) could be obtained at low cost from the bagasse itself; and the other component (phenol, melamine, urea) added as monomer, and the resin generated in-situ. This composite concept is based on utilizing the generated chemical without recovering it from the bulk of bagasse. Minor amounts of additional chemicals can be added to produce a composite.

While the hemicellulose is converted into furfural, lignin will also undergo changes under acidic conditions. Lignin remains unaltered when HCl is used. However, H_2SO_4 and H_3PO_4 acids are known to produce condensation leading to advancement in molecular weight. Condensed lignin can also serve as a binder for bagasse fibers.

The wet process, in-situ generated resin bonded bagasse composite process is shown schematically in Figure 3.

Bagasse, acid and water are refluxed for 16 hours to produce furfural. Phenol, melamine or urea are added and then allowed to condense under basic conditions (CaO) at reflux temperature for 4 hours. The water is then removed by forced evaporation. The residue, consisting of bagasse fibers and the generated resin is

Table 1. Analysis[a] of Whole Bagasse From Several Sources[7]

	Louisiana, Fresh (Houma, 1941)	Florida, Fresh, Dry Screened (Clewiston, 1948)	Hawaii, Variety 8560 (1952)	Hawaii, Variety 1933 (Ewa Plantation)	Puerto Rico (Aguirre, 1951, 1952)	Mexico[b] (San Cristobal, 1953)	Philippine (Negros Island, 1952)
Moisture (%)	4 9	7 3	13 2	7 7	7.5	6.2	10.2
Extractives in							
Alcohol-benzene (%)	6.0	10 8	3 2	3 6	5 4	2 3	3 0
Hot water (%)	8 8	11 2	5 7	4 0	8 0	7 6	2 8
1% NaOH (%)	35 9	39 9	33 9	31 3	27 3	40 1	31 3
Ash (%)	2.4	2.2	5 4	2.6	3 9	4 9	2.3
Lignin (%)	18 9	18 1	21 3	19 3	18 1	22 4	22 3
Pentosans (%)	30 0	27 9	27 7	31 3	29 6	29 9	31 8
Cross and Bevan cellulose (%)							
Ash free	53 3	52 0	50 2	55 0	50 9	46 0	56 8
Pentosans in	27.9	26.9	27 4	33.4	31 1	25 5	30 6
Alpha as run							
Ash free	67.3	68 0	69 6	o2 6	63.7	61.3	70 2
Pentosans in	7 0	4.7	8 8	8.2	7 0	5.5	12 5
Alpha-basis original, ash and pentosan free	33.4	33.7	31 8	31 6	30 1	26 6	34 9

[a](All values, except moisture, on oven-dry basis)
[b]Pith

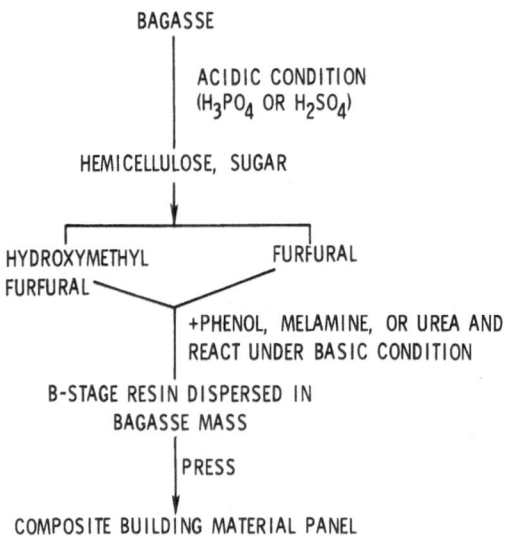

Figure 3.   In-Situ Resin Generation and Composite
            Preparation Method.

then pressed and cured at 390°F for 20 minutes under 625 psi. Water resistance of the "in-situ resin" bonded composite was then tested by immersing specimen in water for up to six weeks.

Most of the research was devoted to study of in-situ generation of phenolic and melamine resins in bagasse.

In-Situ Generated Phenolic Resin

Sulfuric acid hydrolyzed samples were neutralized with CaO and then approximately 10% of monomeric phenol was added. The mass was "cooked", dried and pressed producing a strong board resistant to six weeks water immersion (Table 2). This result was most encouraging since phenol costs less than half the selling price of commercial phenol/formaldehyde B-stage resins. Combinations of nitric and sulfuric acids were at least no better than sulfuric acid alone (Table 2).

Phosphoric acid, at the same concentration as the sulfuric acid, did not produce water resistant panels with about 10% phenol monomer. However, incorporating about 3 weight percent of a water proofing ethylene/vinyl chloride copolymer (Monflex 4530 latex) gave water resistant pressed panels (Table 3).

In-Situ Generated Melamine Resin

Sulfuric acid hydrolyzed bagasse was reacted with melamine at concentrations of approximately 7 to 10 weight percent melamine. Encouraging results were obtained with in-situ generated melamine in that pressed panels successfully passed four week water immersion tests (Table 4).

As in the phenolic samples, the addition of about 3 weight percent of ethylene/vinyl chloride copolymer (Monflex 4530) into the formulation appeared to significantly improve water resistance of samples containing the lower (7 weight percent) concentrations of melamine. The ethylene/vinyl chloride copolymer alone (no melamine) did not produce water resistant panels (Table 4).

Phosphoric acid hydrolyzed bagasse reacted with melamine also yielded samples with inferior water resistance, unless 3 weight percent ethylene/vinyl chloride copolymer emulsion was added into the formulation (Table 5).

In-Situ Generated Urea Resin

The in-situ generation of urea bonding resin was evaluated using phosphoric acid only; with and without incorporation of ethylene/vinyl chloride copolymer into the formulation. At the same or higher monomer concentration, the in-situ generated urea

Table 2. In-Situ Generated Phenolic Resin Bonded Bagasse Panels

| Sample Number | Composition(a,b) | | | | | | General Remarks |
	Bagasse (g)	H2O (l)	H2SO4 (ml)	HNO3 (ml)	Phenol (g)	CaO (g)	Other	
1	200	1.5	10	--	--	12	--	Control sample without phenol or melamine; after 6 weeks immersion; this sample is highly swollen.
2	200	1.5	10	--	20	12	--	After 6 weeks immersion in water, this pressed sample is rated very good.
3	200	1.5	5	10	20	8	Z6020, 1 ml	Tensile strength of the dry panel is 1300 psi.
4	200	2.0	5	20	20	12	Z6020, 1 ml plus Camphor, 1 g	First panel pressing at 225°F was flexible. Repressed same panel at 390°F. Good panel but surface cracks apparent.

(a) Legend

H2SO4   98% sulfuric acid
Z6020 Silane   Coupler (Dow Corning)
HNO3   Concentrated nitric acid
H2O   Water
CaO   Calcium oxide

(b) Process - Bagasse, acid and water refluxed overnight to produce furfural; add phenol, along with CaO and condense at reflux temperature for 4 hours. Remove water from the mass in an oven at 50°C. Press and cure for 20 minutes at 390°F under 625 psi.

Table 3. Evaluation of Phosphoric Acid

Sample Number	Composition(a,b)						General Remarks
	Bagasse (g)	H₂O (l)	H₃PO₄ (ml)	Phenol (g)	CaO (g)	Monflex 4530 (g)	
1	200	1.5	10	15	12	---	Pressed and cured panel swells in water and therefore no good.
2	200	1.5	28	10	55	10	Good water resistance.
3	200	2.0	28	10	45	10	Panel O.K. in water for 6 weeks.
4	200	1.5	20	15	15	15	Sample looks O.K.
5	200	1.5	6	15	10	15	Compares with 4 in overall panel evaluation.

(a) Legend

H_3PO_4 85% phosphoric acid
H_2O Water
CaO   Calcium oxide
Monflex 4530   Ethylene/vinyl chloride emulsion (Monsanto)

(b) Process – Bagasse, acid and water refluxed overnight to produce furfural; add phenol along with CaO and condense at reflux temperature for 4 hours. Remove water from the mass in an oven at 50°C. Press and cure for 20 minutes at 390°F under 625 psi.

Table 4.   In-Situ Generated Melamine Resin Bonded Bagasse Panels

Sample Number	Composition (a,b)						General Remarks
	Bagasse (g)	H_2O (l)	H_2SO_4 (ml)	Melamine (g)	CaO (g)	Other Ingredients (g)	
1	200	1.5	5	--	12	Hexamethylene-tetramine, 1 g	Swollen and almost no strength after 6 weeks immersion in water.
2	200	1.5	10	15	12	Hexamethylene-tetramine, 1 g	Better than 1 indicating melamine enters into resinification.
3	200	1.5	10	20	10	---	Tensile strength of dry panel--1000 psi.
4	200	1.5	10	15	10	---	Panel O.K. in water for 4 wks.
5	200	1.5	5	--	12	Hexamethylene-tetramine, 1 g Monflex 4530, 1 g	Sample looks O.K. after 6 wks. immersion in water
6	200	1.5	10	--	12	Monflex 4530, 1 g	Monflex 4530 does not seem to contribute to properties.

Table 5.   In-Situ Generated Melamine Resin: Evaluation of Phosphoric Acid

Sample Number	Composition (a,b)						General Remarks
	Bagasse (g)	H_2O (l)	H_3PO_4 (ml)	Melamine (g)	CaO (g)	Other Ingredients	
1	200	1.5	25	20	25	---	Partial resinification takes place
2	200	1.5	10	20	12	---	Panel swells in water and therefore no good
3	200	1.5	25	20	20	Z6020, 1 ml	No inorganic filler and therefore the silane does not couple; surface characteristics not improved
4	200	1.5	15	15	20	Monflex 4530, 15 ml	Sample looks O.K.

(a) Legend

H_2SO_4	98% sulfuric acid
H_3PO_4	85% phosphoric acid
H_2O	Water
CaO	Calcium oxide
Monflex 4530	Ethylene/vinyl chloride emulsion (Monsanto)
Z6020 Silane	Coupler (Dow Corning)

(b) Process - Bagasse, acid and water refluxed overnight to produce furfural; add melamine along with CaO and condense at reflux temperature for 4 hours.
Remove water from the mass in an oven at 50°C.
Press and cure for 20 minutes at 390°F under 625 psi.

resins were more water sensitive than the phenolic or melamine counterparts (Table 6).

In-Situ Generated Urethane

In a single experiment, sulfuric acid hydrolyzed, base neutralized, dried whole bagasse was tumble blended with about 10 weight percent of 4,4'-diphenyl methane diisocyanate (MDI) and compression molded. An apparently well bonded panel was obtained, but not tested for strength or water resistance. This approach should be further investigated since the isocyanates can be reacted with hydrolyzed bagasse chemicals to form urethanes at near room temperature. The urethanes also show less mold sticking than the phenolic, melamine or urea resins.

In-Situ Resin Generation from "Other" Agricultural Residue

Sulfuric acid, and combinations of sulfuric acid and nitric acids, were used in attempts to hydrolyze pine needles, corn cobs, water hyacinth and rice hulls for subsequent in-situ phenolic resin generation. Under the conditions investigated, results were not very encouraging since good, water resistant panels were not obtained (Table 7).

It may be that a part of the bagasse is easier to hydrolyze for in-situ resin generation and the other part is a better reinforcing filler (for composites) than the other residue materials tested.

Table 6. In-Situ Generated Urea Resin Bonded Bagasse Panels

Sample Number	Composition (a,b)						General Remarks
	Bagasse (g)	H2O (l)	H3PO4 (ml)	Urea (g)	CaO (g)	Monflex 4530 (ml)	
1	200	1.5	25	25	2	--	Liquor and bagasse look separated after cooking
2	150	1.2	9	15	10	--	Slightly better than 1
3	150	1.2	18	15	16	10	Resinification occurs with urea but panel is more water sensitive when compared with the phenol counterpart

(a) Legend

H_3PO_4	85% phosphoric acid
H_2O	Water
CaO	Calcium oxide
Monflex 4530	Ethylene/vinyl chloride emulsion (Monsanto)

(b) Process - Bagasse, acid and water refluxed overnight to produce furfural; add urea along with CaO and condense at reflux temperature for 4 hours. Remove water from the mass in an oven at 50°C. Press and cure for 20 minutes at 390°F under 625 psi.

Table 7. Evaluation of Other Agricultural Residues

Sample Number	Residue (g)	Composition (a,b)						General Remarks
		H2O (l)	H2SO4 (ml)	HNO3 (ml)	Phenol (g)	CaO (g)	Other (g)	
1	Pine Needles (100) Bagasse (100)	2.5	10	--	--	15	--	Panel is not completely rigid.
2	Pine Needles (100) Bagasse (100)	2.5	10	--	--	10	Hexamethylene-tetramine (10)	Poor resin flow in mold. Considerable swelling after 2 days soaking in water.
3	Pine Needles (140)	2.5	5	20	15	8	--	Somewhat water-sensitive and "noncompacted".
4	Corn Cobs (250)	1.5	10	--	20	13	Cunilate 2440 (10)	Panel does not look strong; corn cobs are good source of furfural but are not good filler.
5	Water Hyacinth (200)	1.5	10	--	10	--	--	Panel is rigid, but not resistant to water immersion.
6	Rice Hull (127)	1.0	4	15	20	10	Z6020 (1)	Panel is extremely water-sensitive and disintegrated rapidly.

(a) Legend

H_2SO_4	98% sulfuric acid
HNO_3	Concentrated nitric acid
H_2O	Water
CaO	Calcium oxide
Z6020 Silane	Coupler (Dow Corning)
Cunilate 2440	Fungistat (Venton Corporation, Beverly, Massachusetts)

(b) Process – Residue, acid and water refluxed overnight to produce furfural; add phenol along with CaO and condense at reflux temperature for 4 hours. Remove water from the mass in an oven at 50°C. Press and cure for 20 minutes at 390°F under 625 psi.

CONCLUSIONS

This work is preliminary and points to the potential of preparing useful building composites from bagasse by use of minor amounts of phenol, melamine, urea and isocyanate chemicals and in-situ resin generation. Formulation optimization was not undertaken and we recommend such a study. The advantage of our concept is that bagasse supplies both the filler as well as one of the two resin intermediates. Another advantage is that we do not have to recover the generated resin chemicals. The advantage of lower cost raw materials (monomers) makes this process economically attractive and justifies further research and development leading to possible commercialization.

ACKNOWLEDGEMENT

The work reported herein was part of a large five year research program sponsored by the U.S. State Department (USAID). Besides the authors, important team members whose major contributions should be recognized include Mr. George L. Ball III (Monsanto), Mr. Dennis W. Werkmeister (Monsanto), Prof. J. P. R. Falconer (Washington University), and Dr. Ben S. Bryant (University of Washington). Valuable help of Mr. D. Robert Askins and Mrs. Jeanne Drake both of the University of Dayton was greatly appreciated.

REFERENCES

1. I. S. Goldstein, Science, 189, 847 (1975).
2. M. R. Ladisch, M. C. Flickinger, and G. T. Tsao, Energy, 4, 263 (1979).
3. A. M. Usmani, G. L. Ball, I. O. Salyer, and D. W. Werkmeister, J. Elastomers and Plastics, 12, 18 (1980).
4. A. M. Usmani, I. O. Salyer, G. L. Ball, and J. L. Schwendeman, J. Elastomers and Plastics, 13, 46 (1981).
5. H. F. Wenzel, "The Chemical Technology of Wood," Academic Press, New York, 1970 .
6. M. Chang, J. Polymer Sci., Part C, 36, 343 (1971).
7. E. C. Lathrop and S. I. Aronovsky, Tappi, 37(23), 24A (1954).

STRUCTURAL IDENTIFICATION OF THE CONDENSATION PRODUCT OF SUCROSE WITH ORGANOSTANNANE DIHALIDES

Charles E. Carraher, Jr.*, Philip D. Mykytiuk*,
Howard S. Blaxall*, Raymond Linville*,
Thomas O. Tiernan*,** and Shelley Coldiron**

*Department of Chemistry
**The Brehm Laboratory
Wright State University
Dayton, Ohio 45435

INTRODUCTION

The term carbohydrate literally means a compound that is a hydrate of carbon. The names sugars and saccharides are synonyms and apply to simple carbohydrates, typically consisting of units containing two to seven carbon atoms (usually five or six). Sugars are typically more or less sweet, water soluble, odorless, optically active, colorless substances.

Almost all plants manufacture sugar through photosynthesis using energy from sunlight, water from the soil or sea, and carbon dioxide from the air or dissolved in the water. The amount of sugar produced yearly through photosynthesis is an item of disagreement but is of the order of 60 to 400 billion tons, depending on the quantity attributed to formation in the oceans[1], with 3 billion tons produced under cultivation.

Sucrose is a disaccharide in which an alpha-D-glucosyl residue in the six-membered ring form is combined with a beta-D-fructoside residue in the five-membered ring form. Sucrose is unstable to acid, converting to invert sugar (equal parts of dextrose and levulose) which is utilized to decrease the tendency of sucrose to crystallize in candy, soft drinks, etc. Sucrose is fairly stable in alkali solutions permitting its esterification allowing sucrose to be converted to plastics, plasticizers, surfactants and surface-coating materials.

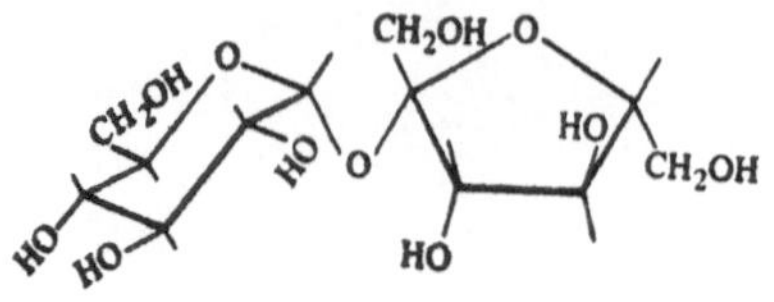

<u>1</u>

Sucrose is known to English-speaking peoples as sugar. It is found in the juice of every land plant examined for it. It is commercially derived from sugar beets and cane.

Organostannane dihalides have been condensed, through the hydroxyl groups, with a variety of alcohols including ethylene glycol, hydroquinone, polyvinyl alcohol and more recently cellulose derived from cotton and dextran[2-5]. An extension is the use of other naturally occurring hydroxyl-containing Lewis bases. Here we report the synthesis of crosslinked networks derived from organostannane dihalides and sucrose; the product microscopically being a heterogeneous structure containing units such as <u>5</u> - <u>7</u>.

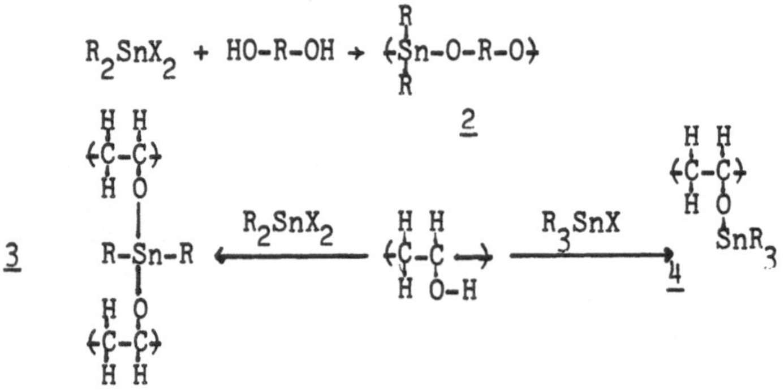

EXPERIMENTAL

Sucrose (Fisher Scientific, Fairlawn, N.J.), dibutyltin dichloride (Fisher Scientific, Fairlawn, N.J.), dilaurytin dichloride (Metallomer Labs., Fitchburg, Mass.), diphenyltin dichloride (Ventron, Beverly, Mass.) diethyltin dichloride (Ventron), dioctyltin dichloride (Ventron), terephthaloyl chloride (Aldrich, Milwaukee), sebacyl chloride (Aldrich), and dimethyltin dichloride (Ventron) were used as received.

Interfacial reactions were carried out using an assembly described in ref. 6. In a typical procedure sucrose, along with added sodium hydroxide, is dissolved in water. The organostannane dihalide is dissolved in a suitable organic liquid. The organic phase is added to a one pint Kimex Emulsifying jar. The metal lid was screwed onto the jar and stirring begun; then the aqueous phase is added. Stirring was stopped and the reaction mixture filtered using a Buchner filter with suction. The filtered solid, a white material, is washed with water and the organic liquid to remove unreacted monomer. The product is left to dry at room temperature.

Infrared spectra were obtained using a Perkin-Elmer 457 Grating Infrared Spectrophotometer employing KBr pellets. Mass spectrophotometry and elemental analyses were carried out as described elsewhere [5,7] utilizing a double-focusing DuPont 21-49 Mass Spectrometer equipped with a modified Hewlett-Packard, HP-2116C computer, having a disc-oriented data system specially developed for the DuPont 21-491 MS. The MS system is under the control of the HP 2116C computer operating with a dual 2.5M byke disk drive, a Hewlett-Packard Cathode Ray Tube terminal, a Tektronix storage scope, driven by a dual 12 bit digital-to-analog converter, and a Versatec printer/plotter. Data is acquired using a 14 bit analog-to-digital converter. The system can operate and process data at rates to 8 KHz.

The MS was calibrated from m/e 12 to m/e 1000 by using a special data processing system which employs a Hall probe and which gives typical scan to scan reproducibility of 200 to 500 ppm.

RESULTS AND DISCUSSION

The preceding paper dealt with the modification of dextran assuming that the reactivity of hydroxyl groups contained on dextran was similar to those contained on other polysaccharides thus permitting information learned from dextran to be directly applicable to other polysaccharides. The same logic is applicable for the current study, that is the reactivity of hydroxyl groups contained on sucrose is similar to those contained on other sugars, thus information learned about the chemical reactivity of sucrose is directly applicable to other water soluble sugars. A point of departure from the preceeding paper is the sole use of dihaloorganostannane since difunctionality is required to achieve polymeric products.

This paper concentrates on the identification of the structural unit of the condensation product between sucrose and organostannane. As in the case of dextran modification, the product contains a mixture of units including those depicted as 5 to 7.

Historical

As previously noted, the reaction of alcohol (both aromatic and aliphatic) containing reactants with mono- and dihaloorganostannanes to form tin ethers is well known (for instance 2-5) and the reaction with sucrose with organostannane dihalides should proceed in an analogous manner.

Control Reactions

The term "control reactions" describes conducting reaction sequences in the usual manner except omitting one of the reactants. Thus the reaction sequence was carried out omitting the organostannane dihalide for one set of reaction attempts and sucrose for other reaction attempts. None of these reaction attempts produced precipitate consistent with the precipitate, obtained when both reactants are present, containing components of both the stannane and sucrose.

Elemental Analysis

Elemental analyses were conducted on samples to determine the amount of tin present. Tin content was typically in the range of 20 to 40%. Thus for the product derived from reaction of sucrose (1.00 mmole) with sodium hydroxide (3.00 mmoles) in 30 ml of water

added to rapidly stirred (18,500 rpm, no load stirring rate)
carbon tetrachloride (30 ml) solutions containing dibutyltin
dichloride (1:00 mmole) the tin content was 37.3% tin. The per-
centage-tin calculated for structure 5 is 28%, 6 is 36%, 7 is 20.1%
and for a structure containing the maximum number (eight) organostannane moieties, the %-tin is 41%. Thus the tin analyses are
reasonable and are consistent with the product containing the
organotin moiety.

Infrared Spectroscopy

Analysis of the infrared spectra is consistent with a product
containing units such as 5-7 with the formation of the Sn-O-R ether
linkage with hydroxyl groups contained on the sucrose. Formation
of the ether linkage is indicated by the presence of bands about
660 to 690 cm^{-1} (Figure 1) characteristic of the asymmetric
stretch of tin ethers and a doublet about 550 to 600 cm^{-1} attributed to the symmetric stretch of tin ethers.

Bands characteristic of the organostannane moiety are
present. For instance, for products from the dibulyltin dichloride (Figure 1) bands characteristic of methylene (CH_2) deformation are present at 1470 and 1150 cm^{-1}; bands characteristic of
methyl groups are present at 1420 (asymmetric stretching) and 1380
(symmetric stretching) cm^{-1}; bands characteristic of the C-H
stretch present in n-butyl groups are present at 2910, 2880 and
2860 cm^{-1}; and bands at 668, 598 and 565 cm^{-1} attributed to the
presence of the Sn-O-R moiety. Bands attributed to the symmetric
C-O stretch are present at 1008, 888 and 865 cm^{-1} and bands
attributed to asymmetric C-O stretching are present at 1070 cm^{-1}.
A broad band from 3600 to 3200 cm^{-1} is present and attributed to
the presence of unreacted hydroxyl groups and/or water trapped
during the condensation reaction.

Further, Figure 1 contains a companion infrared spectrum of
the condensation product of dibutyltin dichloride and dextran.
This spectrum appears strikingly similar to that of the product
derived from sucrose. This is reasonable since the major structural difference is the presence of five membered rings in sucrose
compared to only six membered rings present in dextran.

Mass Spectroscopy

Mass spectral data is also consistent with a product containing units derived from the organostannane and sucrose with frag-
mentation patterns assignable to both moieties. For the product
from diphenylstannane dichloride all of the major ions (normalized
intensities of one and greater) are assigned derived from sucrose
and the diphylstannane moiety (Table 1). Table 2 contains a

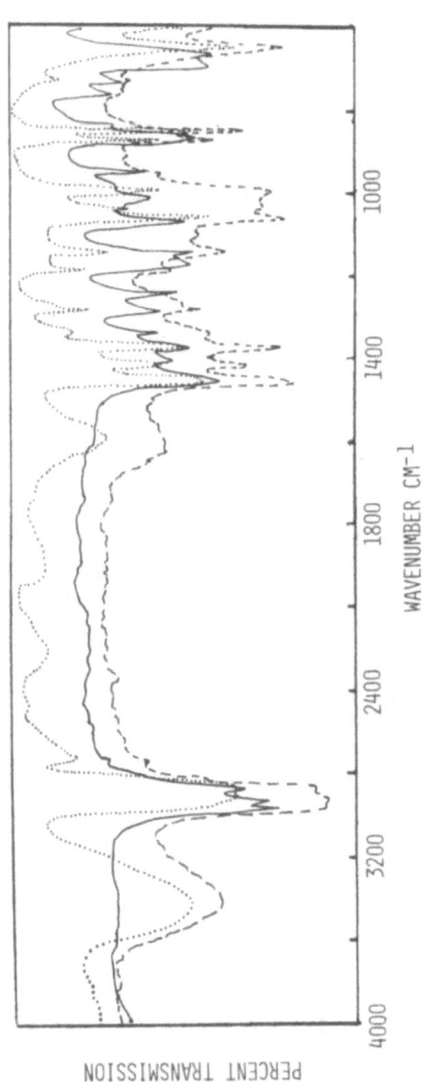

Figure 1.   Infrared spectra of dibutyltin dichloride (———), and the condensation products of dibutyltin dichloride with cellulose derived from dextran (– – –) and sucrose (....).

comparison of the mass spectrum of benzene with that obtained for the degradation product identified as originating from the diphenyltin moiety. The comparison is good and supports the assignment of these mass fragments as being from the diphenyltin moiety. Such comparisons are routinely done where possible.

Table 1.   Major ion fragments for the gasses derived from the thermal degradation of the condensation product of sucrose and diphenyltin dichloride

m/e	Normalized Intensity*	Parent**
17	5.8	S(OH)
18	12.5	S(H_2O)
22	1.8	S
27	4.8	S
29	1.7	S
39	5.3	ϕ,S
41	2.1	S
42	3.4	S
43	6.9	S
50	1.3	ϕ
51	2.6	ϕ
52	2.7	ϕ
67	1.7	S
76	1.0	ϕ
77	5.8	ϕ,S
78	8.3	ϕ
82	2.3	S

* Based on N_2 28=100; all ion fragments with normalized intensities of 1.0 and greater.
**S = sucrose, ϕ = phenyl.

No ion masses are found attributable to tin or a tin-containing moiety consistent with tin remaining in the residue.

Mass spectral results for thermal pyrolysis of the condensation product of sucrose and dibutyltin dichloride are given in Table 3. Again no ion masses are found attributable to tin or a tin-containing moiety consistent with tin remaining in the residue. For both spectra, the presence of only small amounts of m/e=35,37 derived from chloride are present consistent with a high amount of crosslinking.

Table 2. Mass spectrum observed for the degradation product
identified as phenyl from the condensation product
of sucrose and diphenyltin dichloride compared with
generated data of the mass spectrum of benzene

m/e	McLafferty-Stenhagen*	Normalized Intensity	Relative Intensity**
78	1000**	8.3	1000
51	205	2.6	310
52	196	2.7	320
50	179	1.3	160
39	141	5.3	640[a]
77	140	5.8	700[a]
79	65	0.75	90
76	62	1.0	120
38	61	0.13	16
74	48	0.15	18

* E. Stenhagen, S. Abrahamsson and F. W. McLafferty, Eds., "Atlas
of Mass Spectral Data," Wiley-Interscience, N.Y., 1969.
**Normalized intensities relative to the most abundant ion
fragment of benzene, in this case m/e 78.

[a] Also a major ion fragment from sucrose.

Table 3. Major ion fragments for the gasses derived from the
thermal degradation of the condensation product of
sucrose and dibutyltin dichloride

m/e	Normalized Intensity	Parent
17	13.8	S(OH)
18	38.7	S(H_2O)
22	1.1	S
27	14.0	S,Bu
29	40.0	S,Bu
36	1.8	S
38	3.1	Bu
39	24.9	S, Bu
41	100.0	Bu, S
42	17.1	S, Bu
43	91.0	S, Bu
49	1.1	S
51	2.7	Bu
52	1.4	S
55	23.0	S
56	35.0	S

Table 3. (cont.)

m/e	Normalized Intensity	Parent
57	11.5	Bu
58	6.7	Bu
65	1.4	S
67	3.1	S
69	5.7	S
70	7.3	S
77	3.1	S
82	1.5	S
84	1.6	S
85	2.9	S
97	1.3	S
100	1.7	S
114	1.0	S

Solubility

Solubility tests were carried out employing about 1 mg of sample to 3 ml of liquid. A wide variety of liquids were employed including chloroform, benzene, 1,4-dibromobutane, cyclohexane, butenediol, 2-propanol, DMSO, n-heptane and diethyl chlorothiophosphate. The product is insoluble in the afore liquids consistent with the product containing crosslinks. It swells in HMPA and small amounts dissolve in HMPA after several days.

Summary

Crosslinked networks are formed from the interfacial condensation of sucrose with organostannane dichlorides. The presence of the sucrose moiety is indicated by results obtained utilizing infrared spectroscopy, control reactions and mass spectroscopy; the presence of the organostannane moiety is indicated by infrared spectroscopy, elemental analysis, control reactions and mass spectroscopy; the presence of tin-ether bonding is indicated by infrared spectroscopy; and crosslinking is indicated by the general insolubility of the products and low amount of chloride atom found through mass spectroscopy.

REFERENCES

1. N.L. Pennington, "World Book Encyclopedia," Field Enterprises Ed. Corp., Chicago, Vol. 18, p. 766 (1975).
2. C. Carraher, Interfacial Synthesis, Vol. II, Chpt. 21, F. Millich and C. Carraher, ed., Dekker, Inc., N.Y. (1977).
3. C. Carraher, and J. Piersma, Angew. Makromolekulare Chemie, 28, 153 (1973).

4. C. Carraher, J. Schroeder, C. McNeely, D. Giron and J. Workman, Org. Coat. Plastics Chem., <u>40</u>, 560 (1979).

5. C. Carraher and T. Gehrke, Org. Coat. Plastics Chem., in press.

6. C. Carraher, J. Chem. Ed., <u>46</u>, 314 (1969).

7. C. Carraher, H.M. Molloy, T. Tiernan, M.L. Taylor and J. Schroeder, J. Macromol. Sci. - Chem., <u>A16</u>(1), 195 (1981).

8. C. Carraher, D. Giron, D.R. Cerutis, S. Tsuji, T. Gehrke, R.S. Venkatachalam and H.S. Blaxall, Org. Coat. Plastics Chem., <u>44</u>, 1 (1981).

MODIFIED LIGNAN POLYMERS FROM THE RESIN OF PARANÁ PINE TREE KNOTS

Eloisa Biasotto Mano

Instituto de Macromoléculas
Universidade Federal do Rio de Janeiro
Rio de Janeiro, RJ, Brazil

Since 1973, with the petroleum crisis, it has been fully
demonstrated that a search for more reliable, renewable and economical sources of new materials was needed as an answer to the
ever-growing demand of the modern, highly technological world.
This has led man to the origin of the large fossil deposits of
raw materials, especially to the plants.

In this regard, the forest-based industry constituted by far
the most important activity. The pulp and paper as well as the
timber industry leave large amounts of residues, half of which are
unused, representing an economic waste and a source of pollution
that is still a problem to be solved by the chemical industry.
The utilization of the other half is limited to the production of
fiber products, charcoal and ethyl alcohol, by fermentation of the
wood hydrolyzates. Nowadays, a better integration of the forest
products industry and the chemical industry has become increasingly
important.

Besides the main cellulosic components, there is in wood
about 30% weight of lignin and related intermediate structures.

In Brazil, where the tropical and sub-tropical forests constitute the largest green reserve of the world, there is a natural
source of lignin and its precursors: the knots of a pine tree which
is native to Paraná State in the South of the country. This tree
belongs to the class Coniferae, family Araucariaceae, named
Araucaria angustifolia O. Kuntze (or A. brasiliensis Lamb)(Fig.1).
It provides a cheap and abundant raw material for the production
of timber, cellulose and paper. In 1967, the number of Araucaria
pine trees was estimated at 47 million and 50.000 new seedlings

113

were planted.   Thirty or more years are necessary for the trunk of
the tree to reach 0,50 m diameter which is an adequate size for
the timber industry.   The trees may reach as much as 40 meters in
height.

Fig. 1 - The Paraná pine tree may reach as much as 40 m height

Not much has been written on this tree[1-5]; although about
80% of the production of timber in Brazil comes from this species.
The dry distillation of $1m^3$ of pine tree branches gives 120 kg of
charcoal, 20 kg of tar, 17 kg of calcium acetate and about 3,5 kg
of acetone [3,6,7].   Besides timber, wood paste and cellulose,
taken from the trunk, the Paraná pine tree is also a raw material
for a terpene resin, edible fruits and knots.

The terpene resin exudates from the plant, as happens with
other pine trees, as a result of the injuries provoked by insects,
birds or men.   It is a viscous liquid, composed of terpene products, which becomes hard in contact with air.   The analysis of
the Paraná pine tree terpene resin shows moisture, 13,1%; mucila-
ginous material, 39,7%; rosin, 37,2%; turpentine, 5,0%; and miscellaneous, 5,7% [8]

The chemical composition as well as the industrial uses and
the biosynthetic origin of terpene resins are well known [9,13].

The fruit ("pinha") is composed of 40 to 100 seeds ("pinhão"), as seen in Fig. 2, and is used as a nutritive source for men as well as animals. The analysis shows moisture, 53,6%; ether extract, 1,1%; protein, 4,2%; fiber, 0,6%; glucoside, 38,9%; and ash, 1,8%[8].

Fig. 2 - Paraná pine tree fruit ("pinha") showing 2 seeds ("pinhão")

The Paraná pine tree has the peculiar characteristic of pre-senting rather large knots which constitute the supporting base for the branches. These knots, shown in Fig. 3 and 4, may reach as much as 30 cm length and 3 kg weight; they are very hard due to the high content of lignin. These natural by-products have been employed as fuel for house heating.

The successive extraction of the powdered pine knot with benzene, water, ethyl alcohol and the treatment with 72% sulfuric acid lead to the composition presented in Table 1 [14-17]. It can be observed that the major component is lignin (32%) followed by cellulosic materials (21%) and water-soluble carbohydrates (20%). The next largest fraction is composed of lignans and related resins (13%). A small fraction (3%) includes mainly waxy materials.

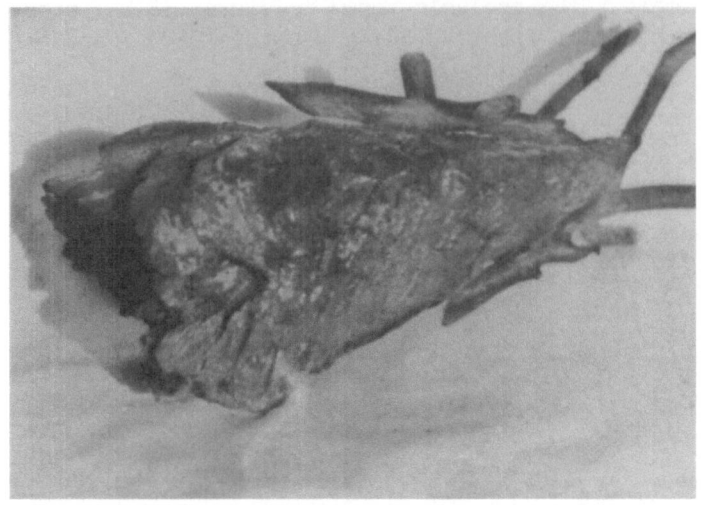

Fig. 3 - Paraná pine tree knot - average length: 30 cm; average
         weight: 3 kg

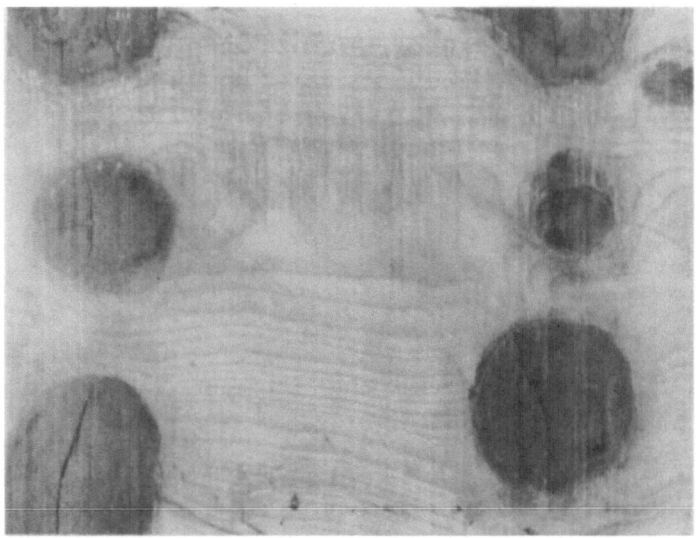

Fig. 4 - Paraná pine tree commercial board (3rd quality) showing
         the knots (black spots)

Table I

Composition of the lignan resin of Paraná pine tree knots [14, 17]

Fraction	Component	Content (%)	Characteristics
Benzene extract (*)	Hydrocarbons waxes, fats	3	Waxy, reddish material, sweet odor
Aqueous extract (*)	Glicosides, sugars tannins, etc	20	Dry, yellowish-brown, powdery material, odorless
Ethanol extract (*)	Lignans, resins, etc.	13	Dry, reddish-brown, powdery material, slight sweet odor
Solubles in 72% H_2SO_4	Cellulose, etc.	21	---
Insoluble in 72% H_2SO_4	Lignins, etc.	43	Dry, reddish powdery material
Ash	Mineral residues	Negligible	White powder
Total		100	

(*) Successive extractions, following the order

(*) Commercialized by CARBOMAFRA Indústrias Químicas, Curitiba, PR, Brazil.

In industry (*) the powdered knots are extracted directly
with 95% ethyl alcohol, obtaining a dry, reddish lignan resin, the
weight of it amounts to about 1/3 of the original weight. This
represents roughly the sum of the benzene, water and ethanol
extracts referred to before. The knots as well as the residue of
the ethanol extraction are used for the production of active
charcoal.

        Anderegg and Rowe [18] analyzed a sample of the knot resin;
the results are presented in Table II. It can be seen that
this resin was mostly made of lignans (88%), of which nearly all
are composed of secoisolariciresinol and its derivatives. This
resin is completely different from the terpene resin already
mentioned.

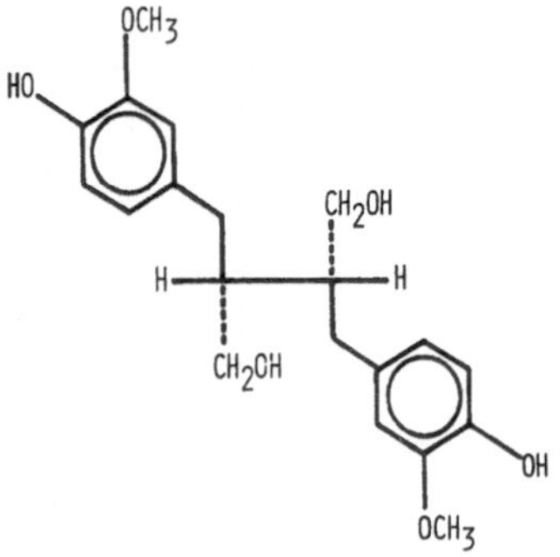

Fig. 5 - Secoisolariciresinol - Principal lignan of the resin of
         Paraná pine tree knots

        Lignans and lignins are components largely found in several
parts of the plant. Lignans are small, soluble molecules, with a
dimeric phenylpropane skeleton; the formula of secoisolariciresinol
is represented in Fig. 5. Lignins are insoluble macromolecules,
related to the lignans. The lignins are much more complex and
have different structures, depending on the origin and the process
of separation; a segment of the macromolecular structure of the
lignins is shown in Fig. 6. The lignins may be defined as a copo-
lycondensate of the dehydrogenation products of the phenolpropane
starting structures [19,20].

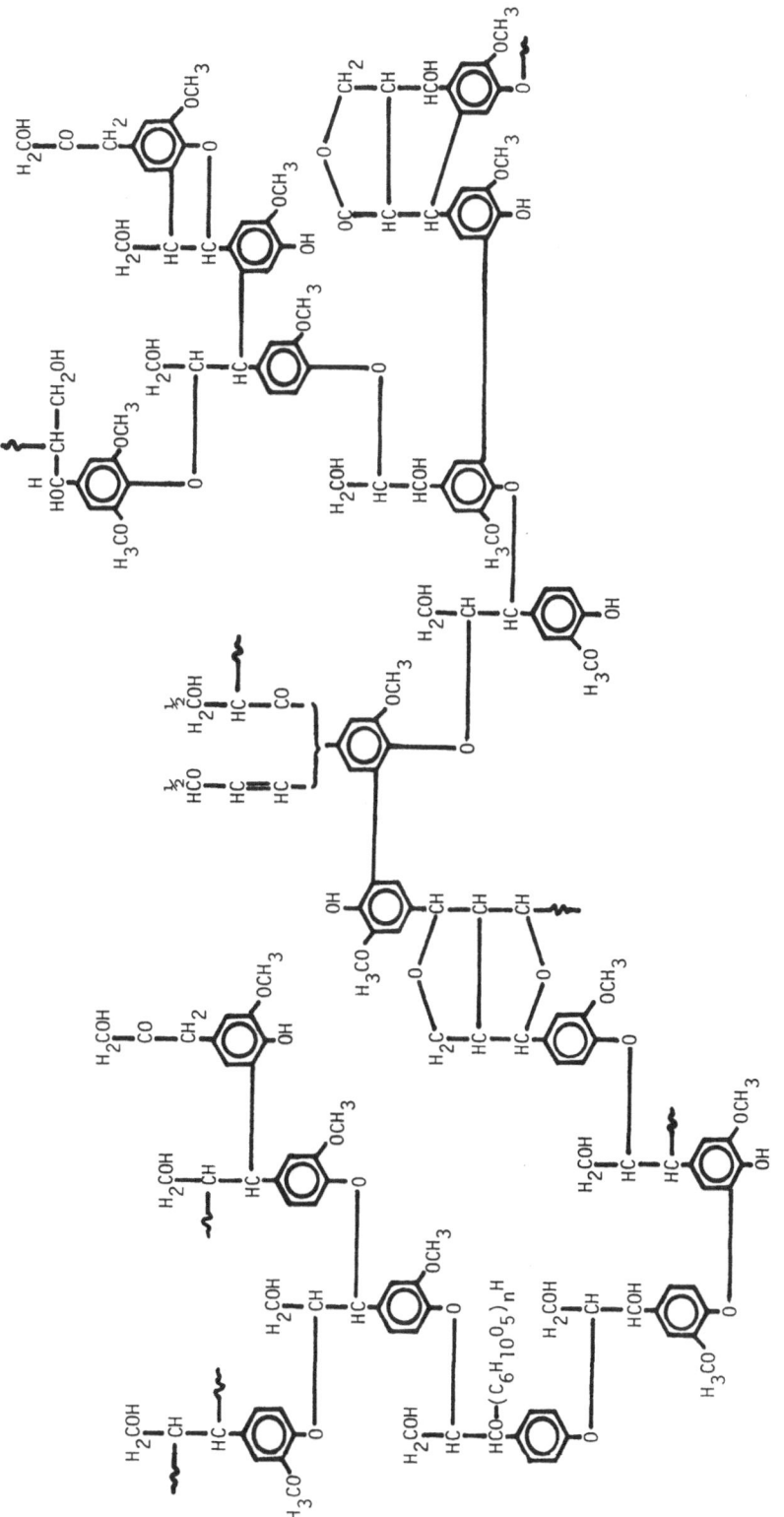

Fig. 6 - Representative section of the lignin structure after Freudenberg

Table II

Chemical compostion of the ethanol extract of the lignan resin of
Paraná pine tree knots [18]

Component	Content (%)
Diterpenes, alkaloids, ash and precipitable with HCHO	none
Total reducing sugars	3.1
Tannins	1.4
Fatty acids	1.0
Fatty esters	0.9
Hinokiresinol	4.1
(+) Pinoresinol dimethyl ether	2.0
(+) Pinerosinol monomethyl ether	1.2
(+) Pinoresinol	3.2
(+) Isolariciresinol	0.8
(-) Secoisolariciresinol	7.2
Lignans related to secoisolariciresinol	~ 70

The term lignan, created by Haworth, defines a family of
natural compounds which presents hydroxylated, methoxylated or
methylenedioxylated aromatic rings, with the propane chain either
non-oxidized as in guaiaretic acid, or oxidized in diversified
ways, forming either lactone or furan, or bifuran condensed rings
or else benzylcyclohexane derivatives [22,23]. More recently, the
term neolignan was proposed to include all dimers formed by the
association of two monomeric precursors comprising the propenyl
and/or allyl derivatives [24]. The Haworth lignans refer mainly to
association by oxidative coupling of acid and/or alcohol structures,
while the neolignans refer to propenyl and/or allyl derivatives
which are not so frequently found in nature. In several cases,
the neolignans show oxygen bridges, a fact which indicates that
neolignans and lignans have different biogenetic origins.

Lignans occur in nature in resins exudéd from plants; as happens with the lignins, lignans may exist in roots, trunks, leaves and fruits [25].

In secoisolariciresinol, there are two phenolic hydroxy groups para to aliphatic substituents of the ring, and two methoxy groups ortho to the phenolic hydroxyls. Thus, each lignan molecule of this type presents two ortho positions still available for reactions of the electrophilic substitution type.

So, lignan molecules may be considered a bifunctional monomer for purposes of a reaction of the phenol-formaldehyde condensation type.

The preliminary technological characterization of the modified lignan resins was carried out using for purposes of comparison a commercial sample of the phenolic Novolac type product and the unmodified lignan resin from the knot of the Paraná pine tree.

The degree of cure was estimated by the soluble fraction in acetone [26] and reported as the percentage of insoluble products due to the crosslinking of the incorporated resin. The water absorption was determined by the difference in weight after two hours of immersing the test sample in boiling water [27]. Figure 7 shows coherent information: the less cured the product, the higher the water absorption. The best performance is shown by the commercial phenolic resin.

Considering the filler content the water absorption follows the order phenol-formaldehyde, lignan-formaldehyde, crude lignan and lignan-furfural resins. In all cases, the water absorption becomes higher as the wood-flour filler increases (Fig. 8). The specific gravity procedures are very close for phenolic and modified lignan resins, with lower results for the unmodified lignan molded specimens (Fig.9) [28].

The Rockwell hardness [29] do not show any significant difference among the crude lignan, the lignan-furfural, the lignan--formaldehyde and the phenol-formaldehyde resins, with any wood--flour content within the studied range (0-150 phr filler) (Fig.9).

The stiffness modulus [30] as well as the modulus of elasticity [31] do not show a great change for the 4 resins, though the results are lower for the furfural-lignan molded pieces (Fig.10).

The impact resistance [32] by the Izod method shows similar behavior for the resins under study. The same can be said for the deflexion temperature under flexural load except for the unmodified lignan resin composition[33] (Fig. 11).

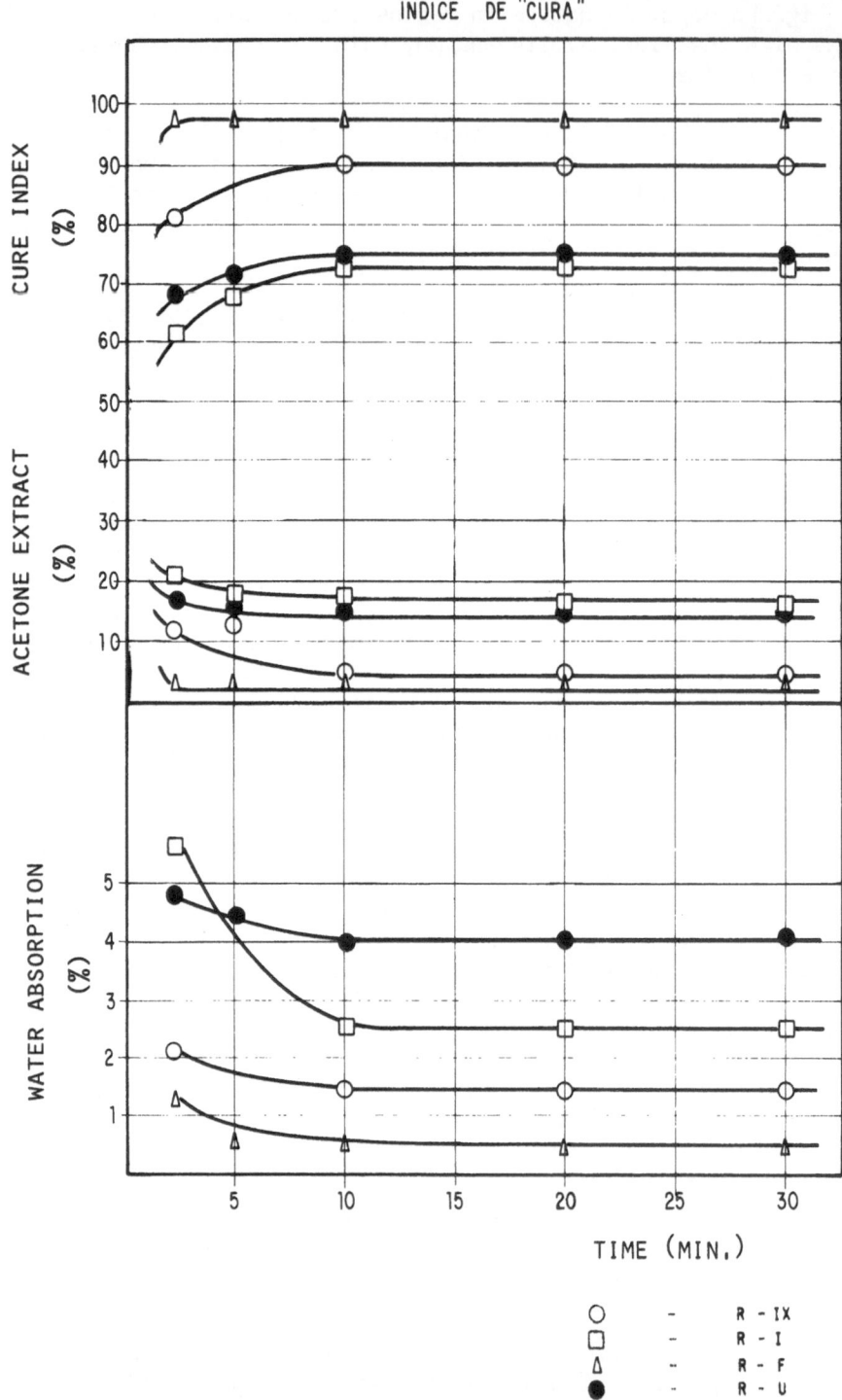

Fig. 7 - Cure index and water absorption in lignan resins

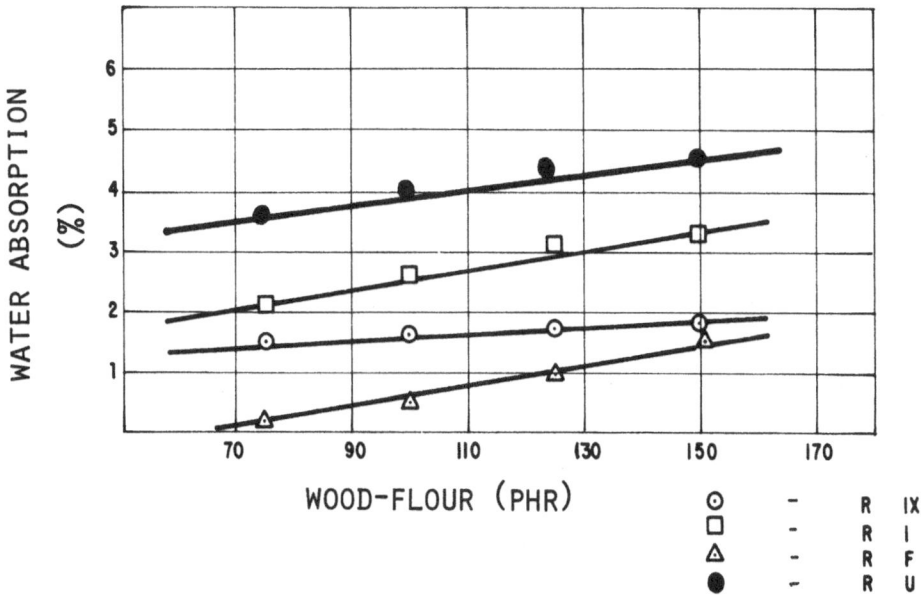

Fig. 8 - Water absorption _versus_ filler content in lignan resins

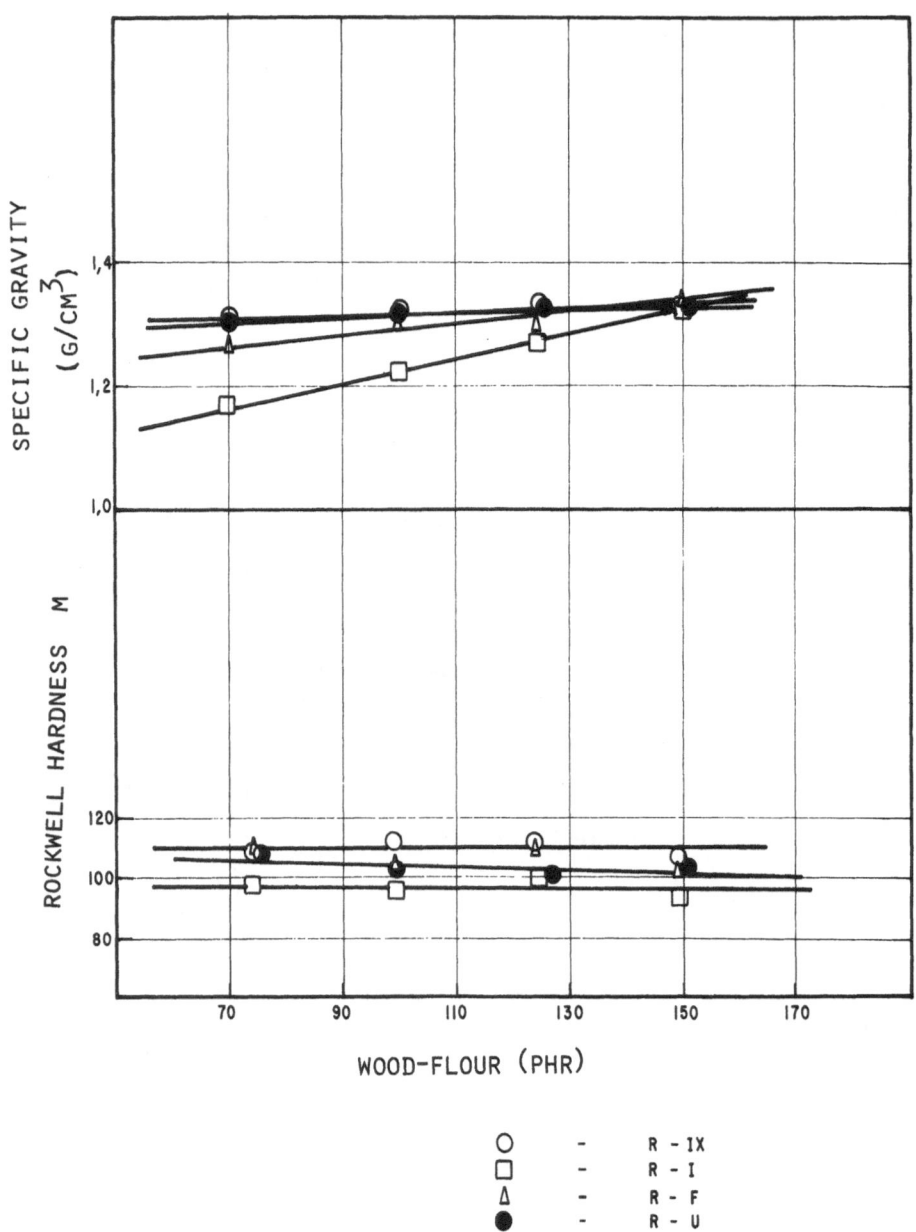

Fig. 9 - Specific gravity and Rockwell Hardness <u>versus</u> filler
content in lignan resins

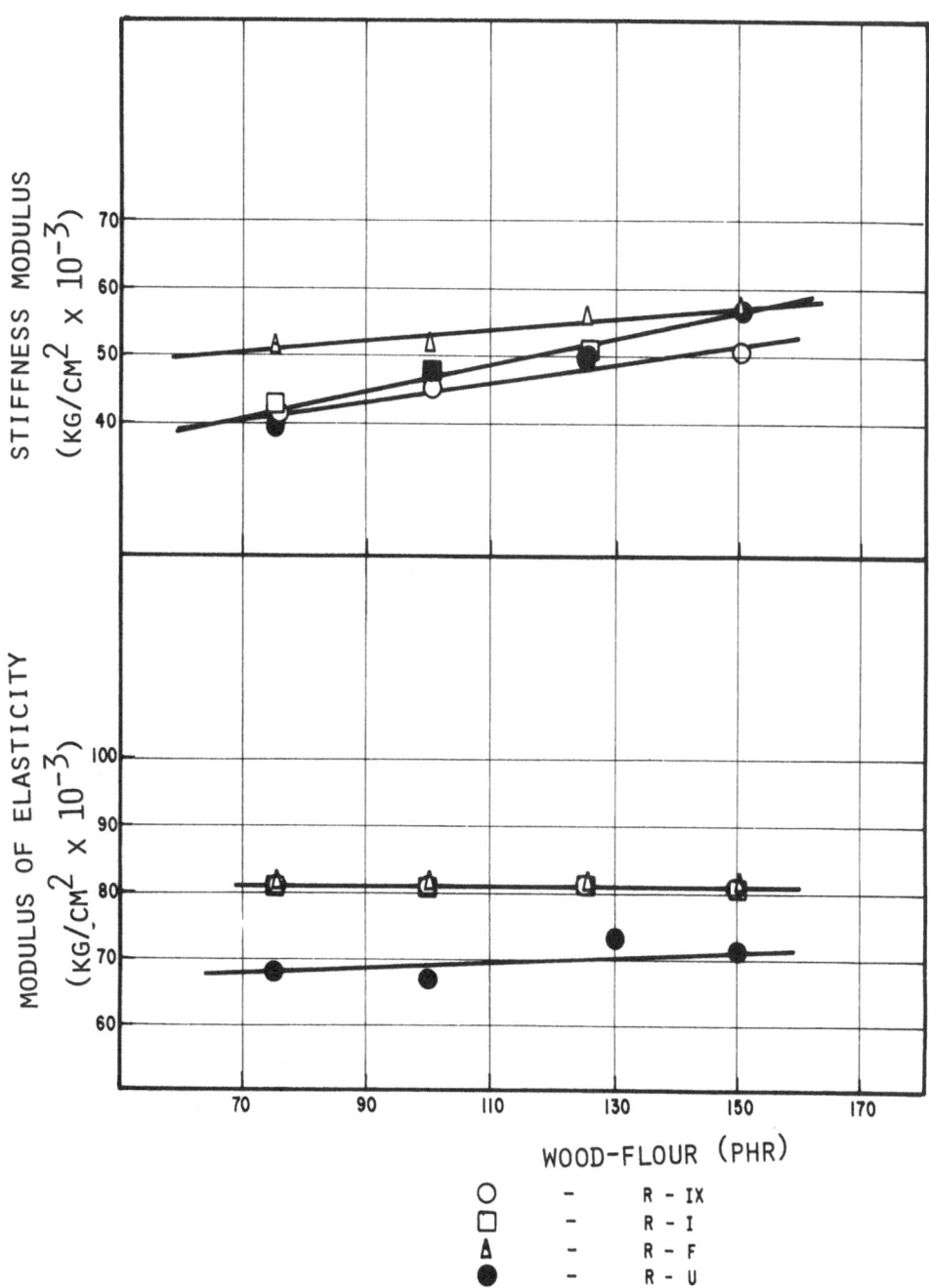

Fig. 10 - Stiffness modulus and modulus of elasticity versus
filler content in lignan resins

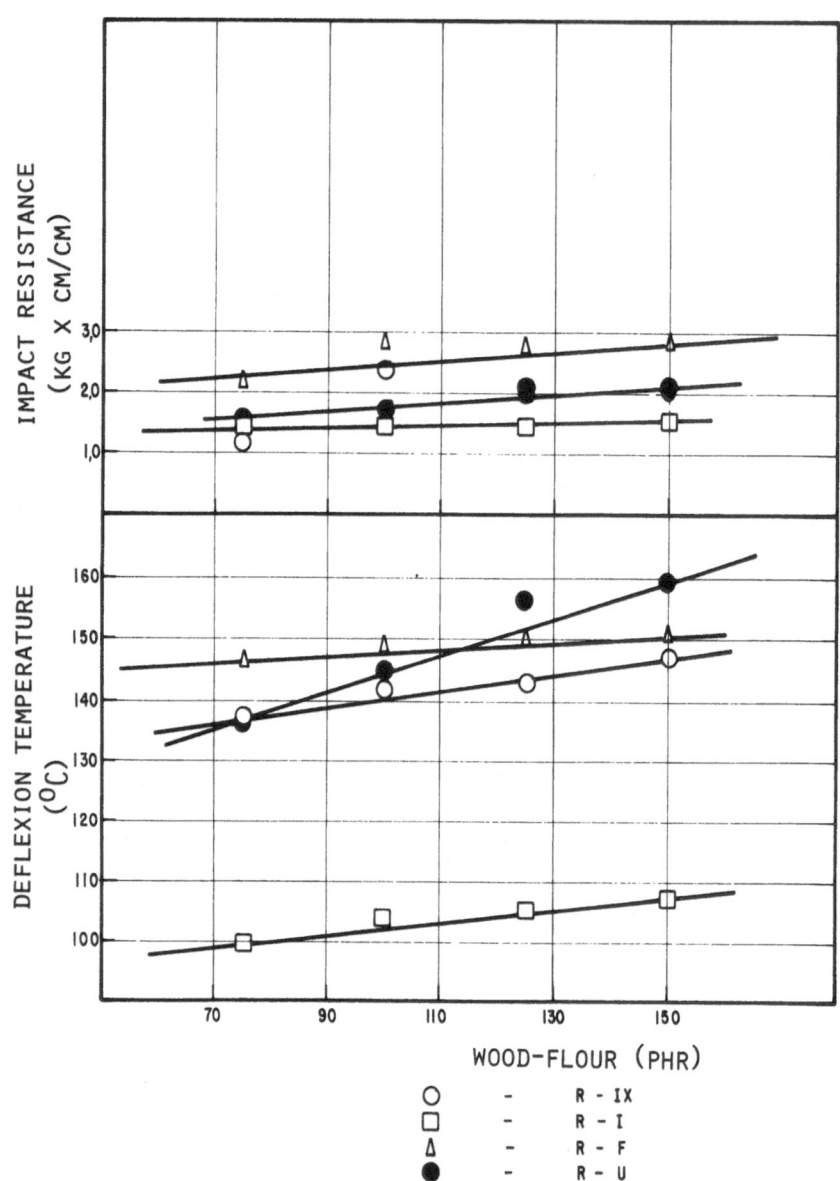

Fig. 11 - Izod impact resistance and deflection temperature
versus filler content in lignan resins

As a general remark, the best set of results is obtained, for each resin, for molding compositions with 100 phr wood-flour field. This preliminary evaluation shows that, taking the phenol-formaldehyde resin as the standard for comparison, the lignan-formaldehyde resin has about the same characteristics.

Concluding from these experimental data, it is possible to use the lignan components of the Paraná pine knot as a substitute chemical for phenol in the preparation of resins of the phenol-formaldehyde and phenol-furfural types. These renewable-source lignan resins may be used in the same field of the petrochemical phenolic resins.

References

1. J.R. Mattos - "O Pinheiro Brasileiro" - Grêmio Politécnico, DLP, São Paulo, Brazil, 1972, 620 pgs.
2. "Anuário Brasileiro de Economia Florestal", Ministério da Agricultura, IBDF, ano 18, nº 18, Rio de Janeiro, Brazil, 1967.
3. S. Martinho - "O Pinheiro Brasileiro", IBDF, no date.
4. R. Descartes de Garcia Paula - Boletim do INT, 2, nº 2, pg.38-43 (1951).
5. T. Peckolt and G. Peckolt - "História das Plantas Medicinais e Úteis do Brazil", Rio de Janeiro, Brasil, Lemmert, 1888, Fascículo 1, pg. 71-78.
6. H.Tarkow - in "Kirk-Othmer Encyclopaedia of Chemical Technology", Interscience, New York, 1970, vol. 22, pg. 384.
7. D. McNeil - in "Kirk-Othmer Encyclopaedia of Chemical Technology", Interscience, New York, 1970, vol. 19, pg. 653.
9. H.I. Enos, G.C. Harris and G.W. Hedrick - in "Kirk-Othmer Encyclopaedia of Chemical Technology", vol. 17, Interscience, New York, 1968, pg. 474.
10. C.T. Gozenbach, M.S. Jordan and R.P. Yunick - in "Encyclopaedia of Polymer Science and Technology" Interscience, New York, 1970, vol. 13, pg. 575.
11. L.Ruzicka - Proc. Chem. Soc., 341 (1951).
12. D.E. Wolf, C.H. Hoffman, P.E. Aldrich, H.R. Skeggs, L.D. Wright and J. Fokers - Am. Chem. Soc., 78, 4499 (1956).
13. F. Lynen, H. Egger, U. Henning and I. Kessel - Angew. Chem. 70, 738 (1958).
14. M.M. Jacob - "Resinas Furfural-Lignânicas de Nó-de-pinho do Paraná", M.Sc. Thesis, Instituto de Macromoléculas, Federal University of Rio de Janeiro, RJ, Brasil(1980).
15. L.C.F. Barbosa - "Nó-de-pinho do Pinheiro do Paraná como Matéria Prima de Resinas Termorrígidas". M.Sc. Thesis, Instituto de Macromoléculas, Federal University of Rio de Janeiro, RJ, Brasil, 1981.
16. M.M. Jacob, L.C.F. Barbosa and Eloisa B. Mano - In press.

17. M.M. Jacob, L.C.F. Barbosa and Eloisa B. Mano - In press
18. R.J. Anderegg and J.W. Rowe - Holzforschung 28 (5), 171-5 (1974).
19. K. Freudenberg - Science 148, 595-600 (1965).
20. K. Freudenberg - in "Lignin structure and reactions", Adv. in Chem. Ser. nº 59, Washington, DC, 1966, pg. 1-37.
21. W.M. Hearon and W.S. McGregor- Chem. Revs. 55, 957-1068 (1955).
22. H. Erdtman - in K. Paech and M.V. Tracey, "Modern Methods of Plant Analysis", vol. III, Springer Verlag, Berlin, 1955, pg. 428-32.
23. T. Geissman and E. Hinreiner - Bot. Revs. 18, 77-244 (1952).
24. O.R. Gottlieb - Fortschritte der Chemie organischer Natur-stoffe 35, 1-72 (1978).
25. M.A. Buchanan - in "The Chemistry of Wood", B.L. Browning, Interscience, New York, 1963, pg. 313-67.
26. American Society for Testing and Materials - D 494.
27. ASTM D 570.
28. ASTM D 792.
29. ASTM D 785.
30. ASTM D 747.
31. ASTM D 790.
32. ASTM D 256.
33. ASTM D 648.

RENEWABLE RESOURCES FROM FOREST PRODUCTS

FOR HIGH TEMPERATURE RESISTANT POLYMERS

Sukumar Maiti, Sajal Das*
Manoranjan Maiti** and Atanu Ray

Polymer Division, Materials Science Centre
Indian Institute of Technology
Kharagpur 721302, India

INTRODUCTION

Recently the price of crude oil and various petroleum fractions has been increased by about 20 times, beyond all expectations. The availability of oil and other petroleum fractions has also become very uncertain. The shortage of oil has led to the severe competition between fuel for energy and feedstock for petrochemicals. Although the entire petrochemicals including fertilizers and synthetic polymers consume less than 10% of the supply of crude oil, the demand for fuel for energy makes the prospect of assured supply of feedstock for petrochemicals at a reasonable price increasingly gloomier. The cost advantage of polymer material in general over other coventional materials such as metals and alloys, glass and ceramics etc. has already been eroded considerably by the price raise of curde oil. Never before have plastics been faced with such a tough competition from conventional materials. Oil as a feedstock is also not renewable and one day, sooner or later, it will dry out.

---

* Present address: Department of Chemistry
  Wright State University, Dayton, Ohio 45435 U.S.A.

** Department of Chemistry, Uluberia College
   Uluberia, Howrah

This has prompted research activities all over the world to find alternative feedstocks for polymers. Agricultural by-products and forest products have become more attractive these days as renewable resources for polymers and plastics additives. Plant exudates such as gum arabic, gum ghatti, gum karaya, gum tragacanth etc. have been used in textiles, cosmetics and pharmaceuticals mainly as emulsifiers and thickners[1]. Other types of gums from seed such as guar gum have similarly been used. But these have not been utilized as the raw material for polymer manufacture.

Lignin, available as a waste product from the pulp and paper industry, has recently been modified and its degradation products were used to produce thermostable polymers. Polyphenylene sulfides have been prepared from the degradation product of lignin such as guiacol and its chloroderivatives[2]. Aromatic polyesters have also been synthesized starting from the degradation products of lignin[3].

Attempts have already been made to develop adhesive substitutes for phenol-formaldehyde resin from the bark extract of pine trees[4]. Bark accounts for about 15% by weight of pine trees. The phenolics from bark extracts can be used as substitutes for phenols. The bark phenolics, having a molecular weight of about 1500, may be reacted with formaldehyde under mild conditions to obtain adhesives[5,6]. Another approach to the use of bark phenolics is to react the bark extract with low molecular weight methylol phenols[6].

Trimellitic anhydride (TMA) is an important raw material for high temperature resistant polymers such as polyesterimide, polyamideimide etc. The cyclic anhydride group in the TMA molecule offers a reaction site for imide formation and the carboxyl group for synthesis of ester or amide linkages. The polycondensation of TMA with glycol and/or diamine results in the formation of semi-ladder structures in the polymer, which is responsible for high temperature resistance.

Since TMA is a petroleum based chemical, its price and supply are likely to be affected adversely in the future. In an attempt to find a suitable substitute for TMA, we have developed gum rosin, the exudate of conifer trees such as pine, as a source of raw materials for polyesterimide[7-9] and polyamideimide resins. Our present program aims at the development of gum rosin as cheap, dependable and renewable source of raw material for polymers that are expected to be a substitute for TMA-based polymers and other thermostable polymers.

Rosin is classified into three main types viz. (a) gum rosin, (b) wood rosin, and (c) tall rosin[10-12]. Gum rosin is obtained as the residue in the spirit of terpentine from crude terpentine pitch by distillation. Wood rosin is produced by naphtha extraction of waste pine wood after recovery of pine oil and terpentine. Tall rosin is obtained from the distillation of tall oil[13]. Among the various rosins the most important is common rosin or colophony obtained from various species of pine trees.

EXPERIMENTAL

Purification of Rosin

Commercially available gum rosins containing 90% abietic acid were purified by the following methods.

(a) 50.0 g of rosin were dissolved in a minimum quantity of hot ethanol and filtered to remove insoluble impurities. It was then kept overnight at room temperature for complete separation of the product from the solution. It was filtered again and the final product was dried under vacuum. The yield and the melting point of the product are 76% and 7-75°C, respectively.

(b) 50.0 g of rosin were dissolved in hot ethanol and filtered to remove insoluble impurities. The solution was poured over excess of 1N hydrochloric acid and the solid material was separated by filtration. It was finally recrystallized from hot methanol and dried under vacuum. The yield of the product was 84%, m.p. 80-85°C.

Conversion of Resin Acids to Levopimaric Acid

Various methods have been reported for the isomerization of resin acids and the isolation of levopimaric acid from the mixture[11-17]. These methods are time consuming and the yield of levopimaric acid is not very satisfactory. We have modified the procedure of Palkin and coworkers[16] for the isomerization of resin acids to levopimaric acid as follows.

120.0 g rosin were dissolved in a mixture of 250 ml ethanol and 2.1 ml conc.HCl. The mixture was refluxed for 45 min and filtered hot to remove insoluble impurities. The filtrate was cooled and treated with 1.6 ml 16N NaOH to neutralize the mineral acid used. Next, 10.3 ml 8N NaOH, 10 ml neutralized rosin solution and 15 ml ethanol were mixed together to have a clear solution. This solution was added to the remaining portion of the neutralized rosin solution with stirring and allowed to stand at room temperature. Almost instantaneous precipitation of the complex Na-salt of levopimaric acid began and was completed within a few minutes. The precipitate was filtered and dried at 60 - 70°C. The Na-salt was treated with 83 ml 1N H_2SO_4 in ethanol to obtain a clear paste, which was poured into water with stirring to precipitate free levopimaric acid. It was filtered and dried in air at 100°C for 3 hrs. and finally under vacuum at 70°C. The yield of levopimaric acid is about 84%, m.p. 145-150°C.

Preparation of Rosin-Maleic Anhydride Adduct

Rosin-maleic anhydride adduct (RMA) was prepared from rosin
obtained from a commercial source following the method of Gosh
et al[9,18]. The details òf the procedure have already been reported[9].

RMA has also been prepared from levopimaric acid. 10.2 g levopimaric acid and 2.95 g maleic anhydride were melted together at
150°C under nitrogen atmosphere in a three-necked flask fitted with
a mechanical stirrer, a thermometer and a nitrogen purge system. The
reaction was allowed to continue for 2 hrs. and was then cooled to
room temperature, both under the blanket of nitrogen. The solid mass
thus obtained was made into a clear solution in diethyl ether. It
was purified from the diethyl ether solution by precipitation with
petroleum ether, filtered, washed with petroleum ether and dried
under vacuum at 100°C.

Preparation of Rosin Maleic Anhydride
Imidodicarboxylic Acid

Imidodicarboxylic acid from rosin maleic anhydride adduct (RMID)
has been synthesized from RMA and amino benzoic acids by following
the procedure reported earlier[8,9,19].

Preparation of Rosin Maleic Anhydride
Imidodicarboxylic Acid Chloride

51.9 g(0.1 mole) of RMID in a 500 ml two-necked flask fitted
with a condenser and a thermometer were mixed carefully with 66 ml
(0.9 mole) freshly distilled and dry thionyl chloride and the mixture was refluxed for 15 hrs. After the reaction was over, excess
thionyl chloride was then removed by distillation under reduced
pressure and the solid residue was extracted with 500 ml dichloromethane. The crude rosin maleic anhydride imidodicarboxylic acid
chloride was obtained after removal of the solvent. The crude product was further pruified by crystallization from chloroform and
dried in vacuum at 100°C. The yield of the product was 65-70%,
m.p. 186-187°C.

Polymer Synthesis

Both polyesterimides (PEI) and polyamideimides (PAI) were synthesized from RMID and/or its dichloro derivatives.

Synthesis of Polyesterimides

In one of the methods polyesterimide was prepared by one-step
process. 4,4'-Diaminodiphenyl methane was dissolved in excess ethylene glycol and rosin-maleic anhydride aduct was added to this solution. The mixture was heated to 190-220°C in the presence of a mix-

Table 1. Physical Properties of Various Rosins

Properties	Gum rosin[a]	Wood rosin[a]	Tall rosin[a]	Our sample
Melting point,°C	76	73	77	56
Specific gravity at 20°/4°C	1.07	1.07	1.07	1.06
Acid number	164	163	165	162.8
Acid content,%	90	90	90	87.6
Saponification value	172	168	174	-

[a]Data from Ref. 20.

ture of Sb_2O_3 and Zn-acetate (1:1 W/W) under nitrogen atmosphere[7,8].

In the two-step process RMA was first converted into RMID, which was isolated and purified. Purified RMID was reacted with excess diethylene glycol (DEG) to form polyesterimide following the procedure of Maiti and coworkers[7-9].

Synthesis of Polyamideimides

Polyamideimides from rosin were synthesized by reacting the dichloro derivative of rosin maleic anhydride imidodicarboxylic acid (RMIDC with various amines. A typical run may be described as follows: 5.56 g (0.01 mole) were dissolved in 30-35 ml DMF in a three-necked flask fitted with a stirrer, a thermometer and a nitrogen purge system. To this solution 1.09 g (0.01 mole) m-phenylene diamine in 10 ml DMF and 0.78 g (0.01 mole) pyridine were added. The mixture was stirred at room temperature for 3 hrs. and at 60-70°C for next 5 hrs. At the end of reaction the mixture was poured into ice/water and the precipitated polymer was isolated by filtration. The crude polymer was purified by dissolving in DMF and reprecipitated by methanol. Appropriate quantities of different amines were taken to prepare various polyamideimides.

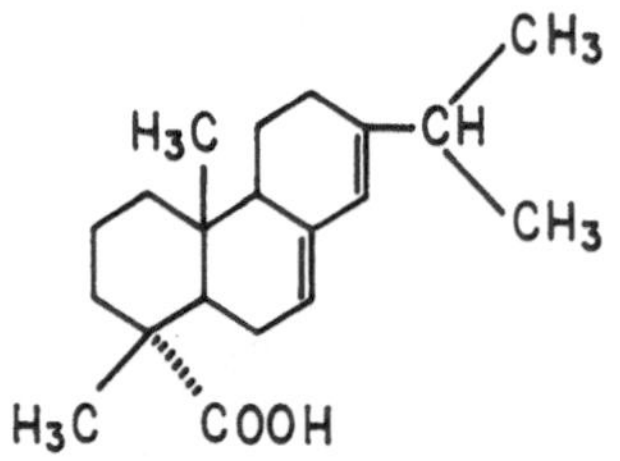

Abietic acid

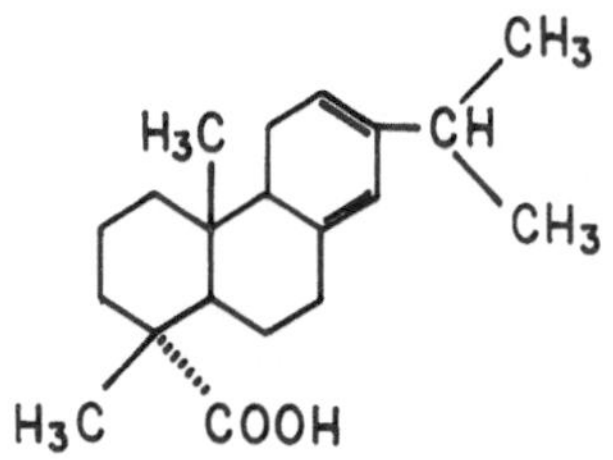

Levopimaric acid

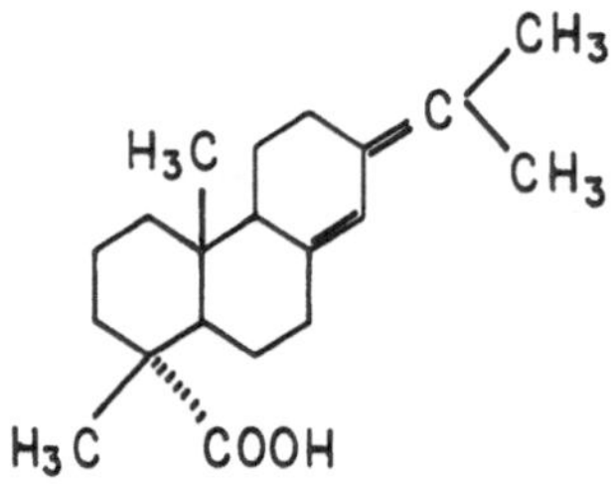

Neoabietic acid

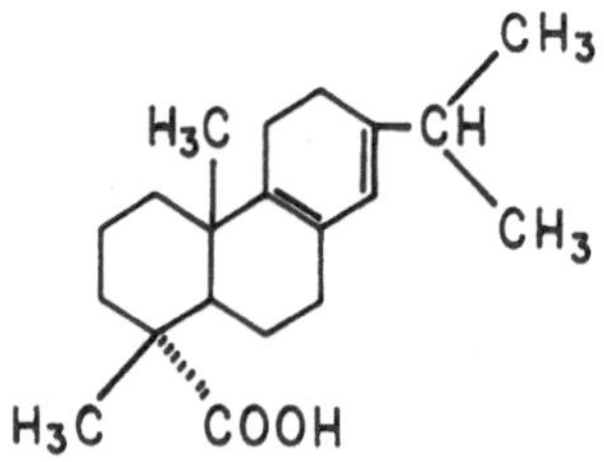

Palustric acid

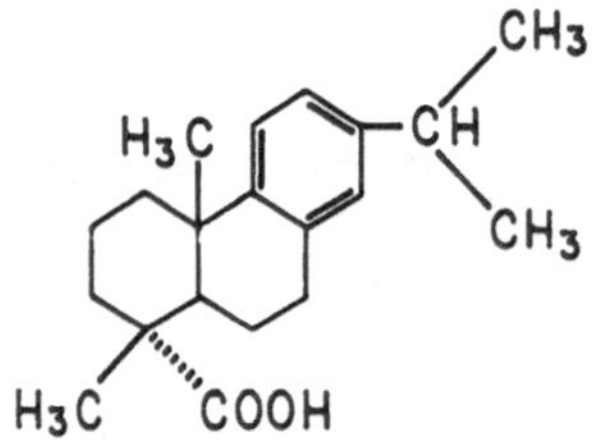

Dehydroabietic acid

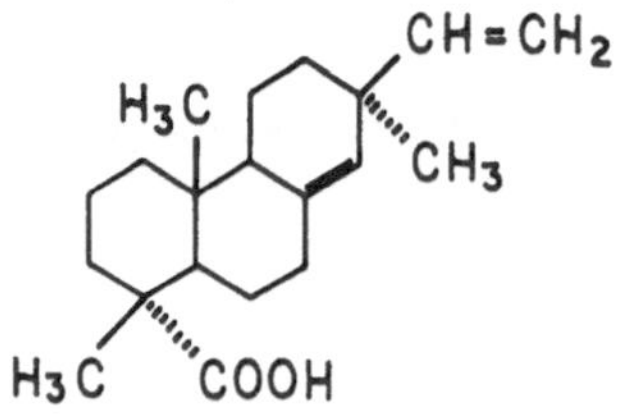

Pimaric acid

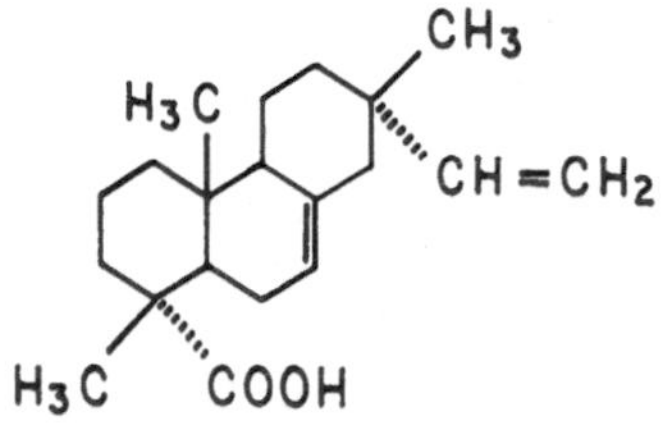

Isopimaric acid

RESULTS AND DISCUSSION

Rosin is a complex mixture of naturally occurring high molecular weight organic acids and related materials. Rosin obtained
from various fractions of exudates of pine trees, known as gum
rosin, wood rosin and tall rosin, is composed of about 90% resin
acids and 10% neutral materials[20] (Table 1). The resin acids are
isomeric monocarboxylic acids of alkylated hydrophenanthrene nu-
clei[21,22].

Of these isomeric acids levopimaric acid has the suitable diene
structure which readily forms a Diels-Alder adduct with maleic anhydride[23-25]. Other isomeric acids are unable to form the adduct
as such with maleic anhydride due to the absence of suitable diene
structure in their molecule. However, under the influence of heat
and/or strong acid they are isomerized to levopimaric acid. When
rosin is used as the raw material for RMA, it is first converted
into levopimaric acid, which actually forms the Diels-Alder adduct.
The increased yield of rosin-maleic anhydride adduct by using levopimaric acid as the starting material (Table 2) corroborates the
fact.

Characterization of Monomers and Polymers

Rosin-maleic anhydride adduct is soluble in alcohol, acetone
and other polar solvents. Characteristic properties of RMA are
shown in Table 2.

Rosin-maleic anhydride imidodicarboxylic acid is soluble in
chloroform, methyl ethyl ketone, m-cresol and other polar solvents
and insoluble in methanol, hexane, chlorobenzene. The physical
properties of RMID have been reported earlier[9].

The dichloroderivative of RMID is soluble in chloroform, dichloromethane, cyclohexane and insoluble in hydrocarbon solvents.
It has been characterized by nitrogen analysis (calc. 2.51%, found

Table 2. Yields and Physical Properties of RMA

Starting material	Yield %	Melting point,°C	Acid number
Rosin	63	205	278.28
Levopimaric acid	89	205	278.50

2.30%) and chlorine analysis (calc. 12.77%, found 12.32%).

Polyesterimides from rosin are found to be soluble in highly polar solvents like THF, cyclohexanone, 1,4-dioxane, m-cresol, DMF, DMSO, DMAC, NMP but insoluble in methanol, ethanol, chloroform, benzene, toluene, cyclohexane etc. The structures of the repeat unit of polyesterimides and polyamideimides are shown in Fig. 1.

Polyamideimides from rosin are found to be soluble only in highly polar solvents such as m-cresol, DMF, DMSO, DMAC and NMP but not soluble in THF, cyclohexanone and 1,4-dioxane. This indicates that the solubility of polyamideimides is less than that of polyesterimides. Physical characteristics of both the polymers are shown in Table 3 and elsewhere[8,9].

Polymers from rosin studied in this investigation have been characterized by IR spectra, and nitrogen analysis (PEI-1: calc. 2.94%, found 2.90%; PEI-2: calc.2.41%, found 2.36%; PAI-1: calc. 7.11%, found 7.17%; PAI-2: calc.7.11%, found 7.03%; PAI-3: calc. 7.01%, found 7.56%). General IR characteristic bands of rosin maleic anhydride imidodicarboxylic acid and polyamideimides are shown in Figs. 2-6.

Characteristic IR absorption bands of RMA (Fig. 2) were observed at 1830 cm^{-1} and 1780 cm^{-1} due to C=O stretching of anhydrides and a sharp band at 1695 cm^{-1} and at 1315 cm^{-1} due to C=O stretching of -COOH. A broad band near 3350-2150 cm^{-1} was also due to -OH stretching of -COOH.

Characteristic IR absorption bands of RMIDC (Fig.3) were observed at 1785 cm^{-1}, 1710 cm^{-1} and 720 cm^{-1} due to imides[19]. The C=O stretching of -COCl were also observed at 1805 cm^{-1}.

Characteristic IR absorption bands of PAI-1(Fig.4), PAI-2 (Fig. 5.) and PAI-3 (Fig. 6) were observed near 1785 cm^{-1}, 1715 cm^{-1} and 725 cm^{-1} due to imide groups and near 1630 cm^{-1} and 1530 cm^{-1} due to amide groups. Because of the complex structure of the polymer, characteristic IR bands observed were not so sharp due to overlapping of different bands of various chromophore groups present in the rosin polymer molecule.

Table 3. Yields and Characteristic Properties
of PEI and PAI from Rosin

Polymer[a]	Color	Yield %	Inherent viscosity dl/g	Density g/c.c.
PEI-1	Yellow	74	0.17[b]	1.20
PEI-2	Light Yellow	88	0.21[b]	1.16
PAI-1	Black	62	0.06[c]	1.12
PAI-2	Black	68	0.14[c]	1.20
PAI-3	Deep brown	63	0.12[c]	1.27

[a]Structure of the repeat unit of polymers is
shown in Fig. 1.

[b]Inherent viscosity has been measured in 0.5%
W/V in cyclohexanone at 35°C.

[c]Inherent viscosity has been measured in 0.5%
W/V in DMF at 30°C.

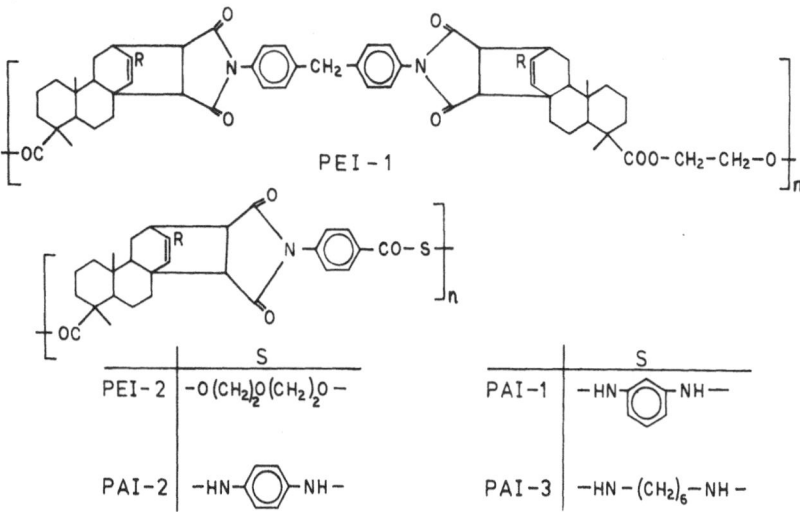

Figure 1. Structure of Polymers from Rosin

Polymers from Rosin

The molecule of RMA, like trimellitic anhydride, has got a carboxylic acid group and a cyclic anhydride group. The former may
be used for the synthesis of ester or amide linkage whereas the
latter for imidization reaction. This principle has been utilized
for the synthesis of polyesterimides and polyamideimides. Thus,
RMA may be regarded as a suitable substitute for TMA and related
petrochemicals such as PMDA etc. The reaction scheme for the synthesis of polyesterimide from RMA may be shown as follows:

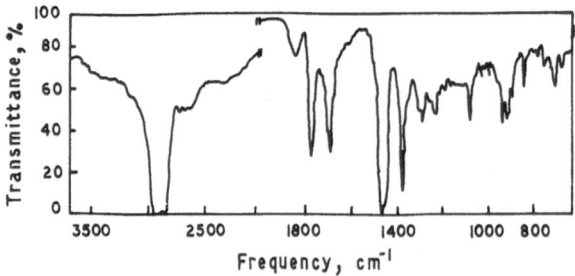

$$RMA + H_2N-R''-NH_2 + HO-R'-OH$$

Polyesterimide

There are two processes for the synthesis of poly-esterimides.
In the one-step process both esterification and imidization reaction
take place almost simultaneously. The kinetics of the reaction are,
therefore, complex and the reaction monitoring is difficult. The
process may be simplified by carrying out the reaction in two stages.

Fig. 2. IR spectrum of RMA

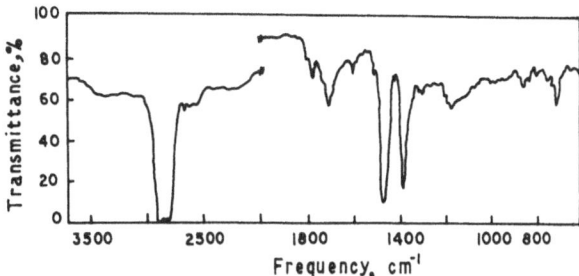

Fig. 3.   IR spectrum of RMIDC

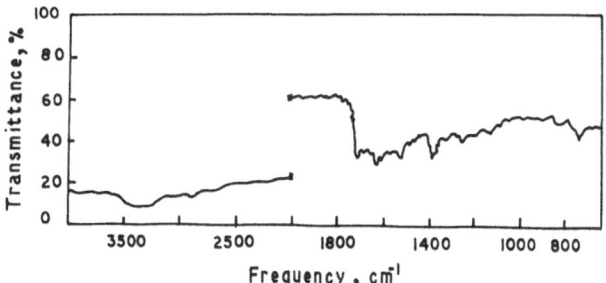

Fig. 4.   IR spectrum of PAI-1

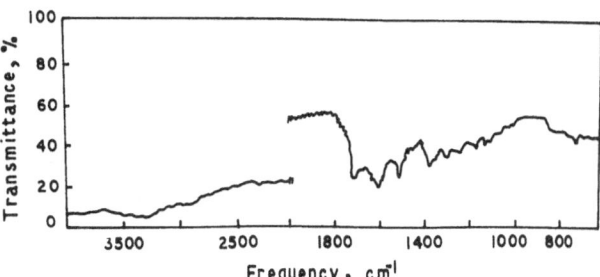

Fig. 5.   IR spectrum of PAI-2

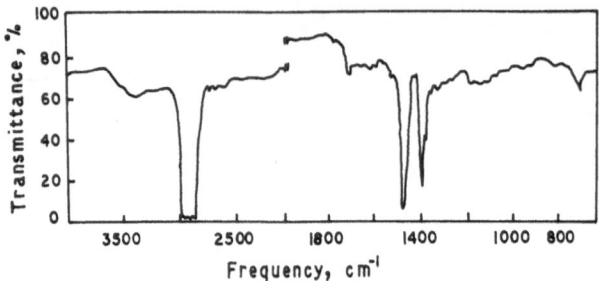

Fig. 6.   IR spectrum of PAI-3

First, the formation of imide group is separately carried out to obtain a dicarboxylic acid containing imide groups. For example, RMA reacts with aminobenzoic acid to produce a dicarboxylic acid containing an internal imide group (RMID).

RMID thus prepared from RMA may be reacted in the second step with a glycol to form polyesterimide.

The above reaction is essentially a polyesterification reaction, and hence can be controlled more conveniently. This has been discussed in detail elsewhere[9].

Aromatic polyamideimides are another class of thermostable polymer[26-32]. Recently we have reported a novel method of synthesis of polyamideimides by low temperature polycondensation of a dicarboxylic acid containing an imide group with a diamine in the presence of thionyl chloride[33]. Following this principle, we have also prepared polyamideimides containing C=C double bonds in the polymer backbone, which will facilitate crosslinking by chain growth mechanism[34].

We wish to follow the same technique of low temperature poly-condensation for synthesis of polyamideimide from rosin. Accordingly, we have tried to react RMID with m-phenylene diamine in the presence of excess thionyl chloride at low temperature (0 -5°C). We failed to get any polymer. This may be due to low reactivity of carboxylic acid groups of RMID because of steric hindrance. The effect of bulky group on the reactivity of -COOH groups of RMID has also been observed earlier in the synthesis of polyesterimide[8].

Recently the use of triphenyl phosphite as a catalyst for poly-condensation reaction of amino acids and dicarboxylic acids with diamines have been reported[35-37]. Similar methods of using this catalyst may be tried to get higher molecular weight polyamideimide from rosin.

However, we have prepared polyamideimides from rosin via the dichloroderivative of RMID. First, the dichloro-derivative of RMID was prepared, which is then made to react with a diamine to produce polyamideimide.

Various amines such as m-phenylene diamine, p-phenylene dia-mine and hexamethylene diamine have been used for the preparation of polyamideimide. It has been found that the molecular weight of the polymers as evident from inherent viscosity values is low. It may be explained on the basis of unequal reactivity of the two acyl chloride groups of RMIDC. The acyl chloride group attached to the benzene ring of p-amino benzoic acid moiety is sterically less hindered than its counterpart attached to its rosin moiety. Unless the molecule behaves like a difunctional monomer satisfactorily, the poly-condensation reaction will be adversely affected.

Low viscosity values of polymers may also be due to the presence of impurities in the dichloride of RMID. The presence of monochloro-derivative of RMID, if any, will terminate the polycondensation process prematurely leading to the formation of low molecular weight products. Although we have purified the dichloroderivative by re-peated crystallization, the purity has not yet been checked by chroma-tographic methods.

The structure of the amine appears to have some influence on the molecular weight of the polymer. It is evident from Table 3 that the higher yield and viscosity values of polymers have been found employing p-phenylene diamine and hexamethylene diamine rather than employing m-phenylene diamine. This may be explained on the basis of steric hindrance, though the reactivity of m-phenylene diamine is higher than p-phenylene diamine[19].

Table 4.   Thermal Stability of Polyesterimides[a]

% weight loss at Temp.°C	TMA-based polymers				PEI-2[d] from rosin
	PEI[b]	Para-polymer[c]	Meta-polymer[c]	Ortho-Polymer[c]	
100	2.0	2.0	3.0	2.0	3.0
200	3.0	3.0	5.0	4.0	3.5
300	8.0	6.0	10.0	18.0	4.0
400	24.0	20.0	24.0	83.0	50.0
500	38.0	80.0	86.0	96.0	80.0
600	48.0	90.0	94.0	97.0	98.0

[a]Calculated from TGA curves; [b]Ref. 38; [c]Ref. 19; [d]Ref. 9

Thermal Stability of Rosin Polymers

TMA-based polymers i.e., polyesterimides, polyamideimides etc. possess better thermal stability than most of the large volume commo-dity polymers. Although their heat resistance is somewhat lower than polyimide or polybenzimidazole, their superior solubility behavior and melt-flow characteristics make them easily processable. Combi-nation of these two desirable properties is responsible for their commercial interest.

We have found that RMA appears to be a suitable substitute for
TMA as a raw material for polymers. Rosin polymers such as polyesterimides via RMA or RMID or its derivative possess similar solubility
and other characteristics to those synthesized from TMA. The thermal
behavior of a polyesterimide (PEI-2) from rosin has been compared
with that of TMA-based polyesterimides[19] (Table 4). It is found
from Table 4 that the thermal stability of the polyesterimide from
RMA is similar to that of TMA. The glass transition temperature,
T_g, of PEI-2 is 180°C, which is also similar to the T_g's of polymers
from TMA. Rosin polymers may, therefore, be regarded as a new class
of processable heat resistant polymer from a renewable resource.

Prospect of RMA and Rosin Polymers

Synthesis of polyesterimides and polyamideimides may also be
tried from RMA as shown in the following reaction schemes.

First, RMA is converted into monoacid chloride derivative, which
may subsequently react with diamines to form polyamideimide.

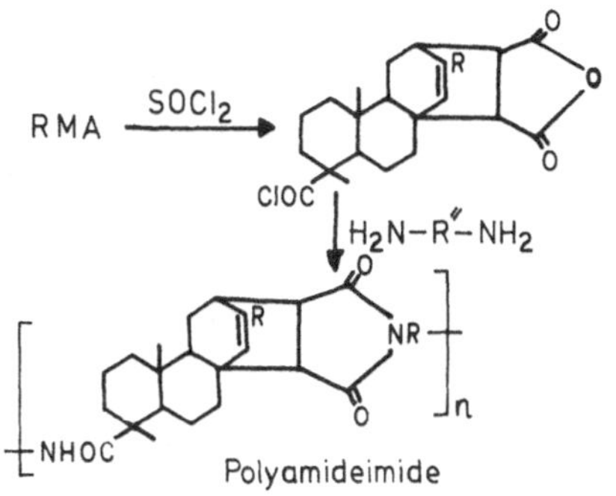

Synthesis of polyesterimide from rosin may also be. tried with a
dianhydride containing an ester group, which may be prepared as shown
in the reaction scheme:

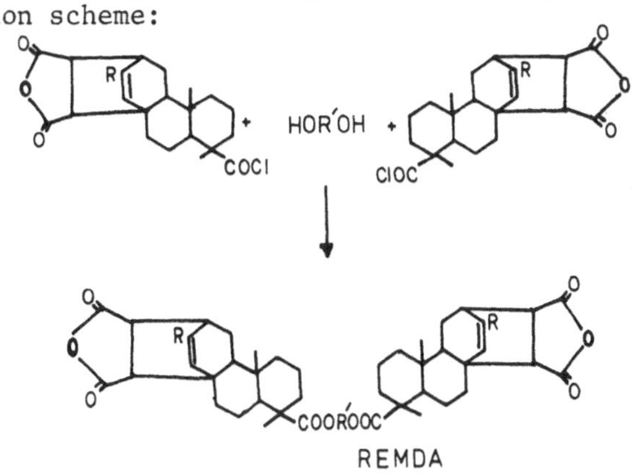

REMDA

The rosin ester maleic dianhydride (REMDA) thus prepared may then be reacted with a diamine to obtain polyesterimide. The reaction is, in fact, a polyimidization reaction

REMDA + H$_2$N - R'- NH$_2$

Polyesterimide

R = -CH(CH$_3$)$_2$; R' = alkyl, cyclo-alkyl; R'' = alkyl, cycloalkyl and aryl.

The properties of such polyesterimides are expected to be different from the polyesterimide reported earlier[9], but these will resemble closely to the polyesterimide prepared by one-step process[7].

Another process to prepare polyesterimides by low temperature polycondensation is the reaction between RMIDC and a diol. This method will produce a similar type of polymer as prepared from the RMID and a diol.

The rosin moiety of the polymer, polyesterimides and polyamideimides alike, offers interesting possibilities which may be explored further. One such possibility is the utilization of the residual unsaturation in the hydrophenanthrene ring structure of the rosin moiety. Although this unsaturation site is sterically very much hindered and therefore stable, attempts may be made to exploit it for crosslinking or other reactions by free radical processes. If this is successful, rosin polymers may be used for fabrication of void-free high temperature laminates and composites with suitable formulation.

Another interesting study may be initiated on the possiblity of oxidizing isopropyl group of the rosin moiety to hydroperoxide, which is subsequently utilized for synthesis of graft copolymers and crosslinking. A schematic of the proposed reaction is shown on the following page.

However, one should not be too optimistic about these interesting possibilities. These two potential reaction sites are very much shielded by the bulky hydrophenanthrene group and the cyclic anhydride ring of the maleic anhydride residue. The probability of reaction of these groups is, therefore, not very high. Ordinary reaction conditions may not be adequate; extreme or unusual conditions may be required to make these reactions possible.

Due to the unsymmetrical nature of the molecule and the presence of bulky groups rosin polymers are found to be amorphous and possess better adhesiveness. The hydrophenanthrene structure of the rosin moiety makes the polymer flexible. It is expected that incorporation of rosin polymers into other polyesterimides or polyamideimides will offer a better cable enamel. By controlling the molecular weight we can easily control the solubility of rosin polymers in commercial solvents. These prepolymers or low molecular weight polymers may be cross-linked subsequently for cable enamel and other coating applications. Studies on the polyblend of rosin

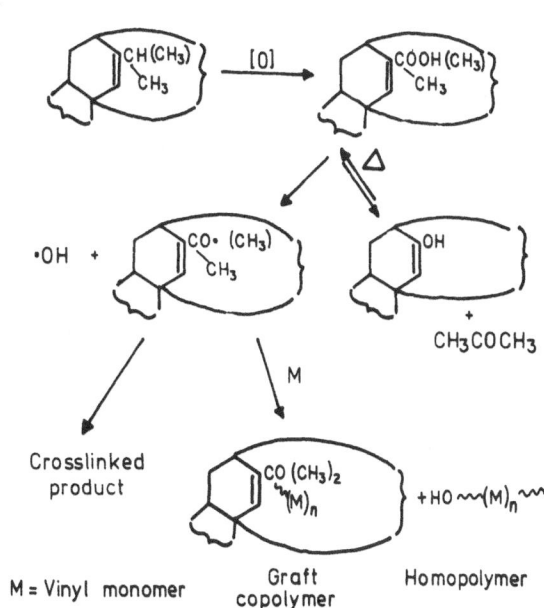

polymers and similar other polymers will produce interesting and useful information. For example, rosin polymers may be compatible with epoxy, phenol-formaldehyde resins, amino resins, etc., due to the presence of similar polar groups. We have found recently that polyesterimides synthesized from TMA form compatible blends with phenol-formaldehyde resin, which have very interesting properties[39]. It has been proposed that the blend of polyesterimide and phenol-formaldehyde resin crosslinks on heating[39].

REFERENCES

1. R.L. Whistler (Ed.), "Industrial Gums", 2nd Ed., Academic Press, New York, 1973, Chpts. 10-14.

2. B. Hortling and J.J. Lindberg, J. Appl. Polym. Sci., Appl. Polym. Symp. 35, 89 (1979).
3. J.J. Lindberg, J. Appl. Polym. Sci., 28, 269 (1975).
4. R.W. Hemingway and G.W. McGraw, Applied Polym. Symp., No. 28, (T.E. Timell, Ed.), Vol. 3, John Wiley, New York, 1976, p. 1349.
5. D.G. Roux, D. Ferreira, H.K.L.Hundt and E. Malan, Applied Polym. Symp. No. 28, (T.E. Timell, Ed.), Vol. 1, John Wiley, New York, 1975, p. 335.
6. R.W. Hemingway and G.W. McGraw, Proc. TAPPI. Forest Biology Wood Chem. Symp., Madison, 1977, p. 261.
7. S. Maiti and S. Das, Polym. Preprint, 21, 190 (1980).
8. S. Das, Ph.D. Thesis, I.I.T., Kharagpur, 1980.
9. S. Das, S. Maiti and M. Maiti, J. Macromol. Sci. Chem., A17#8, 1177 (1982).
10. J.M. Peterson, Ind. Eng. Chem., 24, 168 (1932).
11. A.R. Hitch, Ind. Eng. Chem., 23, 1135 (1931).
12. A. Zinoviev, Masloboino Zhirovoe Delo 24, 10 (1928); Chem. Zentr. 1, 3040 (1929).
13. R.C. Palmer, Ind. Eng. Chem., 26, 703 (1934).
14. L.L. Steele, J. Am. Chem. Soc., 44, 1333 (1922).
15. C.C. Kesler, A. Lowy and W.F. Faraghen, J. Am. Chem. Soc., 49, 2898 (1927).
16. S. Palkin and T.H. Harris, J. Am. Chem. Soc., 56, 1935 (1934).
17. J.S. Stinson and R.V. Lawrence, Ind. Eng. Chem., 46, 784 (1954).
18. G. Ghosh, K.D. Choudhuri and K.J. Balakrishna, Indian J. Technol., 2, 262 (1964).
19. S. Maiti and S. Das, J. Appl. Polym. Sci., 26, 957 (1981)
20. H. Mark and N.G. Gaylord (Eds.), "Encyclopedia of Polymer Science and Technology", 2nd ed., Vol. 12, John Wiley, New York, 1970. p. 145.
21. C.A. Genge, Anal. Chem., 31, 1750 (1959).
22. Kirk-Othmer, "Encyclopedia of Chemical Technology", 2nd ed., Vol. 17, John Wiley, New York, 1968, p. 475.
23. L. Ruzicka, P.G. Ankersmit and B. Frank, Helv. Chim. Acta., 15, 1289 (1932).
24. B. Arbuzov, J. Gen. Chem. (USSR) 2, 806 (1932).
25. L.H. Flett, W.H. Gardner, "Maleic Anhydride Derivatives", John Wiley, New York, 1952, p. 191.
26. "Amoco Amide-Imide Polymers", Technical Bulletin AM-1, Amoco Chemical Corp., USA.
27. H. Lee, D. Stoffey, K. Nevelle, "New Linear Polymers", McGraw-Hill, New York, 1967, p. 173.
28. U.S. Pat. 3,260,691 (1966) to Monsanto Chemical Co., USA; Chem. Abstr. 65, 7412 (1966).
29. U.S. Pat. 3,179,635 (1965) to Westinghouse Electric Corp., USA; Chem. Abstr. 63, 1967 (1965)
30. W. Wrasidlo and J.M. Augl, J. Polym. Sci., Part A-1, 7, 321 (1969).

31. W.M. Alvino and L.W. Frost, J. Polym. Sci., Part A-1, 2209 (1971).

32. K. Kurita, H. Itoh and Y. Iwakura, J. Polym. Sci., Polym. Chem.
     Edn., 16, 779 (1978)

33. S. Das and S. Maiti, Makromol. Chem., Rapid Commun., 1, 403 (1980).

34. A. Ray, S. Das and S. Maiti, Makromol. Chem., Rapid Commun., 2,
     333 (1981).

35. N. Yamazaki, M. Yamaguchi, H. Kakinoki and F. Higashi, Synthesis,
     1979, 355.

36. N. Yamazaki, M. Matsumoto and F. Higashi, J. Polym. Sci., Polym.
     Chem. Ed., 13, 1373 (1975).

37. H. Higashi, M. Goto and H. Kakinoki, J. Polym. Sci., Polym. Chem.
     Edn., 18, 1711 (1980).

38. S. Maiti and S. Das, Angew. Makromol. Chem., 86, 181 (1980).

39. S. Maiti and S. Das, Angew. Makromol. Chem., 1981 (Communicated).

LIGNOCELLULOSE-POLYMER COMPOSITES   II

Ahmed Nagaty, Olfat Y. Mansour, and
A.B. Mustafa

National Research Centre
Dokki, Cairo
A.R. of Egypt

SYNOPSIS

The change of the properties of lignocellulose-polymethyl methacrylate composites with the polymer load is affected by the glass to
lignocellulose ratio. In general, there is a limit, i.e., maximum
or minimum, for the change of the properties with the change of the
glass to lignocellulose ratio. The lignocellulose substrate has been
found to affect the composite properties. Ball-milling of the substrate deteriorates the properties. Composites prepared with the
use of sodium bisulfite-soda lime glass system as initiator showed,
in general, improved properties, compared with ceric ammonium sulfate. The reverse was achieved on prehydrolysing the substrate.
Prehydrolysis leads to composites of inferior properties, in general.
Alkaline pre-treatment of bagasse ground or semichemical pulp, depressed the compression strength. The other properties improved for
composites from bagasse semichemical pulp, while improved or deteriorated for samples from ground bagasse, depending on the kind and concentration of the alkali used in pre-treatment, also, on the particle
size of the ground bagasse. Subsequent mechanical treatment of alkaline treated unground bagasse resulted in improved properties, compared to those due to mechanical nontreatment of the ground one.

INTRODUCTION

In Part I of this series[1] the properties of the lignocellulose-
polymethyl methacrylate composites were found to be affected by both
the polymer load and the mesh size of the ground bagasse. Grafting
in absence of soda lime glass from the initiator system sodium bi-

sulfite-soda lime glass, led to composites which properties differed
greatly to those containing glass. Some of the properties of the
composite containing cupric oxide instead of soda lime glass of the
initiator system were improved, while those of the one containing fer-
ric oxide were deteriorated. A composite from a true grafted sample
(homopolymer free) showed improvement for some of the properties,
while the others were deteriorated. Composites prepared from impregnated bagasse with polymer or homopolymer behaved differently in their
properties and from those of crude grafted samples, depending on the
mesh size of the ground bagasse they are made from.

In the present work, further investigation of the influence of
the polymer load on the properties of the composite was taken into
consideration. It was also aimed to investigate the effect of the
glass to lignocellulose ratio, the substrate, state of cellulose and
initiator, as well as the prehydrolysis and alkaline treatments of
the lignocellulose substrate followed or not by mechanical ones on
the properties of the composite.

EXPERIMENTAL

Bagasses, ground, ball-milled, prehydrolyzed and alkaline treated
were grafted with methyl methacrylate or ethyl acrylate to be used for
the preparation of the composite samples. Semichemical pulp from bagasse of Edfu Company of A.R. of Egypt, ground rice straw, mixture of
spruce and pine sulfate pulp and prehydrolyzed wheat straw were also
similarly grafted and processed into composites. The ground and ball-
milled substrates were pre-extracted with methanol-benzene mixture
1:1 before its use. 2% sulfuric acid based on substrate was used
for prehydrolysis, at liquor ratio 20:1, for 1 hr., at 120°C. Alkaline treatment was affected with 2, 4% sodium hydroxide and 12.5,
25% ammonium hydroxide solutions, at liquor ratio 20:1, for 24 hr.,
at 25°C.

Mechanical treatment, namely, cutting or beating, was carried
out in a Jokro mill beater on unground alkaline treated bagasse, at
3 and 6% consistency, for cutting and beating, respectively.

Grafting was carried out using ceric ammonium sulfate or sodium
bisulfite-soda lime glass system as initiators. Ten grams of the
lignocellulosic substrate are placed in a 500-ml well-stoppered
ground joint glass bottle. In case of using sodium bisulfite-soda
lime glass system as initiator, 0.3-g sodium bisulfite is added in
a ratio to lignocellulose 3:100, at glass to lignocellulose ratio
1:2, liquor to total solid ratio 10:1, liquor to lignocellulose ratio
15:1 and monomer to lignocellulose ratio 1:2. Other glass and monomer
to lignocellulose ratios, also liquor to total solid ratio were used
to obtain various polymer loads. Samples 1, 4 and 3, 6; Table 1,
were prepared at monomer to lignocellulose ratios 1:1, for the two

former and 1:3, for the two latter. Sample 13; Table 2, was prepared at liquor to total solid ratio 20:1 and liquor to lignocellulose ratio 30:1. Purified nitrogen is passed through the mixture for ten minutes. With the use of ceric ammonium sulfate initiator, 100-ml 0.3-g of it in 1% sulfuric acid are added, followed by 50-ml distilled water. Purified nitrogen is passed for ten minutes, then 5-ml of the monomer is added.

The reaction bottle is placed in a water bath at $40^{\circ}C$ and occasionally shaken for twenty four hours when using sodium bisulfite-soda lime glass system as initiator and 4 hours when using ceric ammonium sulfate. The product is thoroughly washed with water, left to dry at $60^{\circ}C$ and then weighed for the determination of the crude grafting yield.

Crude-grafting yield: $C\% = (F-S) - Z/Z$

S, Z and F are the weights of glass, lignocellulose and crude grafted product, respectively.

About ten grams of the product were heated at $170^{\circ}C$ in a disk form (mould), then pressed under 160 kg/cm^2, for ten minutes. The resulting composite samples were investigated for:

1. Compression strength: determined as the weight at which the composite sample is deformed or breaks (ton/cm^2).

2. Percent deformation: calculated from the difference between the length before and after compression divided by the first length X 100.

3. Hardness: measured by a hardness testing machine HPK by Ball Kögel, Leipzig.

4. Compression strength to percent deformation: calculated by the Poisson ratio obtained for the composite (Ref. 2, p. 127).

5. Compressibility K: calculated according to the following equation (Ref. 2, p. 30):

$K = -(1/V) (dV/dP)$

in which V is volume and P is pressure.

RESULTS AND DISCUSSION

Effect of Glass to Lignocellulose Ratio on the Change of the Properties with the Polymer Load

From Part I of this series[1] the way in which the properties of
lignocellulose-polymer composites obtained from crude grafting with
polymethyl methacrylate of ground bagasse changed with the polymer
load was assumed to relate to either one or both of the following:

   i. Differences in the particle size of the ground bagasse.

   ii. a limit, i.e., maximum or minimum for the change of the properties with the polymer load.

   Table 1 of the present work shows a maximum compression strength
with the change of the polymer load, for glass to lignocellulose
ratio 1:2, while values increased for 1:3 ratio. However, it is
noticeable that for the latter ratio, the compression strength
achieved at 30.0% polymer load was practically equal to the highest
(maximum) one obtained at 8.8% polymer load for glass to lignocellulose ratio 1:2. This may therefore indicate the shift of the limit
(maximum) of the compression strength beyond 30.0% polymer load for
the glass to lignocellulose ratio 1:3. In other words, two different
glass to lignocellulose ratios, for the change of the compression
strength with the polymer load were studied.

   Regarding hardness, decreased values accompanied the increase
of the polymer load at glass to lignocellulose ratio 1:2, while a
maximum was achieved at 1:3; Table 1.

Table I

Glass to ligno-cellulose ratio		1.2			1:3	
Sample no.	1	2	3	4	5	6
Polymer load (%)	4.0	8.8	22.2	8.2	15.3	30.0
Density (g/cm^3)	1.42	1.33	1.34	1.36	1.36	1.12
Compression strength (ton/cm^2)	2.48	3.77	1.90	1.96	2.73	3.34
Hardness (HPK)	410	380	210	746	910	470
Deformation (%)	25.2	36.1	38.7	32.0	40.2	55.3
Compression strength to % deformation	0.10	0.10	0.05	0.06	0.07	0.06
Compressibility (K)	0.022	0.021	0.045	0.035	0.026	0.037
Water uptake (%)						
after 24 hr	49.8	37.9	28.8	37.7	31.1	17.7
72 hr	53.2	39.7	30.1	39.3	32.3	19.9
96 hr	53.3	39.8	30.2	39.4	32.4	20.2
240 hr	57.3	46.9	33.5	45.8	39.1	22.1

Increased percent deformation and decreased water uptake followed increased polymer load for both 1:2 and 1:3 glass to lignocellulose ratios.

As the compression strength to percent deformation and the compressibility are products of the two former properties, they therefore changed or behaved differently with the increase of the polymer load for both glass to lignocellulose ratios; 1:2 and 1:3.

In general, the change of the properties of lignocellulose-polymer composites with the polymer load is affected by the change of the glass to lignocellulose ratio.

The results of Table 1 also show that at practically the same polymer load, namely, 8.8 and 8.2%, the compression strength, compression strength to percent deformation and compressibility became worse, while the percent deformation and the hardness improved, on changing the glass to lignocellulose ratio from 1:2 to 1:3, i.e., on decreasing the glass content from 50 to 33.3%. The water uptake was kept only practically the same.

From Part I of this series[1] it has been shown that absence of glass (0%) in the lignocellulose-polymer composites obtained from crude graftings of ground bagasse with polymethyl methacrylate improved the compression strength, percent deformation, compression strength to percent deformation and the compressibility. Only the hardness became worse.

Summing up, it may be concluded that there is, in general, a limit, i.e., maximum or minimum, for the change of the composite properties with the change of glass to lignocellulose ratio.

Effect of Substrate

In Part I of this series[1] it was shown that among the samples of composites of crude graftings containing glass, the least deformation and the highest compression strength and compression strength to percent deformation were obtained from bagasse of 60 mesh of the intermediate polymer load; 15.3%, while the highest hardness from bagasse ground to 40 mesh. These results are tabulated here in Table 2 (samples 12 and 13, respectively) and were compared with the best properties relating to the samples obtained from the semichemical pulp similarly prepared at the same glass to lignocellulose ratio; 1:2 (sample 2; Table 1). The comparison reveals that with the exception of the compression strength, the composite from semichemical pulp is inferior than from ground bagasse. The same result is also achieved if the comparison is carried out at practically the same polymer load (sample 3, Table 1 and sample 13, Table 2).

Table 2

Monomer	MMA			EA		MMA	
Initiator	NaHSO$_3$-SLG			CAS		NaHSO$_3$-SLG	
Sample no.	7	8	9	10	11	12	13
Polymer load (%)	15.0	39.0	4.4	2.0	2.3	15.3	23.3
Density (g/cm³)	1.32	1.46	1.17	0.98	1.07	1.42	1.23
Compression strength (ton/cm²)	1.25	1.80	1.70	0.84	0.69	1.53	0.97
Hardness (HPK)	540	350	135	150	150	430	510
Deformation (%)	10.4	17.0	24.1	17.3	6.8	7.7	12.0
Compression strength to % deformation	0.12	0.10	0.07	0.05	0.10	0.20	0.08
Compressibility (K)	0.016	0.018	0.028	0.036	0.020	0.010	0.024
Water uptake (%) after 24 hr	23.3	17.7	30.2	----	48.0	24.9	----
72 hr	23.9	----	----	----	----	----	----
96 hr	25.4	20.4	29.8	----	49.4	----	----
240 hr	26.9	20.6	30.1	----	49.4	27.2	34.5

Samples: 7, rice straw 60 mesh; 8 & 9, mixture of spruce and pine sulfate pulps; 10, ball-milled bagasse; 11 & 12, bagasse 60 mesh; 13, bagasse 40 mesh.
MMA: methyl methacrylate. EA: ethyl acrylate.
NaHSO$_3$-SLG: sodium bisulfite–soda lime glass. CAS: ceric ammonium sulfate.

The result above achieved relates to partial removal of the lignin and incrustants, such as silica and inorganic salts, present in bagasse, i. e., purification, due to semichemical pulping; Table 6. In this table, ash refers to silica and inorgnanic contents. From Part I[1], absence of glass (silica) leads to inferior hardness, while superior compression strength. Also, as a result of purification, more cellulosic hydroxyl groups become available, hence increased compression strength due to increased hydrogen bonding. It is noteworthy that practically no change in the hemicellulose content, respectively the hydroxyl groups, accompanied the production of the semichemical pulp; Table 6. Removal of lignin may lead to increased elasticity of the fiber walls, respectively, increased percent deformation and compressibility.

A composite obtained from rice straw ground to 60 mesh was compared with that similarly obtained from bagasse having practically the same polymer load (samples 7 and 12, respectively; Table 2). The comparison reveals practically equal values for the water uptake. The other properties were superior for bagasse than rice straw. The hardness was only an exception. Its value was less for the sample from bagasse than from rice straw. This again relates to the lower content of inorganic matter of the former than the latter; ash determination showed 3.85 and 16.23% for bagasse and rice straw, respectively.

Effect of State of Cellulose

A composite obtained through grafting of polyethyl acrylate onto ball-milled bagasse using ceric ammonium sulfate as initiator was of distinctly low polymer load; 2.0% (sample 10: Table 2). This agrees with what was achieved in a previous work that ball-milling leads to the deterioration of the grafting yield[3]. The poorer properties of this sample of compsoite related to its low polymer load and the destruction of the fiber structure due to ball-milling.

Effect of Initiator

Ceric ammonium sulfate and sodium bisulfite-soda lime glass system were used as initiators for grafting polymethyl methacrylate onto bagasse ground to 60 mesh, also onto a mixture of spruce and pine sulfate pulps.

Table 2 shows distinctly decreased polymer loads for the samples prepared with the use of ceric ammonium sulfate than with sodium bisulfite-soda lime glass system, from ground bagasse (samples 11 and 12, respectively) and the mixture of spruce and pine sulfate pulps (samples 9 and 8, respectively). This is consistent with what was achieved in a previous work[4].

Due to the presence of glass in the composite samples prepared with the use of sodium bisulfite-soda lime glass system, their hardness was superior to those of the corresponding samples obtained

Table 3

Prehydrolyzed substrate	Bagasse		wheat straw	
Monomer	EA		MMA	
Initiator	CAS		NaHSO$_3$-SLG	
Sample no.	14	15	16	17
Polymer load (%)	10.0	5.9	2.0	2.0
Density (g/cm^3)	1.04	1.02	1.14	0.76
Compression strength (ton/cm^2)	0.75	0.61	0.46	det.
Hardness (HPK)	120	-----	------	230
Deformation (%)	10.2	10.1	8.7	det.
Compression strength to % deformation	0.07	0.06	0.05	det.
Compressibility (K)	0.028	0.033	0.036	det.
Water uptake (%)				
after 48 hr	60.3	24.9	57.1	170.1
72 hr	126.6	37.9	61.5	-----
96 hr	-----	41.9	------	-----
240 hr	-----	41.9	66.5	-----

EA: ethyl acrylate. MMA: methyl methacrylate
CAS: ceric ammonium sulfate
NaHSO$_3$-SLG: sodium bisulfite-soda lime glass
det.: deteriorated

with the use of ceric ammonium sulfate initiator. The other properites were also, in general, superior. This may be related to the higher polymer load which accompanied the use of sodium bisulfite-soda lime glass system than ceric ammonium sulfate.

Effect of Prehydrolysis

Prehydrolyzed bagasse led to composites of distinctly low polymer load; 2.0%, when using sodium bisulfite-soda lime glass system as initiator for grafting polymethyl methacrylate (sample 16, Table

3), compared with the samples similarly prepared from the semichemical pulp (Table 1) and ground bagasse (cf. Table 1, Part I[1]).
This may relate to the oxidizing effect of sulfuric acid on the polyhydric alcohol structure of lignin and its transformation into the
quinonoid form. Oxidation of the polyhydric alcohol of lignin into
quinone has been achieved on using ceric ammonium sulfate which inhibited graft polymerization reactions on using this initiator[5].
The comparison reveals also, in general, inferior properties for the
sample from prehydrolyzed bagasse. This may relate to its distinctly
low polymer load.

The polymer load of the composite sample similarly prepared from
prehydrolyzed wheat straw (sample 17, Table 3), was also as low as
that from prehydrolyzed bagasse. However, its properties were deteriorated compared with those of the latter. This may relate to the
effect of the substrate.

On the other hand, the polymer load of the composite obtained
from prehydrolyzed bagasse was slightly raised when using ceric
ammonium sulfate as initiator (sample 15, Table 3), compared with the
one similarly obtained from bagasse ground to 60 mesh (sample 11,
Table 2). Here also, the properties of the composite from prehydrolyzed bagasse were inferior, in spite of its relatively higher polymer load. Only, the water uptake was superior, since it is rather
affected by the polymer load[1]. Inferior properties may relate to
hemicellulose removal which on determination was found to be 6.0%
for prehydrolyzed unground bagasse, compared with 23.69 and 27.0%
for bagasse ground to 20 and 40 mesh, respectively, and treated with
methanol-benzene mixture; Table 6. Removal of hemicellulose leads,
on the other hand, in decreased hydrogen bonding, respectively, compression strength, and in increased voids, respectively increased
deformation.

In conclusion, prehydrolysis has, in general, a deteriorating
effect on the properties of the composites, with the exception of
the water uptake which change follows the polymer load.

It has been shown before that composites prepared with the use
of sodium bisulfite-soda lime glass system as initiator showed,
in general, improved properties, compared with the use of ceric
ammonium sulfate. The reverse was observed on prehydrolyzing the
substrate before grafting. Only, the percent deformation was exception (samples 15 and 16; Table 3). The reverse change or properties may relate to the reverse change of polymer loads.

A composite obtained from prehydrolyzed bagasse grafted with
polyethyl acrylate with the use of ceric ammonium sulfate was of a
higher polymer load, compared with the one similarly prepared through
grafting with polymethyl methacrylate (samples 14 and 15, respectively; Table 3). The former sample showed improved compression strength,

Table 4

	20 Mesh	40 Mesh		20 Mesh		40 Mesh	Semichemical pulp	
Bagasse								
Treatment	Ammonium hydroxide				Sodium hydroxide			
Concentration (g%)	12.5		25	4		2	24	4
Sample no.	18	19	20	21	22	23	24	25
Polymer load (%)	3.3	23.3	23.3	7.0	14.0	2.6	29.0	29.3
Density (g/cm^3)	1.05	0.72	1.03	1.15	0.94	0.91	1.36	1.28
Compression strength (ton/cm^2)	1.33	0.70	2.15	0.82	0.59	0.30	0.87	1.17
Hardness (HPK)	140	210	550	120	688	240	339	788
Deformation (%)	24.7	31.3	27.7	21.5	44.8	48.1	18.0	21.4
Compression strength to % deformation	0.050	0.020	0.080	0.040	0.013	0.006	0.048	0.055
Compressibility (K)	0.038	0.071	0.027	0.052	0.161	-----	0.044	0.039
Water uptake (%) after 48 hr	139.8	49.2	60.1	70.2	36.3	71.5	20.2	20.5
72 hr	256.5	58.8	62.5	88.8	38.5	89.6	25.9	26.3
96 hr	-----	-----	-----	91.2	45.0	90.9	29.1	28.3
240 hr	-----	-----	-----	218.0	70.8	220.0	36.1	35.9

compression strength to percent deformation and compressibility, compared with the latter. This may relate to the higher polymer load
of the sample grafted with polyethyl acrylate. However, the percent
deformation was practically the same for both samples, while the water
uptake was less for the sample possessing the lower polymer load.
This may relate to the difference of the type of polymer, which may
also affect the other properties.

Effect of Alkaline Treatment

Ground bagasse was processed into composite after treating with
ammonium or sodium hydroxide solutions. Decreased polymer loads were
achieved (Table 4), compared with the correspondings similarly prepared from bagasse ground to the same mesh size and extracted with
methanol-benzene mixture; cf. Table 1, Part I[1]. This may relate
to insufficient removal of pectins and waxes (Table 6) which retard
or hinder the process of grafting.

For semichemical pulp of bagasse, where no methanol-benzene extraction has taken place, alkaline treatment with sodium hydroxide
solutions resulted in increased polymer loads (samples 24 and 25;
Table 4), compared with the correspondings similarly prepared alkali
untreated ones (Table 1). This may relate to increased accessibility
and surface area of the cellulose produced by swelling in alkali.

On the other hand, relatively higher polymer loads accompanied
increased concentration of alkaline treatment for bagasse ground to
the same mesch size (samples 18 and 20; 23 and 22) or grinding to
finer particle size when treating with the same concentration of alkali (samples 18 and 19; 21 and 22). Increased polymer load may relate to one or both of the following:

   i. increased accessibility and surface area of the cellulose
produced by increased swelling and increased removal of lignin with
the higher alkali concentrations (Table 6), or through grinding to
finer particle size.

   ii. transformation of cellulose I into the more reactive cellulose II, respectively extent of transformation, on treating with
the mercerizing concentrations of alkali, namely, 12.5 and 25% ammonium hydroxide.

   iii. presence of different limits for the polymer loading for
different particle sizes.

For semichemical pulp, practically equal polymer loads related
to the alkaline treatment with both 2 and 4% sodium hydroxide. This
may be contributed either to the achievement of a maximum or to a
difficult penetration of the grafting chemicals through a higher com-

pactness of the fiber layers as a result of a higher swelling with the higher alkali concentration, respectively, the deswelling accompanying washing the alkali with water. This state may be achieved due to the absence or partial removal of the fiber walls of bagasse semichemical pulp.[6,7,8]

Regarding properties, decreased compression strength; Table 4, were achieved for samples of composites obtained due to alkaline pretreatment of bagasse ground to 40 mesh treated with 12.5% ammonium hydroxide (sample 19) and bagasse semichemical pulp treated with 2 and 4% sodium hydroxide (samples 24 and 25), compared with the correspondings due to alkali untreatment, similarly prepared; Table 2, sample 13, and Table 1, samples 1, 2, and 3, respectively. This may relate to hemicellulose removal which accompanies alkaline pretreatment; Table 6.

However, for bagasse ground to the same mesh size or bagasse semichemical pulp, increased concentration of the alkaline treatment resulted in relatively increased compression strength. This may relate to one or more of the following:

    i.   increased polymer load.

    ii. increased available hydroxyl groups, respectively, hydrogen bonding due to increased swelling.

    iii. increased delignification than hemicellulose removal; Table 6.

For bagasse ground to different mesh sizes, treated with the same concentration of alkali, lower compression strengths were achieved for those ground to 40 than to 20 mesh, inspite of the higher polymer load and the higher delignification accompanying treating of the finer particle size with alkali; Table 6. This may reveal the influence of particle size on the compression strength. It is noteworthy that hemicellulose removal due to alkaline treatment was practically the same for both particle sizes, or lesser with bagasse of the finer particle size; Table 6.

As regards hardness, a distinctly lower value; 210, was achieved for the sample from bagasse ground to 40 mesh, treated with 12.5% ammonium hydroxide (sample 19, Table 4), compared with 510, for the one of the same polymer load; 23.3%, produced as a result of alkali untreatment of bagasse ground to the same mesh size (sample 13, Table 2). This may relate to the removal of some of the incrustants through the alkaline treatment.

However, for semichemical pulp, alkaline treatment resulted in distinctly high hardness (samples 24 and 25, Table 4), compared with

samples similarly obtained at glass to lignocellulose ratio 1:2,
due to alkali nontreatment (Table 1); taking into consideration that
decreased hardness accompanies increased polymer load at this glass
to lignocellulose ratio. Also, Table 4 shows highest hardness; 550,
688, and 788, for samples from bagasse ground to 20 mesh, treated
with 4% sodium hydroxide, respectively (samples 20, 22 and 25, respectively.

In previous work[6,7,8] it has shown that treatment of cellulose with sodium hydroxide solutions, even at concentrations as
low as 5%, led to transformation of the fiber layers into higher
states of order or/and strengthening of the fiber wall. In partial
removal or absence of the latter, the outer layers of the fibers
are strengthened through being highly compacted. These structural
changes have been related to the swellability which accompanies the
alkaline treatment, followed by deswelling on washing with water.
The extent of the changes depends on the rate of deswelling which
in turn depends on the extent of swelling, respectively, sodium hydroxide concentration of the alkaline treatment and presence or absence of the fiber wall, also, the extent of elasticity of the latter
if present.

Based on this, the highest hardness achieved due to treatment
with relatively higher alkali concentrations, namely, 25% ammonium
hydroxide and 4% sodium hydroxide, may relate therefore to fiber
wall strengthening or strengthening of the outer fiber layers in
absence or partial removal of the fiber wall, which may be the case
with the semichemical pulp.

However, it may be due to the coarser particles of bagasse
ground to 20 mesh, the penetration of sodium hydroxide through the
cellulosic material is difficult, leading to slight swelling, respectively, structural changes. A low hardness; 120 (sample 21),
is thus achieved.

Also, bagasse ground to 40 mesh or the semichemical pulp, when
treated with the lower concentrations of alkali; 2% sodium hydroxide
and 12.5% ammonium hydroxide, lesser swelling, respectively, lesser
strengthening of the fiber walls may be achieved, leading to relatively lower hardness; 210, 240 and 339 (samples 19, 23 and 24, respectively).

Again, it may be due to the coarser particles of bagasse ground
to 20 mesh and the lower concentration of ammonium hydroxide; 12.5%,
used for treatment, still lesser structural changes are achieved, respectively the hardness; 140 (sample 18), compared with 210, for the
sample similarly produced from bagasse ground to the finer particle
size; 40 mesh (sample 19).

In conclusion, the differences in the degree of hardness which

follow the alkaline treatments of bagasse, may relate to lignin removal and differences in the structural changes of the fibers which
accompany such treatments, and which depend on the kind and concentration of the alkali used, and the type of bagasse and its particle
size if ground.

On considering percent deformation, the values achieved for composites from ground bagasse treated with alkaline solutions, were in
general higher (Table 4), than those of the similarly prepared ones
of crude graftings obtained without alkali treatment; cf. Table 1,
Part I[1]. This may relate to increased voids and increased elasticity of the fiber walls due to lignin and hemicellulose removal;
Table 6. Kerr and Goring, found that the removal of hemicellulose
from wood increased the average pore size in the cell wall during
all states of delignification and overall delignification rate of
the fiber wall with respect to that of the middle lamella. In addition, for ammonia treated ground bagasse increased deformation may
also relate to the enlargement of the capillaries[10] and the loosening
of the cell walls.[11]

However, for the same alkaline treatment differences in the percent deformation were achieved following the differences in the concentration of the alkali treatment and the particle size of the
ground bagasse.

Thus, Table 4 shows that bagasse ground to 40 mesh possessed relatively lower percent deformation when treated with 4 than 2% sodium
hydroxide solution; samples 22 and 23, respectively, in spite of the
criteria favouring the reverse, namely, increased lignin and hemicellulose removal (Table 6) and the higher polymer load accompanying
treatment with the higher sodium hydroxide concentration. The
achieved result may relate to fiber wall strengthening due to treating with the higher concentration of sodium hydroxide.

Also, bagasse ground to 20 mesh, treated with 4% sodium hydroxide solution (sample 21), showed distinctly lower percent deformation, compared with the one due to bagasse ground to 40 mesh, treated with 2% sodium hydroxide (sample 23) inspite again of the criteria favouring the reverse, namely, increased removal of lignin
(Table 6) and the relatively higher polymer load of the former sample; the change in the hemicellulose content was practically the
same (Table 6). Here, the achieved result may relate to decreased
void contents of the sample ground to 20 mesh, and this again reveals the influence of particle size on the composite properties
It is noteworthy that for bagasse ground to 20 mesh, treatment with
4% sodium hydroxide did not result in fiber wall strengthening as
previously indicated on dealing with hardness.

However, the decrease in percent deformation was only slight
due to bagasse of 20 and 40 mesh, when treated with 25 and 12.5%

ammonium hydroxide, respectively (samples 20 and 19) inspite of the
practically equal polymer loads and the slight differences in the
removal of lignin and hemicelluloses (Table 6); also, the fiber wall
strengthened, accompanied treatment with 25% ammonium hydroxide, as
previously indicated by the value of hardness. The slight decrease
in percent deformation may therefore relate to easier deformation
of cellulose II than cellulose I, where more of the former will be
produced as a result of treating with the higher concentration of
ammonium hydroxide.

For semichemical pulp, alkaline treatment led to composites
(samples 24 and 25, Table 4), of percent deformation less than those
obtained due to alkali untreatment (Table 1); taking into consideration that increased percent deformation accompanies increased polymer load. This may relate to compactness of the fiber layers which
accompanies treatment of the semichemical pulp with alkaline solutions, then washing with water, due to the absence or partial removal of the fiber walls during the production of this type of pulp,
as previously stated.

On conclusion, one may summarize the factors influencing the
percent deformation of an alkali treated lignocellulosic substrate
as: particle size, extent of removal of hemicellulose and lignin,
transformation of cellulose I into II, compactness of the outer fiber layers and fiber wall strengthening.

The change of compression strength to percent deformation and
the compressibility of the composite samples produced from alkali
treated ground bagasse and semichemical pulp, followed the change
of both the compression strength and percent deformation (Table 4).

As regards the water uptake of samples of alkali treated ground
and semichemical pulp of bagasse, it is observed from Table 4 that
treatment with 2 and 4% sodium hydroxide solutions led to composites
which water uptakes run in accordance to the polymer load; decreased
values for the water uptake with increased polymer load (samples
21-25).

The same also applies to samples of ground bagasse treated with
ammonium hydroxide solutions. However, it is observed that at the
same polymer load, different values for the water uptake were
achieved on treating bagasse with different concentrations of ammonium hydroxide; 12.5 and 25% (samples 19 and 20, respectively).

Also, at practically the same polymer load for samples of bagasse ground to 40 mesh, treated with 2% sodium hydroxide and to
20 mesh, treated with 12.5% ammonium hydroxide (samples 23 and 18,
respectively), contrary to expectation, the higher value for the
water uptake related to bagasse of the decreased surface, i.e.,
ground to 20 mesh.

The results above achieved may be attributed to partial mercerization which accompanies treatment of bagasse with high concentrations of alkali, namely, 12.5 and 25% ammonium hydroxide and which
may lead to the transofrmation of cellulose I into the more hygroscopic cellulose II. Increased mercerization is known to accompany
increased mercerizing alkali concentrations.

Alkali-Mechanical Treatment

Unground bagasse treated with 2 or 4% sodium hydroxide was subjected to mechanical treatment, namely, cutting or beating, before
processing into composite. Distinctly high polymer loads were
achieved (Table 5), compared with those (samples 21, 22 and 23; Table 4) similarly prepared from ground bagasse treated with the same
concentrations of sodium hydroxide and mechanically untreated. This
may relate to increased surfaces through cutting or beating.

Increased surfaces may also contribute for the increased polymer load from 23.3 to 26.6% due to the increase of time of cutting
from 3 to 10 minutes or beating till 10 minutes; samples 29, 28,
and 27, respectively; Table 5. It is a known fact that increasing
the time of cutting decreases the fiber length, consequently increased surface area.

Again, increasing the sodium hydroxide concentration of the
alkali treatment from 2 to 4% resulted in increased polymer load
from 26.6 to 33.3%, on cutting till 10 minutes (samples 28 and 26,
respectively). This may relate to increased surface on cutting bagasse treated with the higher sodium hydroxide concentration due
to an accompanying increased swellability.

The compression strength showed also distinctly higher values
(Table 5), compared with those (samples 21, 22 and 23, Table 4),
for samples similarly prepared from ground bagasse treated with the
same sodium hydroxide concentrations and mechanically untreated.
The increase in compression strength may correlate with the increase
of polymer load. Increased surfaces, respectively available hydroxyl
groups and hydrogen bonding, obtained as a result of the mechanical
treatment may also contribute for the increased compression strengths.

For samples subjected to mechanical cutting (samples 26, 28 and
29; Table 5), the results of the compression strength refer to the
existence of a maximum at 23.3% polymer load (sample 29). The sharp
increase of compression strength (sample 27) accompanying beating
till 10 minutes may also contribute to a sharp increase in the available hydroxyl groups through fibrillation brought about by the
beating process.

As regards hardness, samples of the unground bagasse treated

Table 5

	Sodium hydroxide concentration (g%)			
Alkaline treatment	4		2	
Mechanical treatment	Cutting (10 min)	Beating (10 min)	Cutting (10 min)	Cutting (3 min)
Sample no.	26	27	28	29
Polymer load (%)	33.3	26.6	26.6	23.3
Density (g/cm^3)	1.40	1.19	1.21	1.26
Compression strength (ton/cm^2)	1.50	4.95	1.47	2.37
Hardness (HPK)	400	---	540	780
Deformation (%)	13.1	44.1	30.4	29.6
Compression strength to % deformation	0.11	0.11	0.05	0.08
Compressibility (K)	0.017	0.013	0.045	0.014
Water uptake (%)				
after 48 hr	22.5	40.7	34.2	41.1
72 hr	32.4	41.1	34.2	44.1
96 hr	33.7	44.4	36.4	48.2
240 hr	33.9	74.7	36.7	71.1

with 2% sodium hydroxide, then mechanically cutted (samples 28 and 29; Table 5), showed higher hardness; 540 and 780, respectively, compared with 240 for the sample of bagasse ground to 40 mesh, treated with the same sodium hydroxide concentration and mechanically untreated (sample 23, Table 4). The same was also achieved with respect to treatment with 4% sodium hydroxide, where higher hardness; 400, was achieved for the sample from unground bagasse treated with this sodium hydroxide concentration and mechanically cutted (sample 26, Table 5), compared with 120, for the sample from bagasse ground to 20 mesh, similarly treated with sodium hydroxide and mechanically untreated (sample 21, Table 4). From the achieved results, it is obvious that mechanical cutting prefers grinding with respect to hardness. This may correlate with the influence of particle size, respectively, fiber length, on the hardness.

Among the samples of compsoites prepared from unground, alkali-mechanically treated bagasse, the hardness values followed the polymer loads. The hardness showed decreased values from 780 to 540 to 400, as the polymer load increased from 23.3 to 26.6 to 33.3% (samples 29, 28 and 26, respectively; Table 5).

For percent deformation, unground alkali-mechanically treated

Table 6

Treatment	Waxes & resins (g%)	Ash (g%)	Lignin (g%)	Extract.[§] hemicell. (g%)
		Bagasse 40 Mesh		
Methanol/ benzene (1:1)	0.0	4.85	19.58	27.0
2% sodium hydroxide (g%)	2.98	----	11.19	13.5
4% sodium hydroxide (g%)	2.90	----	6.36	10.5
12.5% ammonium hydroxide (g%)	2.25	3.92	18.85	25.8
25% ammonium hydroxide (g%)	3.22	----	11.61	23.6
		Bagasse 20 Mesh		
Methanol/ benzene (1:1)	0.0	----	21.62	23.69
2% sodium hydroxide (g%)	1.67	----	12.14	12.73
4% sodium hydroxide (g%)	2.06	----	7.07	10.30
12.5% ammonium hydroxide (g%)	1.06	----	21.58	22.55
25% ammonium hydroxide (g%)	3.37	----	17.45	19.13
		Semichemical Pulp		
Untreated	----	0.82	2.63	26.94
2% sodium hydroxide (g%)	----	----	2.55	11.52
4% sodium hydroxide (g%)	----	----	1.06	8.88

[§]Extractable hemicellulose

bagasse showed, in general, lower values (Table 5) than the corresponding ground bagasse similarly treated with sodium hydroxide solutions and mechanically untreated (Table 4). This may relate to
decreased voids accompanying the former mechanical treatment, compared with grinding.

However, among the samples obtained from unground bagasse treated with 2% sodium hydroxide, the one undergoing beating till 10
minutes (sample 27, Table 5), showed a relatively high percent deformation; 44.1. This may relate to the fibrillation brought about
by the beating process and which may therefore lead to increased
voids, respectively, increased deformation.

Again, the change in compression strength to percent deformation and the compressibility of samples of unground bagasse treated
with 2 and 4% sodium hydroxide, then mechanically treated, followed
the changes in the compression strength and percent deformation
(Table 5).

The water uptake of alkali-mechanically treated unground bagasse
followed the polymer load and decreased values accompanied increased
polymer load (Table 5). However, at the same polymer load, higher
values for water uptake accompanied beating till 10 minutes, compared with cutting for the same time (samples 27 and 28, respectively; Table 5). This may relate to increased opening of the cellulose
structure due to fibrillation which accompanies the beating process.

REFERENCES

1. A. Nagaty, A.B. Mustafa, and O.Y. Mansour, Lignocellulose-Poly-
    mer Composite. I, J. Appl. Polym. Sci. 23:3263-3269 (1979).
2. E. Scala, "Composite Materials for Combined Functions," Hayden
    Book Company, Inc., Rochelle Park, New Jersey (1973).
3. A. Nagaty and O.Y. Mansour, Some Aspects of Graft Polymerization
    of Vinyl Monomers onto Cellulose by Use of Tetravalent Cerium.
    VI., J. Appl. Polym. Sci. 23:2425-2434 (1979).
4. O.Y. Mansour and A. Nagaty, Graft Polymerization of Vinyl Mono-
    mers onto Cellulose in Presence of Soda Lime Glass, I, J. Polym.
    Sci., Polymer Chemistry Edition 13:2785-2793 (1975).
5. V. Hornof, B.V. Kokta and J.L. Valade, The Xanthate Method of
    Grafting of Acrylonitrile onto High-Yield Pulp, J. Appl. Polym.
    Sci. 20:1543-1554 (1976).
6. O.Y. Mansour, M. El-Saady and F.A. Mottaleb, On Structure of
    Cellulose. I. Study of Changes due to Alkaline Treatment by
    Infrared Spectroscopy, Indian Pulp and Paper 26:71-84 (1972).
7. O.Y. Mansour, On Structure of Cellulose. II. Influence of Alka-
    li Treatment on Lateral Order Distribution, Indian Pulp and Pa-
    per 26:124-128 (1972).
8. O.Y. Mansour and A. Nagaty, On Structure of Cellulose. III.

Processes Accompanying Fiber Deswelling due to Alkali Treatment and Their Influence on Structural Changes of Cellulose, Indian Pulp and Paper 26: No. 12 (1972).

9. A.J. Kerr and D.A.I. Goring, Role of Hemicellulose in the Delignification of Wood, Can. J. Chem. 53 (6):952-959 (1975).

10. O. Labsky, Hydroreactivity and Plasticity of Wood. II. Effect of Ammonia on Chemical Shrinkage of Wood, Drev. Vysk. 19 (4): 203-214 (1974); C.A., 82:166012u (1975).

11. O. Labsky and L. Suty, Some Chemical Changes in Ammonia-Treated Beechwood, Drev. Vysk. 20 (4):161-177 (1975); C.A. 86:44913e (1977).

FAST CURING COPOLYMER RESINS OF PHENOL, FORMALDEHYDE, AND CHEMICALS

FROM FOREST AND AGRICULTURAL RESIDUES

Chia M. Chen

School of Forest Resources
The University of Georgia
Athens, Georgia 30602

INTRODUCTION

For many years, barks and agricultural residues were regarded as waste materials to be disposed of as cheaply as possible. However, in this era of material shortages and ecology awareness, increasing efforts have been directed away from mere disposal and toward positive utilization of barks and residues.

Bark extracts, lignin, tannin and other polyphenols obtained from renewable resources have been proposed for use as raw materials in resins and plastics for more than thirty years. A great number of reports dealing with the basic characteristics of bark and residues have been published[1,2,3,4,5,6,7,8]. These extensive studies have been reviewed from time to time[9,10,11]. The investigations have ranged from molding compounds and laminating resins to adhesives for wood products[4,5]. Some of the investigations have resulted in the issuance of patents[12,13,14], yet there seems to have been little in the way of industrial application of these discoveries on a commercial scale. Possible reasons for lack of application of this technology are because of (1) the lignin or residue resins seem to require excessive cure times, (2) the reproducibility of resin qualities is often not good, and/or (3) the economic factors in the basic resin cost.

Since the oil embargo of 1974, the price of phenol has almost tripled and the balance of supply and demand has become unfavorable to the users. Consequently, new sources of dependable phenolic materials are urgently needed.

169

After reviewing the structure and chemistry of the main constituents of polyphenols and the possible mechanism of their polymerization, it was believed that some useful phenolic copolymer
resins with better quality as adhesives could be synthesized by
carefully controlling and manipulating the process of raw materials
and synthesis of resins.

A series of studies were carried out aimed at investigating the
practicability for industrial practice of using bark and agricultural
residue components as substitutes for at least portions of phenol in
phenol-formaldehyde resins. This research was designed both to study
the opportunities for waste material utilization and to explore possible new sources of raw materials for resins and glue mixes.

The objectives of the research were to investigate the methods
of processing raw materials and their influence on the relative reactivity of the resultant extracts while reaction with formaldehyde,
to evaluate the copolymer resins made of the extracts, phenol, and
formaldehyde, and to determine what kind of materials and processing
conditions (treatments of materials and resin synthesis) might be
required to obtain resins possessed the most desirable characteristics for use as wood adhesives. The primary criterion was to develop
resins capable of producing acceptable bond quality, as determined by
wood products industries' standard, i.e. vacuum-pressure test of the
American Plywood Association PSI-74 for plywood[15] and ASTM D1037-72
for particleboard[16], while at the same time tolerating a reasonable
range of press and assembly times.

REACTIVITIES OF THE RESIDUE EXTRACTS TOWARD FORMALDEHYDE MATERIAL

Southern pine bark, oak bark, pecan nut pith, and peanut hulls
were selected as candidate materials in the study. Not only is there
reason to believe that these may contain useful and extractable
phenol-like compounds[2,3,7,10,17], but it was felt these are available in significant and feasibly retrievable amounts.

It is estimated that over 20 million tons of green bark will
be produced annually from the cut of southern pine timber by the
year 2000[18].

The total volume of pinesite hardwoods is about 54 billion cu.
ft. and half of them are oaks. These hardwoods are harvested along
with the pines[19]. Oak bark is known to be rich in tannin and was
used extensively by leather tanners in pioneer days.

An average of 95,300 tons of pecan were produced each year in
the southeastern United States during the five-year period of 1975
to 1979[20]. Close to one-third of pecan by-products, which amount

to forty percent of the total pecan nut weight harvested, is the pith material. One commercial pecan shelling plant in Georgia produces from 3,500 to 4,000 tons each year[21]. Pecan nut pith had been previously investigated as a source of tannin for processing leather[17], however, most of the pith is currently being disposed of by dumping. It is known to be acidic and thought to contain substantial quantities of phenol-like extractives.

More than 400,000 tons of peanut hulls are available annually as one of the by-products from peanut industry in the United States[22,23]. Approximately one-half of them are produced in the southern area of Georgia, eastern Alabama, and northern Florida and one-third are in the state of Georgia[20,23].

The contents of lignin and extracts from various extractions such as alcohol-benzene, ethyl alcohol, one percent sodium hydroxide solution, and hot water, in raw materials were determined separately with fresh samples according to the ASTM standard method of tests. The results are given in Table 1.

Table 1. Selective Extractive and Lignin Analysis of Materials

Extraction Method	Material			
	Pine Bark	Pecan Nut Pith	Oak Bark	Peanut Hulls
Alcohol-benzene (D1107-56)	7.51%*	3.84%*	7.23%*	8.52%*
Ethyl Alcohol (D1107-56)	11.66%	32.19%	6.86%	11.56%
1% NaOH Solution (D1109-56)	37.66%	90.76%	34.55%	34.48%
Hot Water (D1110-56)	12.04%	55.95%	12.48%	17.80%
Lignin (D1106-56)	57.10%	36.76%	35.22%	28.63%

*Based on the even-dry weight of fresh samples used for each of the separate analysis.

I. Extraction and Dissolution of the Components

The following sixteen treatments were used to digest the raw materials and to bring their extractable components out into aqueous solution or suspension.

1. T-1, T-2, T-3, T-4: Materials were extracted with a 2% sodium hydroxide solution at temperature of 40°C for T-1, 95°C for T-2, 145°C for T-4, and 185°C for T-4.

2. T-5, T-6, T-7: Materials were extracted with a 5% sodium hydroxide solution at temperature of 40°C for T-5, 95°C for T-6, and 185°C for T-7.

3. T-8, T-9, T-10: Materials were extracted with a 10% sodium hydroxide solution at temperature of 40°C for T-8, 95°C for T-9, and 185°C for T-10.

4. T-11: Materials were digested with the kullgreen sulfite pulping (KSP) liquor, that was composed of 12.3% sodium sulfite and 3.0% sodium bisulfite, for 6 1/2 hours after reaching a maximum temperature of 135°C.

5. T-12: Materials were processed with the neutral sulfite pulping (NSP) liquor, that was composed of 3.9% sodium sulfite and 1.0% sodium carbonate. The digestion was for 6 hours after reaching a maximum temperature of 170°C.

6. T-13: Materials were hydrolyzed to produce lignin-like compounds by means of the "Hokkaido Process" with some modification in the laboratory.

For the so-called pre-treatment in the Hokkaido Process, materials were mixed with water at a ratio of 1 to 4, heated to a temperature of 185°C and held for 2 hours instead of steaming; then the mixture was ovendried and powdered. The main hydrolysis was carried out with 80% sulfuric acid at room temperature.

The hydrolyzed product was reactivated and dissolved by sulfonation with a 5% sodium sulfite solution for 4 hours after reaching the maximum temperature of 190°C.

7. T-14: The process was same as that of T-13, except that the pre-treated mixtures were vacuum filtered before being ovendried. In other words, only residues of pre-treatment were subjected to main hydrolysis process.

8. T-15, T-16: The processes were the same as that of T-13 and T-14, respectively, however, a 5% sodium hydroxide solution was used to reactivate the hydrolyzed product instead of using sodium sulfite.

II. Reaction of the Extract Components With Formaldehyde

The reactivities of different phenols have been compared by measuring the rate of disappearance of formaldehyde[24], and it was felt that such information would be a good first step in obtaining the approximate quantity requirement of formaldehyde and in evaluating the resin making potential of these extracts for practical usage in wood products industries.

The purpose of the research was to seek a practical and feasible technology suitable for industrial practice. Therefore, it was felt that the reaction conditions (such as concentration of reactant, temperature, etc.) in the experimental work needed to be within the ball park of current use in the industry.

The non-volatile content of extracts was approximately 20% so that a higher concentration formaldehyde was needed to bring up the concentration of reactant to approach to the 40-50% range being used by the industrial resin manufacturers. Accordingly, paraformaldehyde was used as the source of formaldehyde. Though the reaction rates of paraformaldehyde and formaline toward phenolic materials are different at the early stage of reaction, paraformaldehyde depolymerizes and dissolves completely well before the reactants reach the meaningful reaction temperature during the synthesis of phenolic resol resins commonly used as wood adhesives. And the performance of phenol formaldehyde resins as wood adhesives does not differ based on the formaldehyde source. That is to say, resins made from paraformaldehyde vs. formalin solution, assuming the solution contains no methanol stabilizer, will behave the same in bonding[25].

To simplify the calculations in future resin developments the molecular weight of the extracts including all chemicals dissolved were assumed as 94, the same as that of phenol to be replaced by the extracts. Based on this assumption, the molar ratio of formaldehyde to extracts was 1.9 to 1.0. However, it is possible that the actual molecular weight of extracts is greater, consequently, the actual molar ratio of formaldehyde to extracts was possibly more than 1.9 to 1.0.

The extracts were charged and mixed with the requisite amount of formaldehyde in the form of paraformaldehyde powder into a reaction kettle. The concentration of reactants were adjusted to 30% by adding distilled water. The reactions were carried out at two temperature levels of 60°C and 80°C, respectively. The reactants were heated by a 450 watt heating mantle at 60 volts. It took approximately 20 minutes to reach 60°C and 30 minutes to reach 80°C.

Approximately 15 c.c. samples were withdrawn from the kettle at each of the elapsed times of 0, 15, 30, 60 and 120 minutes after the reactant reached the prescribed reaction temperature. The reaction was terminated at the end of 180 minutes. Samples which had been withdrawn were rapidly cooled in ice water and then stored in a freezer.

Though the accuracy of hydroxylamine hydrochloride method of
formaldehyde determination normally runs at about 94.4% of the theoretical value regardless of time and temperature (20°C-25°C) of
determination[26], yet it is the most commonly used method by the U.S.
wood adhesive industry to determine the content of free-formaldehyde
in phenol formaldehyde resins. Therefore, the unreacted free formaldehyde content in each sample was determined by this method using a
pH meter to detect the end point (pH = 4.00) of titration rather than
using a bromophenol blue indicator. Two determinations were made
for each sample and the average free formaldehyde content calculated. Then the results were converted into percentages of initially
added formaldehyde for comparison.

The percentage of original formaldehyde that remains free at
different reaction times and temperatures are presented in Tables
2, 3, 4, and 5 separately for each of the four materials.

Table 2. Unreacted formaldehyde (as percent of original amount) for
          different times and temperatures of reaction between para-
          formaldehyde and pine bark extracts of various treatments.

Treatment No.	Reaction time at 60°C (Min.)					Reaction time at 85°C (Min.)				
	0	15	30	60	120	0	15	30	60	120
1	64.0	53.0	49.4	47.7	45.2	42.0	37.0	36.7	32.7	28.5
2	68.2	63.5	61.8	59.3	55.6	57.1	54.3	51.2	47.9	42.7
3	71.4	66.4	65.1	62.2	59.7	61.4	55.9	54.6	52.3	50.2
4	92.5	94.8	94.9	94.7	92.4	94.5	90.7	88.7	84.4	81.2
5	29.7	15.8	10.1	7.26	5.24	6.81	2.69	2.56	2.88	2.13
6	40.2	29.6	26.7	22.9	22.2	22.3	18.5	17.2	15.2	14.3
7	70.3	69.2	69.9	69.3	65.9	65.3	58.5	57.9	53.4	50.2
8	19.3	6.88	5.02	3.71	0.29	1.80	1.00	0.62	1.20	1.07
9	12.2	6.57	4.74	2.83	1.69	3.10	2.27	1.44	1.28	1.34
10	36.2	35.9	31.9	31.9	27.7	38.6	31.7	29.8	28.2	27.0
11	56.3	46.5	45.2	40.4	34.9	39.6	33.0	30.6	29.4	27.7
12	92.9	92.5	91.2	91.5	91.3	91.6	87.6	79.0	81.5	79.2
13	57.3	95.9	98.3	97.4	96.9	93.6	92.9	92.3	90.9	90.7
14	90.7	94.2	94.2	92.7	90.5	93.0	92.1	90.2	87.8	86.4
15	85.8	83.5	87.2	86.7	82.9	89.1	88.7	83.1	80.7	74.2
16	98.0	95.7	96.7	94.9	94.4	97.5	94.1	91.3	90.2	85.7

It was possible that some of the formaldehyde was consumed by
the Cannizzaro reaction to form methanol and formic acid instead of
reacted with the extract components, especially for those extracts
with high alkalinity. Therefore, the unreacted formaldehyde values
should not be taken as the absolute reactivities of these extracts

toward formaldehyde. However, it was proved the relative reactivities were properly indicative by the facts that several fast curing resins were developed by the author based on this information.

Table 3. Unreacted formaldehyde (as percent of original amount) for different times and temperatures of reaction between paraformaldehyde and oak bark extracts of various treatments.

Treatment No.	Reaction time at 60°C (Min.)					Reaction time at 85°C (Min.)				
	0	15	30	60	120	0	15	30	60	120
1	57.7	42.1	37.4	34.7	31.9	36.6	33.2	32.6	30.3	26.8
2	83.4	68.0	65.6	62.2	60.0	61.5	56.9	54.2	52.6	50.6
3	96.2	94.9	93.0	93.5	92.3	92.0	88.6	88.1	86.2	82.7
4	97.6	96.3	96.1	95.6	94.5	79.1	75.7	75.8	73.7	72.3
5	25.9	13.2	8.62	5.32	2.90	1.26	0.69	0.94	0.87	0.90
6	25.0	14.6	10.8	6.44	3.89	2.46	2.57	1.45	1.58	1.61
7	92.1	92.6	89.7	88.5	85.9	85.7	81.9	80.8	78.7	75.3
8	9.75	6.29	3.26	1.32	0.81	0.82	0.72	0.92	1.33	1.65
9	25.2	10.0	5.07	2.21	0.83	1.84	1.12	1.18	0.95	0.72
10	64.7	54.4	51.8	48.3	47.2	49.1	44.9	41.6	40.3	38.9
11	65.7	58.8	56.4	49.7	48.7	46.2	50.9	40.9	38.6	37.3
12	98.4	95.7	93.9	93.0	85.2	91.2	88.5	86.5	84.9	82.3
13	26.1	66.6	87.2	88.6	90.2	91.6	89.5	89.7	88.1	86.4
14	26.3	69.8	83.8	90.2	92.4	92.6	89.5	88.5	87.3	85.1
15	96.4	95.8	95.7	94.9	93.3	91.2	87.9	86.9	85.2	83.1
16	97.4	95.4	95.2	93.5	92.2	89.4	86.6	85.3	80.9	80.1

The unreacted formaldehyde increased with the increase of processing temperature and sodium hydroxide concentration. The KSP (T-11) extracts consumed more formaldehyde than that of NSP (T-12). The reactivation (sulfonation vs. sodium hydroxide) after hydrolysis was statistically significant when it interacted with the material.

The unreacted free formaldehyde decreased rapidly at the first 30 minutes of reaction and then the rate of decrease slowed down gradually. However, there were few exceptions, i.e. peanut hull extracts obtained from 2% sodium hydroxide extraction at 185°C, and pine bark; oak bark and pecan nut pith extracts reactivated with sulfonation. The paraformaldehyde was not dissolved completely even though the reactant had reached the prescribed reaction temperature of 60°C. It seemed that the paraformaldehyde dissolving rate at 60°C was greater than the rate of the reaction of formaldehyde toward the extracts, therefore, the unreacted free formaldehyde in solution increased with increase of the time.

The differences among the materials were significant only after their interaction with the method of treatments. And some of the extracts reacted vigorously toward formaldehyde and with a high degree of exothermic reaction at the beginning. On the other hand, some did not show this type of formaldehyde reaction vigor at all, which indicated that the processes of resin synthesis require carefully manipulating and precisely controlling the reaction conditions specifically and individually designed for each of the combinations of material with the particular treatment.

Table 4. Unreacted formaldehyde (as percent of original amount) for different times and temperatures of reaction between paraformaldehyde and pecan nut pith extracts of various treatments.

Treatment No.	Reaction time at 60°C (Min.)					Reaction time at 85°C (Min.)				
	0	15	30	60	120	0	15	30	60	120
1	61.0	53.1	50.2	48.9	44.9	47.8	43.1	39.4	36.4	35.8
2	81.8	78.8	77.0	72.8	71.7	65.6	57.6	55.5	51.1	45.7
3	79.7	74.0	74.7	71.8	70.3	69.8	63.6	59.5	56.4	52.6
4	83.0	83.2	83.3	83.2	81.9	75.3	71.5	68.3	69.1	66.3
5	39.1	22.9	22.1	16.2	12.8	16.8	11.7	9.28	8.82	5.53
6	39.5	26.4	21.1	18.1	17.1	16.9	13.3	12.4	10.8	8.53
7	82.5	77.7	74.3	72.6	70.5	70.7	63.2	61.1	57.8	53.0
8	78.1	41.4	32.1	23.0	21.9	19.8	12.3	9.53	8.67	4.58
9	15.8	7.18	5.48	3.57	2.63	4.61	2.05	1.51	1.65	1.52
10	21.7	15.5	14.6	13.6	13.1	10.1	8.01	6.91	5.81	5.13
11	51.3	45.1	41.0	37.0	33.7	42.8	32.2	29.8	27.7	25.8
12	87.6	87.7	87.8	86.5	85.1	89.4	84.1	84.7	81.9	78.0
13	82.8	83.6	86.9	82.8	87.1	51.7	49.9	50.8	49.1	47.4
14	77.2	95.1	94.4	94.9	94.8	80.4	79.5	77.3	78.7	74.4
15	92.5	89.6	90.5	85.8	85.9	77.5	73.6	69.0	67.4	65.3
16	76.6	79.8	81.3	81.2	79.7	72.1	70.9	67.5	65.3	63.6

COPOLYMER RESINS OF THE EXTRACTS, PHENOL AND FORMALDEHYDE

A series of copolymer resins were investigated sequentially to examine a range of resin types and gluing conditions. At first copolymer resins with 20% by weight of phenol replacement were evaluated in gluing plywood made from southern pine veneer of low and moderate moisture contents. Then, the amounts of catalyst, sodium hydroxide, in synthesis of copolymer resins were adjusted to a moderate level for extracts with high alkalinity.

Some of the reports from previous studies of bark extracts containing resins indicated that inordinately long cure times were needed for these resins[27],[28]. Also, it was determined if the level of phenol replacement could be increased. Thus, the press time/ assembly time relationships and the acceptability of higher levels of phenol replacement were investigated.

I. Copolymer Resin Synthesis

All copolymer resins were prepared by loading the extracts, phenol and formaldehyde into a resin reaction flask then adding water to the mixture, as needed, to adjust the target non-volatile content of the resin to 40% by weight. This was followed by a slow stepwise addition of sodium hydroxide to effect chemical addition of the formaldehyde to the phenolic ring while incurring a minimum of formaldehyde loss due to the Cannizzaro reaction. The mixture was then heated and copolymerized to a target resin viscosity of around 400 cps., as measured at 25°C using a Brookfield viscometer.

Table 5. Unreacted formaldehyde (as percent of original amount) for different times and temperatures of reaction between para-formaldehyde and peanut hull extracts of various treatments.

Treatment No.	Reaction time at 60°C (Min.)					Reaction time at 85°C (Min.)				
	0	15	30	60	120	0	15	30	60	120
1	75.7	50.5	44.6	40.4	38.1	40.5	34.0	32.1	31.7	28.2
2	84.6	65.1	61.5	56.0	51.6	53.5	49.3	46.6	44.1	40.1
3	94.5	87.6	83.9	80.6	78.6	79.4	74.9	70.5	66.6	61.5
4	16.4	57.4	80.3	96.5	98.5	98.6	97.7	97.5	96.7	91.8
5	29.8	12.6	8.45	4.35	3.40	0.93	1.11	0.72	0.48	0.88
6	25.0	13.1	10.3	5.18	2.97	4.38	0.77	1.09	0.70	0.75
7	95.3	93.5	93.1	91.1	8.88	83.6	79.9	75.9	72.5	67.1
8	11.5	6.64	3.83	1.16	0.56	0.93	0.79	0.86	0.89	0.84
9	20.5	10.7	7.62	4.19	2.87	3.70	1.13	1.22	1.10	0.65
10	77.1	63.5	58.3	56.2	50.1	55.8	50.6	47.8	46.2	42.0
11	74.2	67.6	61.7	57.8	51.7	55.8	49.2	48.9	48.8	45.4
12	96.3	95.1	93.3	91.4	91.1	96.0	90.6	93.6	92.1	87.2
13	95.7	93.1	91.7	91.5	90.5	91.1	90.9	89.3	87.4	85.4
14	93.2	92.4	91.0	91.0	90.2	88.0	84.9	83.6	82.6	80.0
15	92.4	91.3	90.8	88.6	86.4	86.5	81.5	77.2	73.4	69.4
16	89.8	88.5	88.1	86.5	86.0	85.8	80.7	79.5	76.8	71.8

All resins were stored in a freezer immediately after preparation. They were subsequently allowed to thaw out overnight and come to room temperature before use in gluing evaluations.

II. Gluability of Copolymer Resins in Southern Pine Plywood

Thirty-six copolymer resins were prepared with 20% by weight of
phenol replaced by those extracts, which had been found to consume
more than 50% of the available formaldehyde at the end of 120 minutes reaction time at the 85°C reaction temperature.

Commercial southern pine veneer at 1/8 inch thickness was obtained from a middle Georgia mill, cut into 12-inch by 12-inch sheets
and used to make three-ply plywood panels. Prior to panel production, the veneer was checked to assure conformance to a thickness
tolerance of $\pm$ 0.005 inch from nominal and then divided randomly into
two groups. The first group was conditioned to an average of 1.9%
moisture content (MC) and then preheated (in sealed bags) to 120°F
in order to produce the dryout tendencies associated with warm, low
MC veneer. The second group was conditioned to an average of 3.7%
MC and then cooled (in sealed bags) to room temperature. This was
considered to represent more moderate gluing conditions.

A widely used, commercially available pine plywood resin was
employed as a comparative control throughout all tests. The glue
was applied using a (Black Brothers) roller spreader and spread rate
was controlled within the range of 83-87 lbs. per 1000 square feet
of double glueline (lbs./MDGL).

All layups were stored in a gravity convection oven at 100°F
for an assembly time period of either 20 or 60 minutes. No prepres-
sing was done but during assembly time the panels were stored under
a slight dead load to prevent edge lifting. The layups were then
hot pressed, one panel per opening, for 4 minutes at a platen temperature of 300°F and a panel pressure of 200 psi. Immediately upon
removal from hot press, panels were stored in an insulated but unheated oven for an overnight period of simulated hot stacking.

After the overnight hot stacking period panels were brought to
room temperature and cut into three 3 1/4 inch wide strips (as
measured along the face grain axis). The center strip was held in
reserve and the two outside strips each cut to yield eight standard
plywood shear specimens. The grooving of these two strips was such
that when tested, the specimens were balanced with regard to the
effect of opening and closing of lathe checks. A total of 12 specimens from each panel, six selected at random from each strip group
of eight, were tested according to the vacuum-pressure procedure for
exterior gluelines as outlined in the industry standard[15]. Wood
failure results are shown in Table 6.

The analysis of variance indicated that the differences among
the extraction processes, veneer conditions, assembly times and
their interactions were statistically significant at the 1% level.
The differences among the materials were statistically significant

Table 6. Results of Southern Pine Plywood Sheat Test of Copolymer Resins with 20% by Weight of Phenol Replaced by the Extracts Indicated.

	Percentage of Wood Failure for Various Veneer-Conditions and Assembly Times Indicated			
	Hot (120°F) Veneer with 1.90% M.C.		Room Temperature (78°F) Veneer with 3.7% M.C.	
	Assembly Time		Assembly Time	
	20 min.	60 min.	20 min.	60 min.
Commercial P-F control resin	91	46	94	87
Pine bark T-1	52	1	79	19
Pecan pith "	39	2	88	76
Oak bark "	64	1	87	46
Peanut hulls "	31	1	80	40
Pine bark T-2	68	3	90	77
Pecan pith "	41	2	87	36
Oak bark "	64	7	82	61
Peanut hulls "	10	1	91	41
Pine bark T-5	40	3	51	58
Pecan pith "	46	6	87	34
Oak bark "	21	2	84	22
Peanut hulls "	46	2	81	48
Pine bark T-6	44	1	83	30
Pecan pith "	13	1	76	43
Oak bark "	43	3	91	31
Peanut hulls "	32	0	79	35
Pine bark T-8	49	1	79	32
Pecan pith "	7	1	85	38
Oak bark "	22	3	68	37
Peanut hulls "	31	0	85	36
Pine bark T-9	56	0	86	48
Pecan pith "	67	2	86	38
Oak bark "	15	1	92	21
Peanut hulls "	66	2	87	29
Pine bark T-10	59	7	84	56
Pecan pith "	14	2	90	54
Oak bark "	36	4	86	82
Peanut hulls "	66	5	79	40
Pine bark T-11	31	4	78	65
Pecan pith "	43	4	79	67
Oak bark "	67	10	72	82
Peanut hulls "	67	6	66	65
Pine bark T-13	68	64	69	89
Pecan pith "	91	49	88	87
Oak bark "	90	79	83	82

only after interaction with the extraction process. This result indicates that there is no so-called best extraction process for all materials. In other words, each different residue material requires a particular method of extraction for the best results.

After careful inspection of gluelines, it was realized that the glueline dryout was prevalent and was caused by the somewhat higher content of sodium hydroxide resulted in the relatively advanced co-polymer resins made of extracts obtained via sodium hydroxide treatments.

Therefore, the sodium hydroxide content of copolymer resins was adjusted to a more moderate level of 5.7 ± 0.3% in the following investigation. Copolymer resins, involving the extracts of sodium hydroxide treatments, were prepared and evaluated by the same manner as described in the preceding paragraphs. The results are presented in Table 7.

Based on the analysis of variance, the effects of material was significant only in interaction with the extraction conditions as was in the previous experiment.

As Table 7 indicates, bond quality was improved by adjusting the sodium hydroxide content of resins to a moderate level. The higher wood failure percentages noted for the 60 minute assembly time as compared to 20 minute one was considered to be due to the effects of the higher veneer moisture content and lower veneer temperature.

It was noted that the boiling water gel times for the copolymer resins were shorter than those for the commercial P-F resins. It was also noted that, although several of the copolymer resins were significantly different (statistically) from others, many were not.

Then investigations were carried out to examine the cure time requirements of various copolymer resins and to determine whether the amount of phenol replacement could or should be increased. The copolymer resins synthesized from extracts, phenol and formaldehyde, were made with 40% by weight (vs. 20% in previous experiments) of the phenol replaced with extracts. The resin synthesis process as well as the veneer types, method of panel production and glueline evaluation were the same as outlined in preceding paragraphs. However, press times of 2, 3 and 4 minutes were used.

The effects of press time and assembly time on bond quality may be clearly understood by referring to Figures 1-A through 1-F, which show test results for each of the five best copolymer resins and the commercial control. On each graph, percentages of wood failure were plotted as a function of assembly time with data shown separately for each of the three press times.

Table 7. Results of Southern Pine Plywood Shear Test of Copolymer Resins With 20% by Weight of Phenol Replaced by the Extracts Indicated.

Resins, Material and Method of Extraction to Obtain the Extracts for the Copolymer Resins	Percentage of Wood Failure for Various Assembly Times Indicated	
	20 Min.	60 min.
Commercial P-F control resin	75	93
Pine bark T-1	78	93
" " T-2	81	90
" " T-8	94	92
" " T-9	94	89
Pecan pith T-1	79	85
" " T-2	71	95
" " T-8	89	93
" " T-9	87	90
Oak bark T-1	86	91
" " T-2	86	93
" " T-8	85	91
" " T-9	74	92
Peanut hulls T-1	75	88
" " T-2	86	90
" " T-8	86	90
" " T-9	74	90
Coarse Peanut hulls T-1	79	87
" " T-2	73	95
" " T-8	91	92
" " T-9	93	91

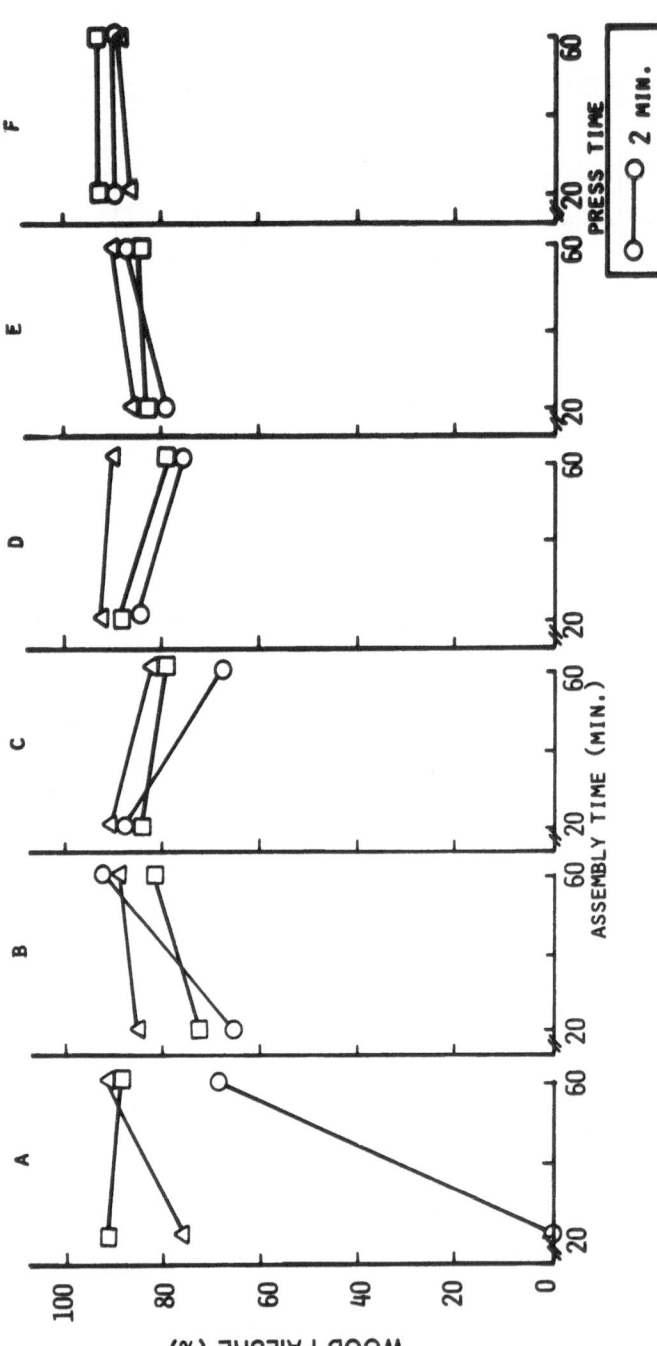

Figure 1. Dependence of Bond Quality on Hot Press Time and Assembly Time for A), Commerical P-F Control Resin, B), Copolymer Resin of Southern Pine Bark Extracts, C), Copolymer Resin of Peanut Hull Extracts No. 1, D), Copolymer Resin of Peanut Hull Extracts No. 2, E), Copolymer Resin of Pecan Nut Pith Extr. No. 1, and F), Copolymer Resin of Pecan Nut Pith Extracts No. 2.

It is obvious that several of the copolymer resins provided superior bond quality in comparison to the commercial control, especially at the short press time of 2 minutes. It was also noted that the copolymer resins containing pecan nut pith or peanut hull extracts seem to be more tolerant of press time and its interaction with assembly time.

Although the commercial phenol-formaldehyde control resin did not perform well with a short press time of 2 minutes, giving average wood failure of 35%, it provided satisfactory bond quality, with an average of 84% wood failure, at the manufacturer's recommended press time of 3 minutes. It did very well (average of 92% wood failure) at the longer press time of 4 minutes.

III. Evaluation of Copolymer Resin as Binder of Particleboard

The copolymer resins of natural products were also evaluated for their bonding qualities in particleboards and in composite panels involving flakeboard cores with veneer faces and backs.

The copolymer resins which had exhibited the best results in gluing southern pine plywood were prepared in the same manner as described in preceding paragraphs, however, the non-volatile content was targeted at 45%. A 45% non-volatile content commercial phenolic resin, specifically tailored for particleboard and catalyzed with three parts of resorcinol resin was used as the control.

The homogeneous southern pine particleboards with a density of 50 pounds per cubic foot were pressed to 5/8 inch stops using a closing pressure of 500 psi with a platen temperature of 360°F for the prescribed press time of 5, 6, or 7 minutes. A resin solids of 6%, based on oven dry weight chips, was used for all boards. No wax or other material was added. Immediately upon removal from the hot press, boards were stored in an insulated, but unheated, oven for an overnight period of simulated hot stacking. Two duplicate boards were made for each hot press time and resin combination.

ASTM D1037-72[16] standard specimen sizes and test procedures were used for evaluation of board properties except for the accelerated aging test where the Canadian standard, CSA standard 0188-1968[29], was used. Before being subjected to the prescribed tests, all specimens were conditioned to an average moisture content of 8% in a room maintained at 20°C temperature and 50% relative humidity. The results of tests were presented in the Figures 2, 3, and 4.

It can be seen from the figures that two of the five copolymer resins, namely those having 40% by weight of their standard phenol replaced by the extracts of (a) peanut hulls and (b) pecan nut pith, exhibited better bonding qualities than the resorcinol resin catalyzed commercial phenol-formaldehyde resin. Specifically, those

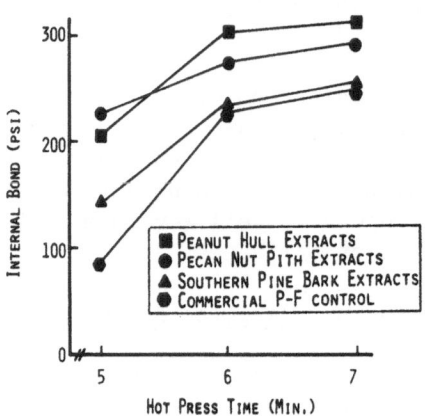

Figure 2.   Dependence of Homogeneous Particleboard Internal Bond on
Hot Press Time for Various Resins.

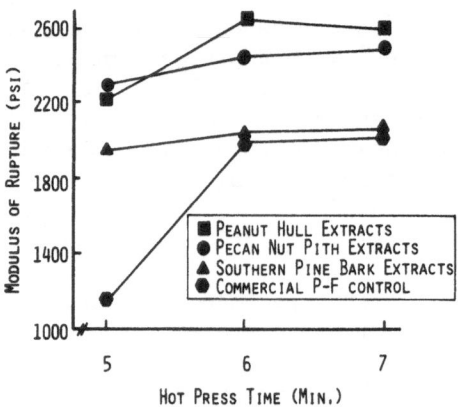

Figure 3.   Dependence of Homogeneous Particleboard Modulus of Rupture
on Hot Press Time for Various Resins.

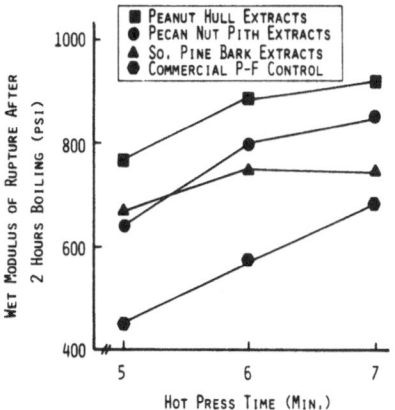

Figure 4. Dependence of Homogeneous Particleboard Modulus of
Rupture (when Tested after 2 Hours Boiling) on Hot Press
Time for Various Resins.

two copolymer resins provided more than 200 psi internal bond and
2200 psi modules of rupture (MOR) with a press time of as little
as 5 minutes at 360°F platen temperature, whereas the commercial
control resin needed longer than 6 minutes press time to achieve
similar board properties. As for the wet bending strength (per
Canadian accelerated aging test) these copolymer resins retained
higher modulus of rupture (MOR) values at a press time of 5 minutes than the commercial resin did at 7 minutes.

IV.   Bonding Composite Panels With the Copolymer Resin

In conjunction with commercial development of the copolymer
resins, two such resins (one tailored for particleboard and the
other for plywood) with 40% by weight of their phenol replaced by
peanut hulls extracts were synthesized as previously described and
submitted to an industrial laboratory for evaluation in bonding
thin (1/4" thick) oak and southern pine flakeboards and in composite
panels where 1/8" southern pine veneer faces and backs were lami-
nated onto oak flakeboard cores.

The particleboard copolymer resin was applied at an 8% rate
based on oven dry weight as the binder for flakeboards. A commercial phenol formaldehyde resin, catalyzed with resorcinol resin
and extra sodium hydroxide was similarly used as the control resin
binder for flakeboard evaluation. The density was 48 pounds per
cubic foot for oak flakeboards and 43 pounds per cubic foot for
pine. The oak flakeboards were pressed at a platen temperature of
350°F, however, the temperature had to be dropped to 335°F for the
pine flakeboards due to a deficiency of steam pressure at the time
they were to be made. The press times were ranging from 60 seconds
to 120 seconds (push button close to push button open).

One day after the flakeboards were made, the oak flakeboards
having a 120 seconds press time were used to laminate with 1/8"
southern pine veneers on face and back to form the 1/2" composite
panel by the plywood resin. After 20 minutes close assembly time,
the panels were hot pressed at a platen temperature of 300°F for 2
or 3 minutes under 200 psi platen pressure.

The American Plywood Association (APA) 6-cycle test was used
to evaluate the properties and durability of the composite panels.
The results of the tests along with other physical properties of
core flakeboards are listed in Table 8. In Figure 5, the dependence
of board thickness (spring back thickness) immediately after hot
pressing on the press time is shown separately for each of the two
wood species and resins.

The most significant and important results from the composite
panel evaluation was that the copolymer resins involving peanut hull
extracts passed the APA 6-cycle test either used in the composite

Table 8. Physical Properties and Test Results for Flakeboard Cores Subsequently used to Form Composite Panels.

Resins	Wood Species	Board Density (pcf)	Mat Moisture Content (%)	Hot Press Push Button Close to Push Button Open Time (sec.)	Temp. (°F)	APA 6-Cycle Test	Internal Bond Before 2 Hour Boil (psi)	After 2 Hour Boil (psi)	Bending Test Modulus of Rupture (psi)	Modulus of Elasticity (psi x 10^5)
Commercial P-F Control	Oak flakes	46	7.0	75	350	--	28	<10	No property test was done	
				90	350	Fail	70	<10	1892	3.18
				105	350	Fail			1142	2.56
				120	350	--			Used to make composite panel	
	Southern pine flakes	43	9.0	75	335	Fail	--	<10	1962	1.65
				105	335	--			No property test was done	
Experimental Copolymer with Peanut Hull Extract	Oak flakes	48	6.5	60	350	--	58	31	No property test was done	
				75	350	Pass	113	9	2823	4.26
				90	350	Pass			2100	2.85
				120	350	--			Used to make composite panel	
	Southern pine flakes	43	9.0	75	335	Pass	--	14	4906	5.00
				105	335	--			No property test was done	

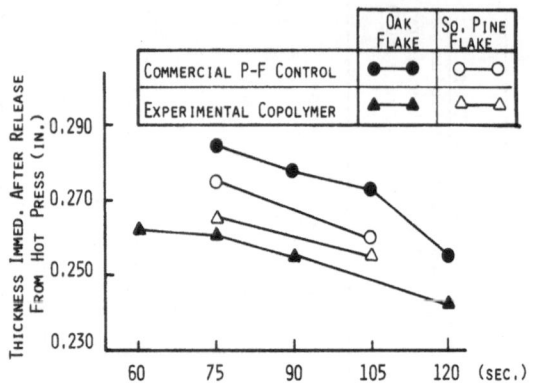

Figure 5. Push Button Close to Push Button Open Press Time.
Dependence of Flake Board Thickness (Immediately after
Release from Hot Press) on Hot Press Time for Different
Resins and Wood Species.

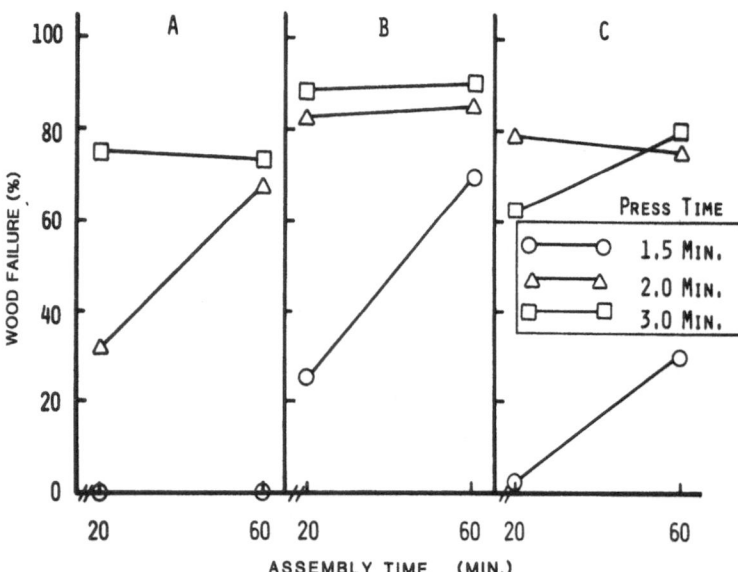

Figure 6. Dependence of Bond Quality on Press Time and Assembly
Time, A) Commercial P-F Resin, B) Copolymer Resin with
60% Phenol Replaced by Peanut Hull Extracts, C) Copolymer Resin with 60% Phenol Replaced by Pecan Nut Pith
Extracts.

bonds of oak flakeboard core with southern pine veneer surfaces or
in the 1/4" oak and pine flakeboard core materials alone. In composite panels, testing involving both the plywood and the particleboard bonds. Press times for the 1/4" thick flakeboard core of as
short as 75 seconds were successful. However, similar 1/4" thick
boards made with the resorcinol catalyzed commercial phenol formaldehyde control resin failed the APA 6-cycle test, even at the longer
press time of 120 seconds.

The thickness of boards immediately after release from the hot
press, including any spring-back in thickness, is a good indicator
of the degree of resin cure. There is less spring-back and the
thickness will be the same as or less than the thickness of stops,
if the resin is cured sufficiently.

As Figure 5 indicates, the copolymer resins of peanut hulls
extract cured much faster than the resorcinol catalyzed commercial
phenol-formaldehyde control resin in bonding these thin flakeboards.
The cure time needed to retain a thickness of 0.250" (the thickness
of stops) can be estimated from the graphs. In so doing, the necessary cure time of the copolymer resin for the 1/4" oak flakeboards
was found to be approximately 104 seconds, whereas, a more than 120
seconds cure time was needed for the commercial control resin.

V. Copolymer Resin With Higher Phenol Replacement

The majority of reports and patents indicated that the substitution of phenol with bark and other residue products was limited
to less than 50% of the phenol. The use of more than 50% natural
products to replace the phenol resulted in resins required substantially increase of cure time[14,27,28].

In viewing the promising results obtained from the gluing of
southern pine and bonding of particleboards with the copolymer resins with 40% by weight of phenol replaced by extracts of peanut
hulls and pecan nut pith. Work was carried out to investigate if
the newly developed copolymer resins could be made of a higher phenol replacement and still possess the advantage of fast curing
property.

The copolymer resins with 60% by weight of phenol replaced by
the extracts of peanut hulls and pecan nut pith were synthesized
and evaluated in gluing southern pine plywood.

The results of the test are shown in the Figures 6A through
6C, which respectively, present the results for each of the experimental copolymer and commercial control resins. On each graph, percentages of wood failure were plotted as functions of assembly time,
and data were shown separately for each of the three press times.

The figures clearly indicated that the copolymer resins cured faster than the commercial one, even though 60% by weight of phenol was replaced by the extracts of natural residues.

REFERENCES

1. P. Fang and G. D. McGinnis, App. Poly. Sym. No. 28, 363 (1975).
2. P. Fang, G. D. McGinnis and E. J. Parish, Wood and Fiber 7, 136 (1975).
3. H. S. Fraser and E. P. Swan, The chemistry of western red cedar bark, For. Tech. Publ. 27, Canadian Forest Service Vancouver, B. C., Canada (1978).
4. J. M. Harkin and J. W. Rowe, Bark and its possible uses, For. Service Res. Note FPL-091, For. Prod. L. USDA (1971).
5. R. W. Hemingway and G. W. McGraw, App. Poly. Sym. No. 28, 1349 (1975).
6. W. D. Ross, Bibliography of Bark, Bibliography Series 6, For. Res. Lab., Oregon State Univ. (1966).
7. L. G. Roth, F. J. Sueger and J. Weiner, Structure, extractives, and utilization of Bark. Bibliography Series 191. Inst. of Paper Chemistry (1960).
8. L. G. Roth and J. Weiner, Structure, extractives, and utilization of bark supplement I. Bibliography Series 191. Inst. of Paper Chemistry (1968).
9. I. Abe, Studies·on the lignin-formaldehyde resin. Hokkaido For. Prod. Res. Ins. Research Report No. 55 (1970).
10. R. W. Hemingway, Adhesives from southern pine bark, Proceeding at Symp. on complete free utilization of southern pine. For. Prod. Res. Soc. (1978).
11. A. Pizzi, J. Macromol. Sci. - Rev. Macromol. Chem. c18 (2), 247 (1980).
12. F. W. Herrick and L. H. Bock, U.S. Pat. No. 2,025,250 (1962).
13. F. W. Herrick and L. H. Bock, U.S. Pat. No. 3,232,897 (1966).
14. J. T. Stephan, U.S. Pat. No. 2,675,336 (1954).
15. American Plywood Association, U.S. Product Standard PSI-74, Tacoma, Washington (1974).
16. American Society for Testing and Materials. Annual Standards Part 22 (1975).
17. C. F. Allen, U.S. Pat. No. 1,078,893 (1913).
18. P. Koch, Utilization of southern pines. Agricultural Handbook No. 420 Vol. 1 USDA (1972).
19. P. Koch, Forest Industries 103 (11), 48 (1976).
20. U.S. Department of Agriculture. Agricultural Statistics 1980.
21. W. P. Reeves, Private communication with author (1976).
22. The Peanut Research Task Force for the Southern Region Agricultural Experiment Stations, Southern region peanut research needs. Ga. Agr. Exp. Stations, Univ. of Ga., Athens, Ga.
23. J. G. Woodroff, Peanuts: Production, processing, products. Second Edition. The AVI Publishing Co., Inc. Westport, Conn. (1973).

24. M. M. Sprung. J. Am. Chem. Soc. <u>63</u>, 334 (1941).

25. C. M. Chen, Ind. and Eng. Chem. Product Research and Development. In press (1982).

26. M. Marasco, Ind. and Eng. Chem. <u>18</u> (7), 701 (1926).

27. F. W. Herrick and L. H. Bock, Forest Prod. J. <u>8</u> (10), 269 (1958).

28. H. McLean and J. A. F. Gardner, Pulp and Paper Mag. of Canada <u>53</u> (8), 111 (1952).

29. Canadian Standard Association, CSA Standard 0188-1968 (1968).

ORGANIC FILLERS FOR THERMOPLASTICS

George R. Lightsey

Department of Chemical Engineering
Mississippi State University
Mississippi State, MS

INTRODUCTION

Historically fillers have been low-cost bulking agents added
to more expensive resins with cost reduction the main objective[1].
Property enhancement, such as improved impact resistance, heat
distortion, etc., is often as important as cost in filler selection. While fillers can be used to improve resin properties, all
successful fillers have the characteristic of not causing an un-
acceptable reduction in end-product properties[2]. The rising cost
of resins and the need to economically enhance certain resin properties has provided an incentive to extend and modify resins with
fillers. The challenge is to match the right filler to the right
job.

Organic vs. Inorganic Fillers

Inorganic fillers such as calcium carbonate, aluminum trihy-
drate, etc., are widely used due to their low cost, abundant
supplies and ability, after surface treatment, to improve selected
resin properties. While inorganic fillers are cheap, the assumption that the lowest-cost filler automatically results in the
lowest-cost composite does not always hold true. High-density
mineral fillers, while cheap on a weight basis, can be quite
expensive on a volumetric basis.

It has been pointed out that "... simply adding filler to a
resin does not automatically assure big cost savings for the
processor"[3]. Mineral fillers because of their high densities
(specific gravities of about 2.5) can actually increase cost on
a volumetric basis as well as make processing more difficult.

193

Organic fillers such as shell flour and wood flour have low specific gravities (about 0.8) and when used to extend resins they offer significant cost savings. They may, however, have a detrimental effect on such properties as flame and chemical resistance, depending upon the degree of polymer loading. Although a number of cellulosic fillers are available, wood flour, shell flour, and cellulose pulp are used the most often[4]. While inorganic fillers will likely continue to dominate the market, there is a definite need to develop more fully low density organic fillers for use in thermoplastics.

Wood Based Fillers

Wood flour, with both low-density and low-cost (3-6 cents per pound), has been used as an extender for thermosetting polymers for many years[5]. In 1975, 40,000 metric tons of wood flour was used in polymer composites, and this is expected to increase to 105,000 metric tons per year by 1985[6]. Published data have indicated that wood flour has potential for use in thermoplastics as well as thermosets[7]. The disadvantage of using wood flour as an extender for thermoplastics is that the resulting composites usually have significantly reduced impact and tensile strengths. The poor properties of thermoplastic composites containing wood flour are due to the low aspect (L/D) ratio, about 2.5[1], of the wood flour and the primarily polar surface of the cellulosic particles. The result is a composite with poor bonding between the non-fibrous filler and the polymer matrix.

An additional problem encountered with resins filled with wood flour is increased water absorption. Wood based fillers that are more hydrophobic than wood flour and that are fiberous would be expected to give thermoplastic composites with significantly improved properties.

A low cost ($12/ton) pulp mill wood residue with L/D ratios of 3 to 19 was tested as a filler in polyethylene and polystrene[8]. The tensile strength and modulus of composites containing the fiberous residue were improved compared to composites containing wood flour. However, the over-all performance of both fillers was relatively poor, probably because of poor compatibility of the non-polar resins and polar fillers.

It has been shown that bark fibers with aspect ratios greater than 15 significantly improve the impact strength of filled polypropylene[9,10]. Since bark contains up to 70% phenolic, non-polar materials compared to about 30% in wood[11] the bark fibers are more compatible with non-polar polymers. The main disadvantages of bark fiber are its relatively high costs and restricted availability.

Shell Flour Fillers

A number of agricultural residues have been studied as possible fillers. One of the most promising is peanut hull flour. When used in a phenolic resin, peanut hull flour, which is high in lignin content, had lower water absorption than wood flour[12]. Also, peanut hulls are an abundant waste product (350,000 tons per year)[13] and are low cost (approximately 4-6¢ per pound)[14]. When peanut hull flour was used as a filler for polyethylene the tensile strength gradually decreased as the percent filler increased so that at 20% filler loading the tensile strength had declined 11%[15]. The tendency of peanut hull flour to evolve gases during processing was noted as a possible processing problem. However, the natural "blowing" action of peanut hull flour could potentially be utilized where some foaming of composites was desired. Commercial grade peanut hull flour was tested as filler in polypropylene at up to 35% loading[16]. Tensile properties were unchanged up to 15% filler, where they began to diminish. SEM photographs indicated good bonding between the peanut hull flour and the polypropylene resin. Pretreatment to remove volatiles from the filler was necessary. A peanut hull flour that had been air classified to remove a fraction that had relatively high water absorbancy was tested as a filler for polypropylene and polystryrene[17]. The air classified peanut hull flour improved composite strength properties relative to whole peanut hull flour. However, the composite properties of both resins were decreased by the addition of the filler.

Driscoll[18] reported that walnut shell flour, which has been commercially successful as a filler for thermosetting resins, had potential as a filler for thermoplastic resins. When added to polyethylene, ABS, or polypropylene, walnut shell flour improved flexural strength and modulus with small to moderate losses in tensile and impact strength. Surface modified rice hull flour was reported to give improved composite properties when added to polypropylene, polystyrene, or polyethylene[19,20]. Rice hull flour without surface modification did not perform well, significantly reducing the impact strength of filled resins.

Surface Chemistry of Organic Fillers

When the data on the performance of organic fillers in thermoplastics is reviewed it is apparent that the surface chemistry of the filler is of major importance. Wood flour is generally a poor filler for non-polar thermoplastics such as polypropylene and polystyrene. The reasons are the polar nature of the wood flour surface (cellulosic) and the non-fibrous nature of the wood flour. Pulp mill residue, which is more fibrous in character than wood flour, was a better filler in non-polar thermoplastic than wood flour. However, the pulp mill residue was still a relatively poor filler, apparently due to its unusually

polar surface causing incompatibility with the resin matrix.
Bark fibers, which would be expected to have a non-polar (lignin-
like) surface gave good results when used as a filler in non-
polar thermoplastics. Likewise, peanut hull flour and walnut
shell flour, which are rich in phenolic, lignin-like material,
are relatively good fillers. Treatment of the surface of rice
hull flour to increase its compatability with thermoplastic resins
has been shown to greatly enhance its value as a filler. None of
the shell flours are fibrous.

It would appear that a low-cost organic filler that had a
non-polar surface and was fibrous would be an excellent candidate for use in thermoplastic resins.

THERMOMECHANICAL FIBERS

Wood fibers are not random mixtures of cellulosic polymers
and lignin but rather these polymers are arranged in a definite
morphological pattern with respect to each other. The major
parts of wood fiber are the middle lamella, which is amorphous in
nature and consists chiefly of lignin and the fiber wall consisting mainly of cellulosic polymers. Fibers can be produced from
wood so that the fiber surface is either primarily exposed lignin
or primarily cellulosic in nature by a process called thermomechanical pulping.

The concept of thermomechanical fiber was introduced in 1929
by Asplun of the Swedish Defibrator Company. The process consists
of breaking down wood into fibers by grinding wood chips under
high steam pressure (30-150 psig) followed by an open stage dis-
charge. The chemical and physical characteristics of the surface of thermomechanical fibers have been shown to depend upon
the pressure and temperature of the steam during the fiberizing
process[21].

In the thermomechanical process lignin behaves as a semi-
glassy solid at temperatures up to approximately 135°C. Above
this temperature fiber areas that are high in lignin are more
susceptible to rupture than fiber areas high in cellulosic
compounds. At higher temperatures the fiber separation occurs
predominantly in zones of high-lignin concentration; i.e., in
the region of the middle lamella. Thus, a high temperature
defibering process produces fibers with high lignin content
on the surface[22].

Evaluation of Thermomechanical Fibers in Plastics

Thermomechanical fibers produced at 175°C were obtained
from a commercial hardboard plant. The moist fibers were dried

at room temperature and the dried cake was broken into fibers
using a hammer mill. The fiber was mechanically screened into
three fractions (-16/+30, -30/+40, and -40/+60 mesh) and the
average fiber length determined as described by Short[23]. Two
commercial thermoplastic resins were tested using the fiber as
filler, e.g., polypropylene (PP) and styrene-acrylonitrile copolymer (SAN), by Gonzalez[24].

Molding and Testing

Polymer composites were prepared using fiber loadings of 3,
6, and 9 percent by weight. The mixing of each polymer with each
filler was accomplished by blending the materials in a two-roll
mill for approximately 10 minutes at 180°C. After the mixing,
the material was sheeted out, granulated, and molded into tensile,
flexure, and impact test specimens by injection molding. The
evaluation of composite properties, i.e., tensile strength, modulus of elasticity, and impact strength, were conducted according
to standard ASTM procedures.

Attenuated Total Reflectance Infrared Analysis

The attenuated total reflectance infrared spectra of the
surface of the thermomechanical fiber was obtained as described
by Kalasinsky[25]. Figure 1 shows the IR spectra of purified wood
pulp which is essentially 100% cellulose with the absorbance
peaks characteristic of lignin and cellulose labled. As shown,
the absorbance peaks for lignin are very weak, indicating an
essentially cellulosic surface. The IR spectra of the surface
of the thermomechanical fibers is shown in Figure 2. The thermomechanical fibers had strong absorbance at the lignin wavenumbers,
indicating a lignin rich fiber surface.

Composite Properties

The tensile strengths of composites containing 3, 6, or 9%
by weight thermomechanical fiber with three fiber lengths corro-
sponding to the three fractions -16/+30, -30/+40, and -40/+60
mesh were determined. The tensile strengths of PP and SAN
composites are shown in Figures 3 and 4 respectively. The tensile
strength of the PP composites declined approximately 10% as the
filler loading increased while the tensile strength of the SAN
composites slightly increased. The shorter fiber tended to give
the lower tensile strength.

The elastic modulus of PP and SAN composites is given in
Figures 5 and 6. In both cases, the elastic modulus significantly
increases with filler loading. The dropped-ball impact resistance of SAN composites is shown in Figure 7. The impact

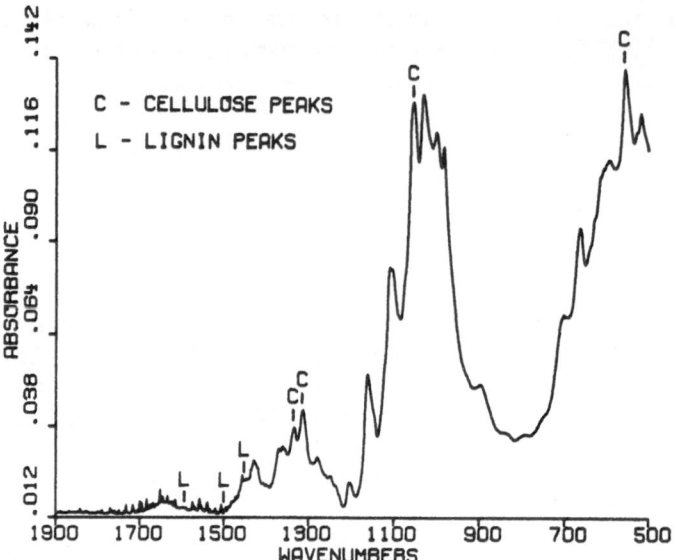

Figure 1: Purified Pulp (100% Cellulose)

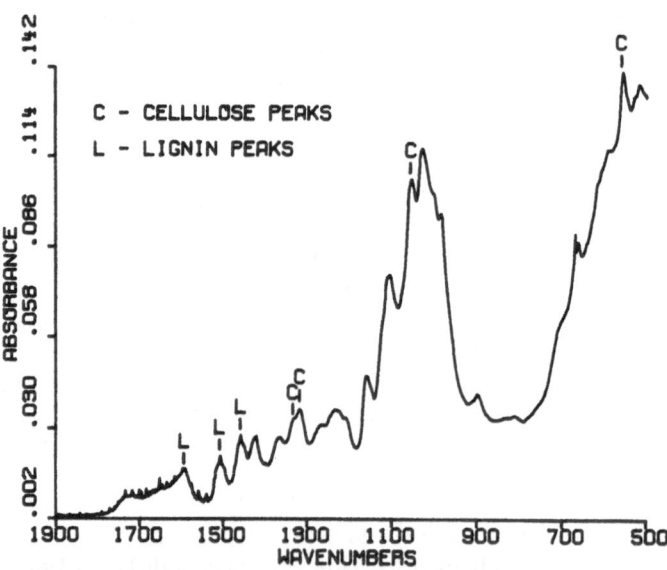

Figure 2: Thermomechanical Pulp Produced at 175°

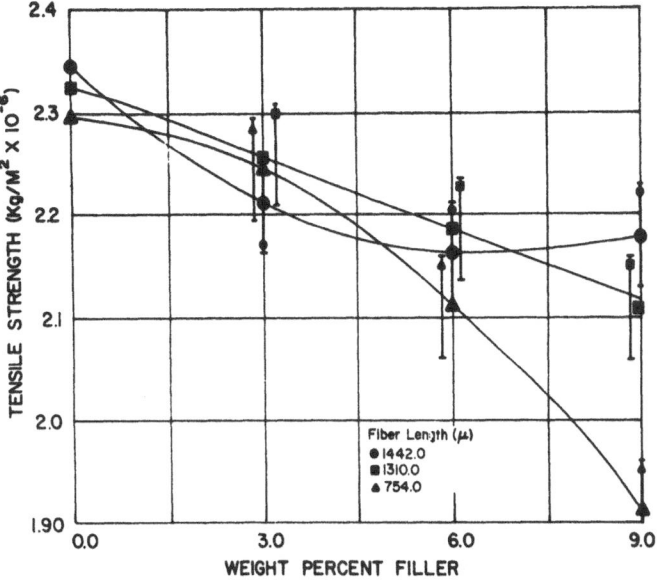

Figure 3.   Tensile Strength of Polypropylene-High Legnin Thermomechanical Pulp Composites vs. Filler Concentrations for
Various Fiber Lengths.

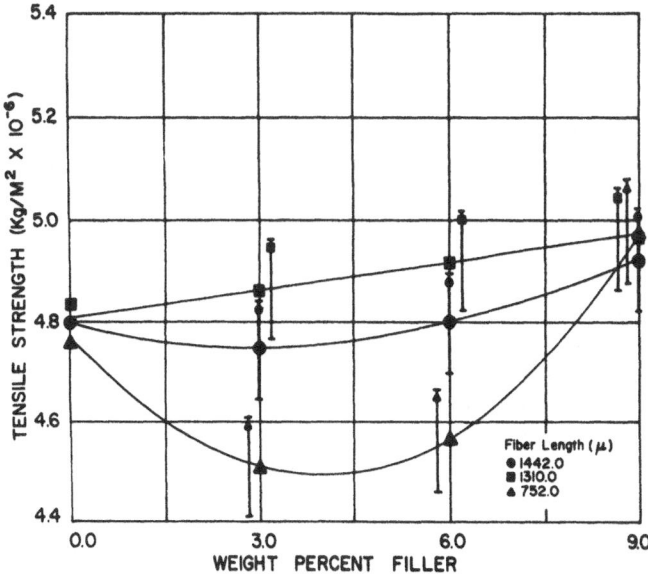

Figure 4.   Tensile Strength of Styrene-Acrylonitrile Copolymer-High
Lignin Thermomechanical Pulp Composites bs. Filler
Concentrations for Various Fiber Lengths.

resistance of the PP composites were not obtained due to the
elastic behavior of the PP composites. Severe deformation of the
test specimens without breakage occurred when the PP composites
were tested using the dropped ball method.

Discussion

The tensile strength and impact resistance of PP and SAN
composites containing thermomechanical fiber are relatively
unchanged from 0 to 9% fiber loading. However, the elastic
modulus of both composites steadily increases with higher fiber
loading. The composites containing thermomechanical fiber had
properties that were significantly better than composites containing wood flour. This indicates that the thermomechanical
fibers, due to their more lignin-like surface, are more compatible
with thermoplastic resins. The thermomechanical fiber composites
had properties that were similar to composites made with peanut
hull flour but did not evolve gases during processing as the
peanut hull flour tends to do.

Thermomechanical Fiber as Filler in Thermoplastic Mixtures

A novel approach to recycling plastics was investigated by
Lowery[26]. When plastics are recycled often the mixture of plastics
has considerably poorer physical properties than the individual
resins in the pure state. This occurs when thermoplastic polymers with greatly differing chemical structures form incompatible
mixtures. Since the surface of thermomechanical fibers is known
to contain both cellulosic and lignin-rich areas it was postulated
that these fibers could act as bonding sites for dissimilar resins,
linking the resins into one general structure. Two resins, polypropylene (PP) and styrene-acrylonitrile copolymer (SAN) were
chosen as having chemical sites on the resins that might be more
compatible with either the cellulosic or lignin-like area on the
thermomechanical fiber surface.

Procedure

Thermomechanical wood fibers produced at 175°C were prepared
as outlined above. The fiber fraction -30/+40 mesh with average
fiber length of 1310 microns was mixed with   50-50 weight
percent mixtures of polypropylene and styrene-acrylonitrile.
Composites containing 0, 4, 8, and 12% by weight fiber were
injection molded into tensile, flexure, and impact test specimens.
A composite containing 8% by weight fiber from the fraction -40/
+60 mesh size (average fiber length of 754 microns) was tested to
determine if fiber length would affect composite properties. All
physical property evaluations were conducted in accordance with
appropriate ASTM procedures. The tensile strength of the 50-50

PP/SAN mixtures containing thermomechanical fiber is shown in
Figure 8. At low filler concentrations an improvement in tensile
strength was obtained indicating that the fibers may indeed be
bonding to both PP and SAN rich regions in the polymer mixture.
The drop in tensile strength at relatively high filler concentrations follows the trend of other low density organic fillers.

The flexural and impact strengths of the mixed resin composites are shown in Figures 9 and 10 respectively. Neither the
flexural nor the impact strength was significantly changed at
the filler loadings investigated. As shown by the properties of
the composite containing 8% +40/-80 mesh filler, the fiber length
of the filler did not appear to greatly affect the performance
of thermomechanical fibers in mixtures of PP and SAN. The impact
and flexural strengths were marginally improved while the tensile
strength was decreased when the fiber length of the filler was
reduced by about 50%.

Discussion

The use of thermomechanical fibers to increase the compatibility of dissimilar resins has promise. While additional work
is needed, initial results indicate that the properties of some
resin mixtures can be improved by the addition of fibers with
both polar and non-polar area on their surface available for
bonding.

SURFACE MODIFICATION OF ORGANIC FILLERS

As noted above, poor results have generally been obtained
when wood flour is added to non-polar thermoplastics. While
commercial wood flour has a low aspect ratio, the primary cause
of the reduction in strength properties of wood flour filled
thermoplastics appears to be incompatibility of the wood flour
and the plastic due to the polar character of the surface of the
wood flour. To determine if modification of the filler surface
would improve composite properties, three types of wood-based
fillers were investigated by molding into composites with polypropylene. One was a commercial grade wood flour with low aspect
ratio. A second was pulp mill residue described elsewhere[11].
The commercial wood flour and the pulp mill residue both were
expected to have primarily polar surfaces. However, the average
aspect ratio of the pulp mill residue was know to be significantly
higher than the aspect ratio of the wood flour. The third filler
was thermomechanical fiber from a hardboard plant produced by
disk refining wood at approximately 175°C. Due to the high
refining temperature the thermomechanical fiber was expected to
have a more non-polar surface than either wood flour as pulp mill
residue.

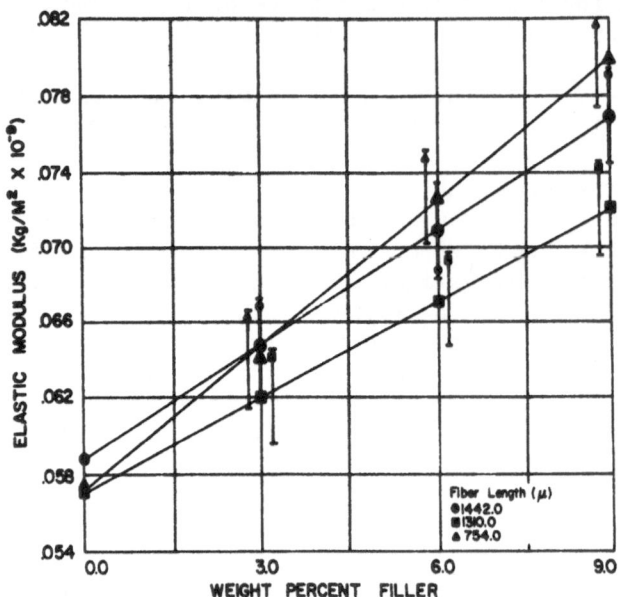

Figure 5.   Elastic Modulus (Bending) of Polypropylene High Lignin
            Thermomechanical Pulp Composites vs. Filler Concentrations
            for Various Fiber Lengths.

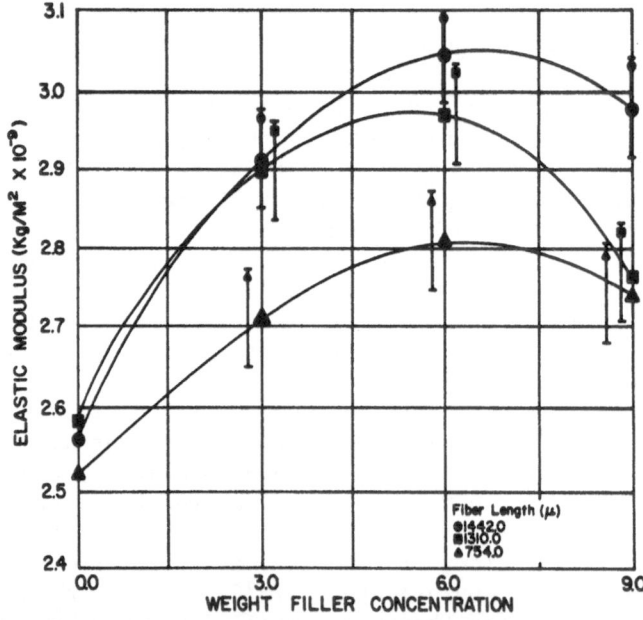

Figure 6.   Elastic Modulus (Bending) of Styrene-Acrylonitrile Co-
            polymer-High Lignin Thermomechanical Pulp Composites vs.
            Filler Concentrations for Various Fiber Lengths.

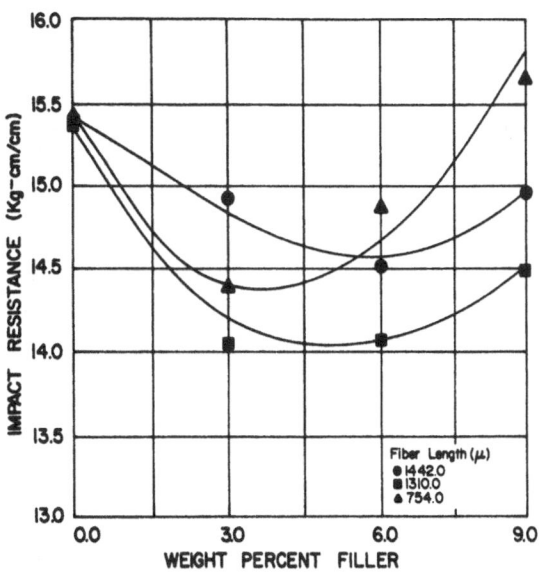

Figure 7. Impact Resistance of Styrene-Acroylonitrile Copolymer vs. High Lignin Thermomechanical Pulp Composites vs. Filler Concentrations for Various Fiber Lengths[1].

Procedure

The aspect ratios of the fillers were determined as described by Short[23]. Ash analysis were obtained using ASTM D 1102-56.

To reduce the surface polarity each of the fillers was coated with a non-polar, organic material commonly used in the paper industry for coating or "sizing" paper. The procedure was to form a slurry of 10% wood particles with water containing 2%, based on wood, paper sizing compound in the sodium salt form. To precipitate the organic sizing onto the surface of the wood particles, alum was added. The excess water was drained from the wood particles and they were dried.

Each of the fillers in both surface modified and as received condition was mixed with polypropylene and then dried at 110°C for 30 minutes just prior to molding. Tensile, flexure, and impact samples were molded using an injection molder. Test specimens were conditioned for a minimum of 72 hours at 23°C and 50% RH prior to testing. All physical property evaluations were conducted in accordance with appropriate ASTM testing procedures.

Results

The physical properties of the cellulosic fillers tested are given in Table 1. The three wood based cellulosic fillers had low ash contents which indicates low abrasion potential during processing. The aspect ratios of the fillers varied considerably with the aspect ratio of wood flour being less than either pulp mill residue or thermomechanical fiber as expected.

Table 1. Physical Properties of Wood-Based Fillers[1]

Filler Type	Aspect Ratio[2]	Ash Content (%)
Commercial Wood Flour	2.56 + .93	0.49
Pulp Mill Residue	7.50 + 2.5	0.53
Thermomechanical Fiber	8.23 + 2.6	1.01

1 - All passing 30 mesh screen
2 - 95% confidence limits

The tensile strengths of composites containing the treated and untreated fillers are shown in Figure 11. At 20% by weight filler loading the tensile strength of polypropylene was reduced by all three fillers. This had been expected since all the

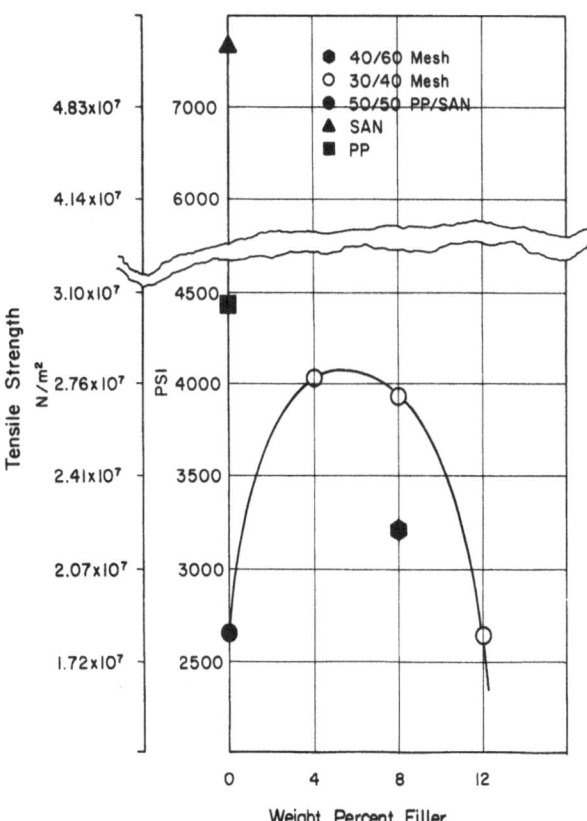

Figure 8.   Tensile Strength of PP/SAN Mixtures with Thermomechanical
            Fiber Fillers.

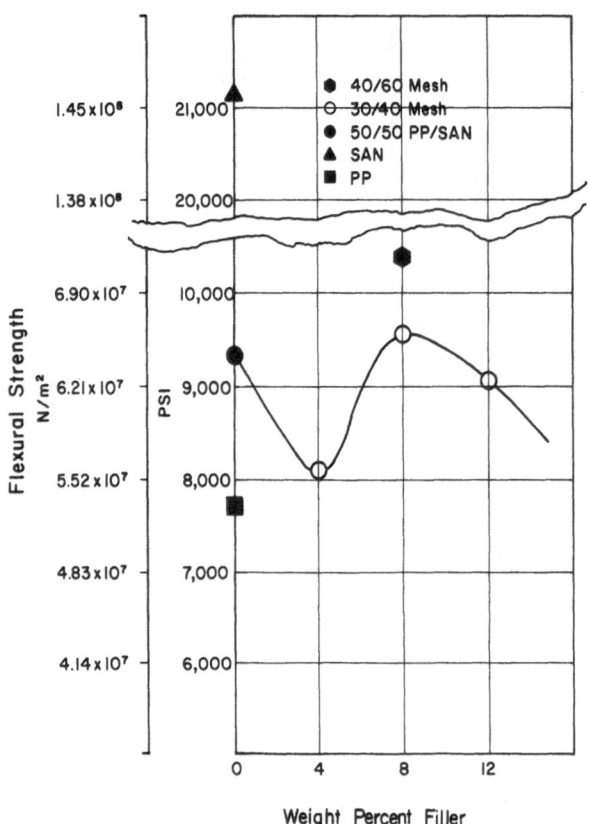

Figure 9. Flexural Strength of PP/SAN Mixtures with Thermomechanical Fiber Fillers.

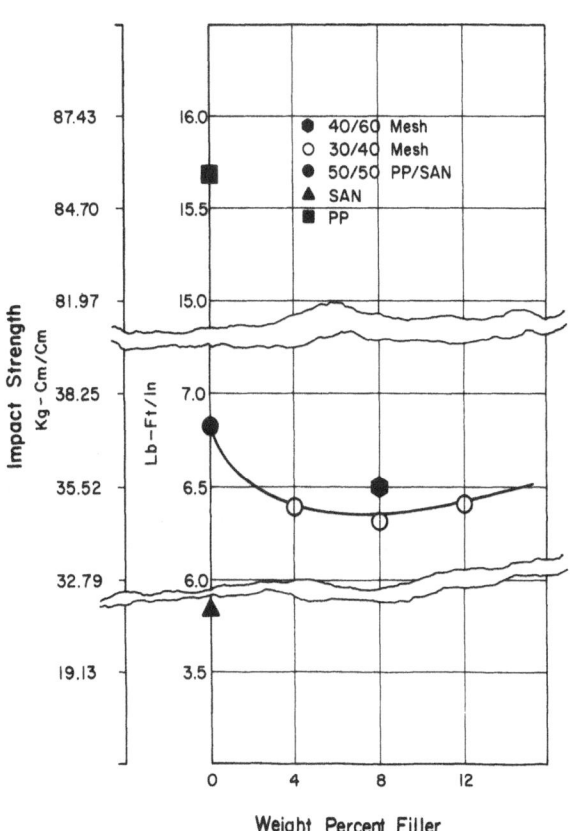

Figure 10. Impact Strength of PP/SAN Mixtures with Thermomechanical Fiber Fillers.

fillers had aspect ratios below 15 which is the minimum aspect
ratio of bark fiber that significantly improved composite.
properties[8]. It should be noted that wood flour with a very low
L/D ratio gave the poorest composite properties when untreated
fillers are compared indicating that even below 15 the L/D ratio
has some influence. Of greater interest was the improvement in
tensile strength of composites containing the surface treated
fillers with the exception of the thermomechanical fiber
polypropylene composite. The small change resulting from the
surface treatment of the thermomechanical fiber supports the
assumption made that the surface of the thermomechanical fibers
was primarily non-polar.

The notched Izod impact strength of the various composites
is shown in Figure 12. The impact strengths of all composites
are reduced at 20% filler loading with thermomechanical pulp
composites giving the smallest decrease when untreated fillers
are compared. As noted with tensile strength, surface treatment
significantly improved the impact strength of composites containing wood flour and pulp mill residue but not thermomechanical
fibers.

Discussion

The improvements noted in composites containing wood flour
or pulp mill residue that had been coating with a non-polar
organic sizing indicates that the surface of organic fillers is
of primary importance in filler-polymer bonding. Coating of
cellulosic fillers with paper sizing is expensive and does not
appear to be feasible. However, by careful control of the
temperature during size reduction, organic fillers with primarily
non-polar surfaces can be produced using the thermomechanical
pulping process.

CONCLUSIONS

Organic fillers have been successful in a range of thermosetting resin formulations. Relatively little work has been
reported on the use of organic fillers in thermoplastics. The
data reported indicates that organic fillers have potential for
use in thermoplastics resins. The low density and low cost on a
volume basis makes organic fillers economically attractive.

In order for organic fillers to compete with inorganic
fillers special care must be taken to select organic fillers that
are compatible with a particular resin system. Shell flour
fillers have been shown to improve certain desirable properties
without   significant reductions in critical composite characteristics in a number of thermoplastics. Thermomechanical fiber

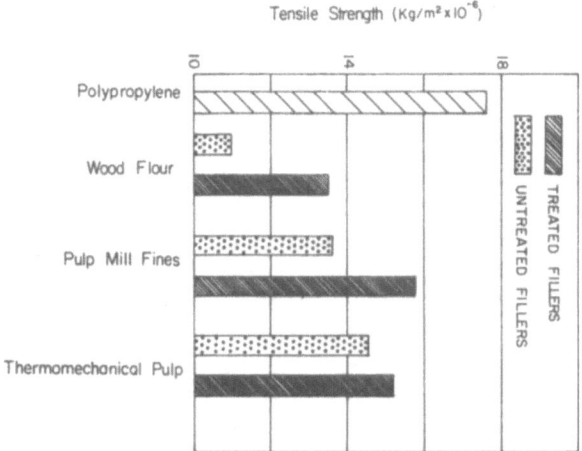

Figure 11. Effect of Treatment of Wood Fillers on Composite Properties (20% Filler in Polypropylene).

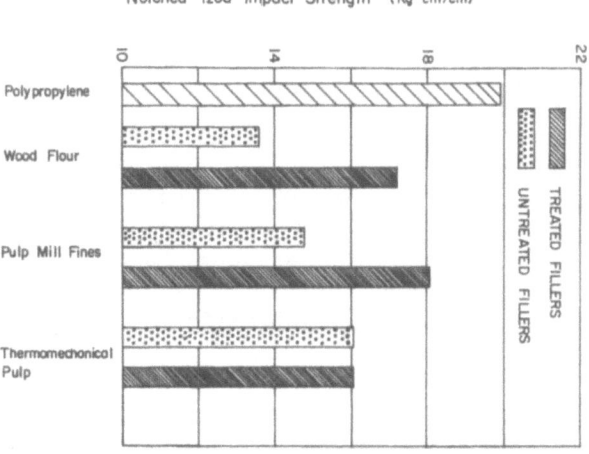

Figure 12. Effect of Treatment of Wood Fillers on Composite Properties (20% Filler in Polypropylene).

offers promise of improvements in composite properties due to its fiberous nature. Also, because the surface chemistry of thermomechanical fibers can be varied by the method of size reduction, the possibility exists that thermomechanical fibers can be "tailor-made" for specific resin systems.

Surface modification of organic fillers can significantly improve their performance as demonstrated with rice hull flour and wood flour. Additional effort to develop surface treatments for organic fillers is needed.

REFERENCES

1. W. C. Wake, "Fillers for Plastics, " ILIFFE Books, London, England, 1971.
2. V. Staymer, "Modern Plastics Encyclopedia, "McGraw-Hill Publications Company, New York, NY, 1978.
3. R. Eller, Plast. World, 32, 70 (1974).
4. P. D. Ritchie, S. W. Critchley, and A. Hill, "Plasticizers, Stabilizers, and Fillers," ILTFFE Books, London, England, 1972.
5. E. E. Halls, Plastics, Lond. 6, 267 (1942).
6. A. S. Wood, Modern Plastics, 51, (6), 42 (1975).
7. Anon., Modern Plastic 51, (6), 42 (1975).
8. D. J. Miller, J. D. Wellons, R. L. Krahmer, P. H. Short, Forest Products J., 24 (8), 18 (1974).
9. E. L. Sould, and H. E. Hendrickson, Forest Products, J. 16, (8), 17 (1966).
10. S. Z. Chow, and K. J. Pickles, Wood and Fiber 3 (3), 166 (1971).
11. G. R. Lightsey, P. H. Short, V. K. K. Sinha, Polymer Engineering and Science 17 (5) 305 (1977).
12. T. F. Clark, R. V. Williamson and E. C. Lathrop, Mod. Plast., 23, 158 (1945).
13. W. J. Albrecht, F. E. Barton, II, and D. Bordick, Trans. ASAE, 16, 650 (1973).
14. Gold Kist Inc., 3348 Peachtree Rd. N.E., Atlanta, Georgia, Personal Communication.
15. G. R. Lightsey, A. L. Hines, D. W. Arnold, and V. K. K. Sinha, Plastics Engineering, p. 40, May (1975).
16. G. R. Lightsey, Plastics and Rubber: Materials and Applications, p. 69, May (1978).
17. G. R. Lightsey, B. D. Herzog, Plastic Design and Processing, p. 28, May (1979).
18. S. B. Driscall, Proceeding of 35th ANTEC of Society of Plastic Engineers, Montreal, Canada, p. 366, (1977).
19. T. S. Connor, Plastics Design and Processing, July (1979).
20. S. Willis and T. S. King. Proceeding of Reinforced Plastics/Composites, 36th Annual Conf., Washington, D.C. Feb. (1981).

21. D. Atack, Svensk Papperstidning, 75, 89 (1972).
22. S. H. Badwin, D. A. Goving, Svensk Papperstidning, 71, 647
     (1968).
23. P. H. Short, Wood Science, 9, p. 37 (1976).
24. G. G. Gonzalez, "Evaluation of Polar and Non-Polar Plastics
     Compounded with Thermomechanical Wood Fibers", Dissertation,
     Dept. of Chem. Engr., Miss. State Univ., Aug. (1979).
25. K. S. Kalasinsky and G. R. Lightsey, "Quanitative Analysis
     of Cellulosic Filler Surface Chemistry Using Infrared Reflec-
     tance Spectroscopy and Computer Spectral Addition (In press).
26. F. Lowery and G. R. Lightsey, Un-Published Data, Dept. of
     Chem. Engr., Miss. State Univ. (1980).

CRYSTALLIZATION OF GUTTA PERCHA NETWORKS

AND ASSOCIATED ELASTICITY

K. J. Smith, Jr.

Department of Chemistry
SUNY College of Environmental Science and Forestry
Syracuse, NY

INTRODUCTION

The force exerted by an amorphous, stretched rubbery network held at constant length varies directly with temperature. But crystallizable networks exhibit deviations from force-temperature linearity once temperature falls below the melting temperature into the crystallization region. Incipient crystallization drives the force downward[1], sometimes sharply so[2]. A select few networks display a resurgence of force as temperature continues to fall[3-5]; polyethylene[3,4] and gutta percha[5] are examples. Gutta percha exhibits the U- or V-shaped stress-temperature profile in a spectacular fashion. Upon cooling its stress falls precipitously to negative values, reemerges in the positive domain at still lower temperatures, and climbs back to a high level[5].

Behavior of this sort has been attributed to a transformation of one crystalline morphology into another[3,4]. Cooling is thought to first promote fibrillar crystallization, which reduces retractive force, followed by conversion of the fibrillar crystallites into chain-folded lamellae at still lower temperatures. Stress regeneration is attributed to lamellae formation. This hypothesis has never been proven; in fact, it is claimed[6] that lamellae generation in rubber networks is actually accompanied by declining stress. And Gent[5], who made the original observations of stress regeneration in gutta percha networks, rejected the transformation hypothesis in favor of a simple explanation involving only thermal contractility of the semi-crystalline structure.

A new explanation of the behavior of stretched gutta percha

networks has recently been proposed[7,8].   It involves only fibrillar
crystallization, but several features previously omitted from
earlier theories are considered. Specifically included are the
points:

1. A network chain, or portion thereof, passing through
    a crystallite may traverse the crystallite in one
    direction relative to the chain displacement vector or
    in the exact opposite direction. The two configurations
    are not equally probable.
2. Changing crystallization continuously dispatches the
    crosslinks to their most probable positions.
3. Lateral crystallite growth is permitted by a contingent
    of amorphous chains.

The first point recognizes that two configurations of crystallization are available to a chain, each generating a different
tension on the two amorphous sub-chains emanating from the
crystallite. A crystallite incorporates chains of each configuration because its growing faces accrete new chains as they
are encountered in the melt[9]. The second point recognizes that
crystallization upsets the balance of forces about a crosslink,
forcing it to a new equilibrium position[7,8]. The third point
simply recognizes that all chains do not crystallize simultaneously; crystallites grow laterally as well as longitudinally.

Incorporation of these features into a theory of stress-
induced crystallization can be done in more than one way. The
results, however, are identical. In this paper a derivation
differing from the original[7,8] is presented. The model adopted
here is more critically defined than before and a little more
tedious mathematically. Following development of the useful
equations a brief analysis of Gent's data[5] on gutta percha is
presented.

Theoretical Development

Since network crystallization is necessarily incomplete, the
free energy of crystallization ΔF_c is customarily taken to be the
sum of two parts. One is the free energy of crystallization in
the absence of restraining forces imposed by amorphous chains and
sub-chains. An amorphous chain is defined here as one spanning
two crosslinks without passing through a crystallite, and a sub-
chain is the amorphous portion of a partially crystalline chain
running from a crystallite to a crosslink. Entry of a chain into
a single crystallite generates two sub-chains as shown in Figure 1.
The second part is the elastic free energy ΔF_e generated in the
amorphous chains and sub-chains in response to the waxing and
waning of crystallinity. That is, the free energy of crystalli-

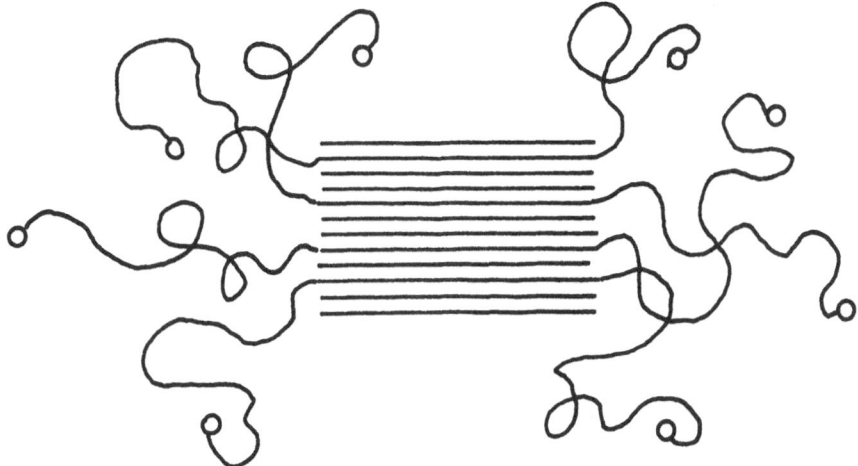

Figure 1. Crystallite model

zation can be written as

$$\Delta F_c = -\omega GN\Delta H_u (1 - T/T_m^o) + \Delta F_e \qquad (1)$$

where G is the total number of network chains running from one
crosslink to another, N is the average number of links per network chain, ΔH_u is the heat of fusion per link, T is temperature,
T_m^o is the melting temperature in the absence of all constraints
(i.e., $\Delta F_e = 0$), and ω is the network degree of crystallinity,
defined as the ratio of the number of crystalline links to the
total number of links

$$\omega = G_c n/GN$$

Here G_c is the number of chains that crystallize and n is the
average number of crystalline links in the crystalline chains.

The elastic free energy is the important quantity to be
reckoned with. To calculate it requires first of all a recognition of an important characteristic of polymer networks,
amorphous or semi-crystalline. And that is that crosslinks in a
network, while not completely immobile, are severely limited in
their fluctuations because the convergence of several chains,
usually four, at a crosslink creates a zone of multiple entangle-
ments with neighboring chains, which obstruct free motion in much
the same manner as a grappling hook is obstructed by a web of

random yarn. If a path is traced along several chains through connected crosslinks, it is reasonable to expect that fluctuations of the crosslinks at the ends of the path will have little effect on interior chains near the center if the path contains enough chains. Interior chains behave as if the path ends are securely anchored at fixed positions. For example, G_j connected chain vectors make up a path $\underset{\sim}{R}_j$ that appears constant to many of its interior chains. If $\underset{\sim}{r}_i$ represents the $\underline{i^{th}}$ network vector, then

$$\sum_i C_{ij} \underset{\sim}{r}_i = \underset{\sim}{R}_j = \text{constant} \qquad (2)$$

The coefficients C_{ij}, which have values 0 or 1, allow the summation in equation (2) to extend over all chain vectors in the network. For chains participating in the $\underline{j^{th}}$ path the coefficients are unity, and zero otherwise. A similar equation holds for each path required for complete network characterization. Constancy of $\underset{\sim}{R}_j$ is established by the condition

$$\sum_i C_{ij} \, d\underset{\sim}{r}_i = 0 \qquad (3)$$

If G_c of the G network chains crystallize, leaving $G-G_c$ amorphous chains, there are $G+G_c$ amorphous elastic elements (amorphous chains plus amorphous sub-chains) that comprise the semi-crystalline network. It is here assumed that a chain enters a crystallite only once or not at all. By and large, multiple entry into crystalline regions is probably rare except in very lightly crosslinked networks. Let each elastic element (chain or sub-chain) behave in a Gaussian fashion so that we may write for the elastic free energy F_e of a specified network

$$F_e = \frac{3kT}{2b^2} \sum_k \frac{r_k^2}{N_k} + \text{constant} \qquad (4)$$

The number of statistical links in the element vector $\underset{\sim}{r}_k$ is N_k and each statistical link is of length b. The summation extends over all elastic elements (amorphous chains and sub-chains) in the network. Assigning the crosslinks to their most probable positions requires that F_e be a minimum, i.e.

$$\sum_k \frac{\underset{\sim}{r}_k \cdot d\underset{\sim}{r}_k}{N_k} = 0 \qquad (5)$$

The conditions (3) still hold. Crystallization perturbs the various vectors $\underset{\sim}{r}_i$, but distant crosslinks still appear immutable to those chains in the middle zone of the path.

A distinction exists between the vectors r_i and $\underset{\sim}{r}_k$. Those denoted by the subscript i are network chain vectors $\underset{\sim}{}_i$ from crosslink to crosslink. Those denoted by k run from crosslink to crosslink and/or from crosslink to crystallite. Upon crystallization the ith network chain partitions itself into a crystallite vector $\underset{\sim}{l}_i$ and two sub-chains $\underset{\sim}{r}_{11}$ and $\underset{\sim}{r}_{12}$:

$$\underset{\sim}{r}_1 = \underset{\sim}{r}_{11} + \underset{\sim}{r}_{12} + \underset{\sim}{l}_i$$

Equation (2) may then be written as

$$\sum_i C_{ij}\, \underset{\sim}{r}_i = \sum_i C_{ij}\, (\underset{\sim}{r}_{11} + \underset{\sim}{r}_{12}) + \sum_i C_{ij}\, \underset{\sim}{l}_i = \underset{\sim}{R}_j$$

By re-indexing with k this becomes

$$\sum_i C_{ij}\, \underset{\sim}{r}_i = \sum_k C_{kj}\, \underset{\sim}{r}_k + \sum_i C_{ij}\, \underset{\sim}{l}_i = \underset{\sim}{R}_j \qquad (6)$$

At constant $\underset{\sim}{l}_i$ the condition expressed by equation (3) is exactly

$$\sum_k C_{kj}\, d\underset{\sim}{r}_k = 0$$

Introducing into equation (5) such a condition for each path (j) gives

$$\left(\frac{\underset{\sim}{r}_k}{N_k} - \sum_j \underset{\sim}{\beta}_j\, C_{kj} \right) \cdot\ d\underset{\sim}{r}_k = 0$$

where $\underset{\sim}{\beta}_j$ is the jth Lagrangian vector. The orthogonal solution can be obviated by imposing the null vector solution in all instances. This simply requires that F_e be a minimum also with respect to the components of $\underset{\sim}{r}_k$. Thus the null vector solution is

$$\underset{\sim}{r}_k = N_k \sum_j \underset{\sim}{\beta}_j C_{kj}$$

The summed terms can be written as $\underset{\sim}{\beta}_k\, G_k$, where G_k is the number of paths containing $\underset{\sim}{r}_k$, and $\underset{\sim}{\beta}_k$ is a resultant vector analogous to $\underset{\sim}{\beta}_j$. But $\underset{\sim}{\beta}_k$ cannot be determined without detailed network information beyond our present knowledge. It is therefore necessary to resort to an assumption that $\underset{\sim}{\beta}_k$ is independent of   k, so that $\underset{\sim}{\beta}_k = \underset{\sim}{\beta}$ and

$$\underset{\sim}{r}_k = \underset{\sim}{\beta}\, G_k\, N_k \qquad (7)$$

This amounts to summing equation (6) over all paths j and holding the resultant constant:

$$\sum_i \sum_j C_{ij} \; \underset{\sim}{r}_i = \sum_j \underset{\sim}{R}_j = \text{constant}$$

A single Lagrangian vector $\underset{\sim}{\beta}$ suffices to handle this condition.

Combining equations (6) and (7) gives

$$\sum_i C_{ij} \; \underset{\sim}{r}_i = \underset{\sim}{\beta} \sum_k C_{kj} \; G_k \; N_k + \sum_i C_{ij} \; \underset{\sim}{l}_i$$

This equation can be better handled if it is summed over all j:

$$\sum_i \sum_j C_{ij} \; \underset{\sim}{r}_i = \underset{\sim}{\beta} \sum_k \sum_j C_{kj} \; G_k \; N_k + \sum_i \sum_j C_{ij} \; \underset{\sim}{l}_i$$

The sums are:

$$\sum_j C_{ij} = G_i \qquad \sum_j C_{kj} = G_k$$

where G_i is the number of paths the ith chain participates in. Thus

$$\sum_i G_i \; \underset{\sim}{r}_i = \underset{\sim}{\beta} \sum_k G_k^2 \; N_k + \sum_i G_i \; \underset{\sim}{l}_i$$

which yields for $\underset{\sim}{\beta}$

$$\underset{\sim}{\beta} = \frac{\sum_i G_i \; \underset{\sim}{r}_i - \sum_i G_i \; \underset{\sim}{l}_i}{\sum_k G_k^2 \; N_k}$$

The value of $\sum_i G_i \; \underset{\sim}{l}_i$ is critically dependent upon the geometry of crystallization. It is specifically assumed that two crystalline configurations are available to a chain, both shown in Figure 2. If P_{ai}, P_{bi} are the respective probabilities that the ith chain crystallizes in the indicated configurations, we may write

$$\underset{\sim}{l}_i = P_{ai} \; \underset{\sim}{l}_{ia} + P_{bi} \; \underset{\sim}{l}_{ib} = \Delta P_i \; \underset{\sim}{l}_{ia}$$

Note that $\underset{\sim}{l}_{ia} = - \underset{\sim}{l}_{ib}$. These probabilities have been previously estimated[9] as

$$\Delta P_i = 3\sigma \; \underset{\sim}{b} \cdot \underset{\sim}{r}_i / N_i b^2$$

The assumption is made that a small number σ of statistical links

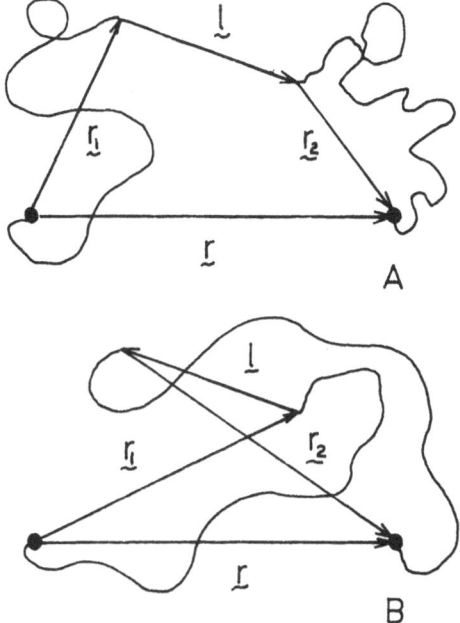

Figure 2. Two configurations of a chain

deposits onto a crystallite face and determines the configuration a or b according to the vector direction at deposition. If the number of links initially deposited is less than σ, configurational fluctuations will possess enough energy to remove the links back into the amorphous domain, but if equal to or greater than σ firm adherence to the crystallite is achieved. The probabilities in question are then the relative probabilities that σ consecutive links are oriented at an angle θ or $\pi-\theta$ with respect to $\underset{\sim}{r}_1$. The vector $\underset{\sim}{b}$ is a statistical link vector located along $\underset{\sim}{l}_{1a}$.

We have then

$$\sum_i G_i \underset{\sim}{l}_i = \sum_i G_i \Delta P_i \underset{\sim}{l}_{1a} = \frac{3\sigma}{b^2} \sum_i G_i \left(\frac{\underset{\sim}{b} \cdot \underset{\sim}{r}_1}{N_1} \right) \underset{\sim}{l}_{1a}$$

Since it is assumed that a chain can pass through a crystallite only once, the ratio r_1/N_1 refers to amorphous chains only and is determined from equations (7,8) when $\underset{\sim}{l}_1 = 0$; i.e.

$$\underset{\sim}{r}_1/N_1 = G_i \sum_i G_i \underset{\sim}{r}_1 / \sum_i G_i^2 N_1$$

We now write for F_e

$$F_e = \frac{3kT}{2b^2} \left[\frac{1}{\sum\limits_k G_k^2 N_k} \right] \left[\sum\limits_i G_i \, \underset{\sim}{r}_i - \frac{3\sigma}{b^2} \sum\limits_i G_i^2 \, \underset{\sim}{b} \left(\frac{\sum\limits_i G_i \, \underset{\sim}{r}_i}{\sum\limits_i G_i^2 \, N_i} \right) \underset{\sim}{l}_{ia} \right]^2$$

$$+ \text{ constant} \tag{9}$$

The sums must be evaluated. Since G_i is the number of paths containing the $i\underline{th}$ chain the sum over G_i is simply the number of chains G times the average of G_i, $<G_i>$. It is then easy to deduce that

$$\sum\limits_i G_i \, \underset{\sim}{r}_i = G <G_i> \, \underset{\sim}{r} \tag{10}$$

$$\sum\limits_i G_i^2 \, N_i = G <G_i^2> N$$

$$\sum\limits_i G_i^2 \, \underset{\sim}{l}_{ia} = G_c <G_i^2> n \, \underset{\sim}{b} = GN <G_i^2> \omega \, \underset{\sim}{b}$$

$$\sum\limits_k G_k^2 \, N_k = G <G_i^2> N \, (1 - \omega)$$

where $\underset{\sim}{l}_{ia} = n \, \underset{\sim}{b}$, n is the number of crystalline units a chain, N is the average number of links in a chain, $< \ >$ represent average values, ω is the degree of crystallization, and r is an average vector characteristic of the network. Substitution of equations (10) into equation (9) yields

$$F_e = \frac{3GkT}{2Nb^2} \left(\frac{1}{1-\omega} \right) \left[\frac{<G_i>^2}{<G_i^2>} \right] \left[\underset{\sim}{r} - \frac{3\sigma\omega}{b^2} (\underset{\sim}{b} \cdot \underset{\sim}{r}) \, \underset{\sim}{b} \right]^2 + \text{constant}$$

The distribution of G_i is unknown, but it is not necessary to know such detail about the network. Without any loss of rigor we can assume a monodisperse distribution so that $<G_i>^2 = <G_i^2>$, which yields the simple result

$$F_e = \frac{3GkT}{2Nb^2} \left(\frac{1}{1-\omega} \right) \left[\underset{\sim}{r} - \frac{3\sigma\omega}{b^2} (\underset{\sim}{b} \cdot \underset{\sim}{r}) \, \underset{\sim}{b} \right]^2 + \text{constant} \tag{11}$$

Limiting deformation to simple elongation along the z axis by amounts great enough to insure that the crystallites have their chain axes lying parallel to the stretch direction, we have for the terms in square brackets above

$$x^2 + y^2 + z^2 (1 - 3\sigma\omega)^2 \tag{12}$$

where x, y, z are the components of $\underset{\sim}{r}$. These components are determined by comparing equation (11) when crystallinity is zero with the traditional result of the kinetic theory of elasticity. For simple elongation along the z axis, such a comparison shows

$$x^2 = y^2 = Nb^2/3\alpha \tag{13}$$

$$z^2 = \alpha^2 Nb^2/3$$

where α is the deformation ratio along z.

Substitution of equations (11, 12, 13) into equation (1) after substracting F_e ($\omega = 0$, $\alpha = 1$) gives ΔF_c

$$\Delta F_c = -\omega GN\Delta H_u \left(1 - \frac{T}{T_m^o}\right) + \frac{GkT}{2} \left(\frac{1}{1-\omega}\right) \left[(1 - 3\sigma\omega)^2 \alpha^2 + \frac{2}{\alpha}\right] - \frac{3GkT}{2} \tag{14}$$

At constant deformation the equilibrium degree of crystallization can be determined from the condition $\partial \Delta F_c / \partial \omega = 0$,

$$\frac{2N\Delta H_u}{k} \left(\frac{1}{T} - \frac{1}{T_m^o}\right) = \left(\frac{1}{1-\omega}\right)^2 \left[(1-3\sigma\omega)(1+3\sigma\omega-6\sigma)\alpha^2 + \frac{2}{\alpha}\right] \tag{15}$$

The melting temperature T_m is obtained at zero crystallinity

$$\frac{1}{T_m} = \frac{1}{T_m^o} - \frac{k}{2N\Delta H_u} \left[(6\sigma-1)\alpha^2 - \frac{2}{\alpha}\right] \tag{16}$$

Equation (15) may be rearranged to

$$\omega = 1 - \left[B/(A+B)\right]^{\frac{1}{2}} \tag{17}$$

where

$$A = \frac{2N\Delta H_u}{k} \left(\frac{1}{T} - \frac{1}{T_m}\right)$$

$$B = (3\sigma-1)^2 \alpha^2 + 2/\alpha$$

And the retractive force at equilibrium crystallization derives from the condition $\partial \Delta F_c / \partial \alpha = L_1 f$, where L_1 is the initial, un-stretched length:

$$f = \frac{GkT}{L_1} \left(\frac{1}{1-\omega}\right) \left[(1-3\sigma\omega)^2 \alpha - \frac{1}{\alpha^2}\right] \tag{18}$$

Discussion

Equation (18) predicts a U- or V-shaped force-temperature profile. In cooling a stretched network, force initially falls sharply with increasing crystallization, declining to zero when

$$\omega = \frac{1}{3\sigma}\left(1 - \alpha^{-3/2}\right)$$

It then turns negative but climbs back to zero when

$$\omega = \frac{1}{3\sigma}\left(1 + \alpha^{-3/2}\right)$$

Further cooling causes the force to increase sharply in the positive domain.

Gutta percha networks behave in this general fashion[5]; Gent[5] reported force-temperature behavior for several samples. From his data the value of σ is estimated to be approximately unity, which is close to the value of 1.24 found for cis-polyisoprene networks. Using this estimated value of σ, taking ΔH_u to be 3000 cal/mole of isoprene units, N equal to 100 isoprene units, $\alpha = 1.21$, and $T_m = 54°C$ the profile given by the solid lines in Figure 3 is predicted. The experimental curve follows the dotted line. Several discrepancies are evident, but overall the agreement between theory and experiment is remarkably good. A serious problem, however, lies with the degree of crystallinity. Gent estimated about 25% crystallization whereas the theory predicts nearly 70%. It is also pointed out that the number of units between crosslinks was reported[5] to be 176 (rather than 100 used here).

Gutta percha, like most polymers, crystallizes in a head-to-tail fashion; neighboring chains in the unit cell traverse the crystal in opposite directions. But this has no effect on the relative probability that a chain enters the crystal in a certain direction. If the chain is wrongly oriented for crystallization at one location - that is, it cannot be accommodated by the unit cell without reversing its direction - it is simply displaced to the next adjacent lattice row, which accommodates it without difficulty and without important statistical consequence because of the small distance of displacement. As a result, the equations herein are little affected by crystallization of this kind.

Obviously the theory is not quantitatively correct. But it does indicate a need to rethink current ideas about stress-induced crystallization. It shows that stress regeneration at low temperatures need not involve a morphological change. It demonstrates that crosslink mobility and wrong-way chains can have extremely important consequences. And it demonstrates a necessity for extensive, reliable experimental data on a number of polymer networks.

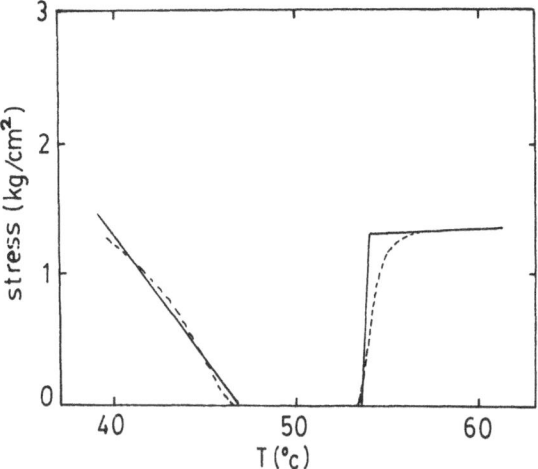

Figure 3.   Comparison of theory with experiment.
Stress-temperature of gutta percha

REFERENCES

1. K. J. Smith, A. Greene and A. Ciferri, Kolloid-Z, <u>194</u>, 49
    (1964).
2. A. N. Gent, J. Polymer Sci., <u>A2</u>, 3787 (1965).
3. J. T. Judge and R. S. Stein, J. Appl. Phys., <u>32</u>, 2357 (1962).
4. A. Keller and M. Machin, J. Macromol. Sci., <u>B1</u>, 41 (1967).
5. A. N. Gent, J. Polymer Sci., <u>A2</u>, <u>4</u>, 447 (1966).
6. D. Luch and G. Yeh, J. Appl. Phys., <u>43</u>, 4326 (1972);
    J. Macromol. Sci., <u>37</u>, 121 (1973); J. Polymer Sci., <u>11</u>, 467
    (1973).
7. K. J. Smith, J. Polymer Sci. Submitted.
8. K. J. Smith, J. Polymer Sci. Submitted.
9. K. J. Smith, Polymer Sci. and Eng., <u>16</u>, 168 (1976).

JAPANESE LACQUER-A SUPER DURABLE COATING

(PROPOSED STRUCTURE AND EXPANDED APPLICATION)

Ju Kumanotani

Institute of Industrial Science, The University of Tokyo

7-22-1 Roppongi, Minatoku, Tokyo, Japan 106

INTRODUCTION

Mankind is indebted to nature with respect to food, clothing and shelter since time immemorial. Though the origin of Japanese lacquer, a naturally occuring phenolic coating material from the lacquer tree, Rhus vernicifera, is not clear, historically, it is known to have been used in continental China for 5,000 to 6,000 years[1], and is said to have come to Japan from China in the 6th Century along with Buddhism. Lacquer from other trees (Rhus succedanea, Melanorrhorea usitata, et al) have been used in Asian countries. An outstanding property of Japanese lacquer wares in man's everyday living, there being many items of many cultures, exhibited in museums in U.S.A. and Europe, some from ancient excavations. At the Shosoin, a temple in Japan, a great number of cultural treasures coated with Japanese lacquer are exhibited, all having been preserved for more than a thousand years without having lost their original elegant beauty.

What gave Japanese lacquer this durability ? This intiguing question launched the author on a quest which started some 25 years ago. We knew of this super-durability but no one .... nothing told us why. Our studies started out to learn what it is.. .... then what happened and why ....

The sap of the lacquer trees is the material for Japanese lacquer and is collected from the lacquer trees, Rhus vernicifera, as a latex as a water in oil (urushiol) type emulsion. This sap consists of urushiol (65-70%)(see Fig. 1), plant gum (5-7%), laccase (less than 1%) and water (20-25%).

225

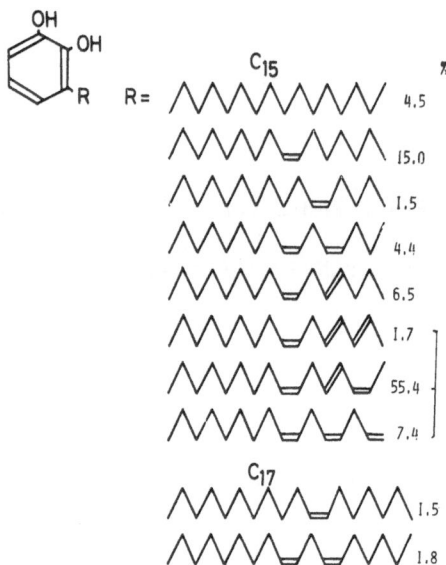

Fig. 1. Heteroolefinic side chain urushiol of the
sap lacquer trees (Rhus vernicifera).

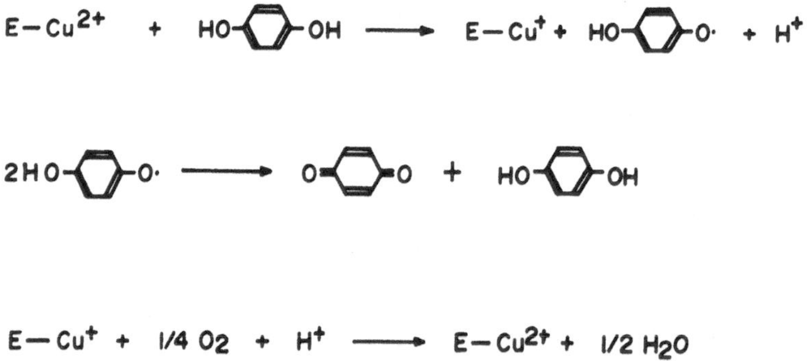

E—Cu^{2+} + HO—◯—OH ⟶ E—Cu$^+$ + HO—◯—O· + H$^+$

2HO—◯—O· ⟶ O—◯—O + HO—◯—OH

E—Cu$^+$ + 1/4 O$_2$ + H$^+$ ⟶ E—Cu^{2+} + 1/2 H$_2$O

Scheme 1. Laccase-catalyzed reaction of hydroquinone.

Urushiol, the first constituent mentioned, is a mixture of 3-substituted catechol derivatives with a 15 or 17 carbon chain having a olefin number 0-3[2,3,5]. The average number of olefins in the side chain of urushiol is 2.0-2.5[4]. Constituents of this Japanese lacquer urushiol are seen to be the same as those of urushiol in poison ivy of U.S.A.[5]

The next mentioned plant gum is made up of water soluble(oligomeric and polymeric) gum and water insoluble gum (8:2), both existing in a salt form with counter cations of Ca, Mg and Na. The water soluble gum indicated a gel permeation chromatogram with two maximum peaks corresponding to M_w of 22,000 and 77,000, respectively[6]. Sugar units of the acid form polysaccharides, derived from the above water soluble salt form plant gum fractions by its acidification, were analyzed to consist of 5 and 5 mole% of L-rhamnose, 7 and 12 mole% L-arabinose, 26 and 22 mole% of 4-O-methyl-D-glucuronic acid and 57 and 56 mole% of D-galactose[7,8], respectively.

The lastly mentioned laccase is a copper glycoprotein, p-quinol $-O_2$ oxidoreductase with M_n of 120,000 and has four atoms of copper per molecule. Scheme 1 is established for hydroquinone as substrate[9]. Explaining on Scheme 1, one electron transfer takes place from hydroquinone to $E-Cu^{++}$, producing semiquinone radical and $^-E-Cu^+$. The formed semiquinone radical undergoes a disproportionation reaction, giving hydroquinone and p-benzoquinone. The reduced $E-Cu^+$ is spontaneously oxidized to $E-Cu^{++}$ by oxygen due to its high affinity toward oxygen, which takes part in the subsequent oxidation of hydroquinone. In these laccase-catalyzed reactions, each of the four copper atoms in laccase participates differently in the electron transfer pathway between hydroquinone and $E-Cu^{++}$ or $E-Cu^+$ and O_2 [10,11].

Let us now turn to how Japanese lacquer is made. Japanese lacquer is made from the sap of lacquer trees as mentioned. The sap is put in a specially designed wooden vessel and stirred at room temperature for 30 min and then at 20-45°C for 6 to 8 hrs under sun rays in the old days and now with an electric heater or charcoal burner heating from overhead. The optimum end point was and is now determined through experience from a darkening of color and increasing viscosity while stirring. The thus made lacquer contains about 3% water and has a viscosity suitable for coating of articles, which, after coatings, are dried in a moist air atmosphere with a relative humidity of 70-80% for one to two days. Most articles coated, now, are made from wood. The dried surface of the wood articles is made smooth and polished with charcoal and water, and recoated, the same procedure being repeated 20 to 30 times untill the finished,polished coating of Japanese lacquer wares is attained.

PROOF OF DURABILITY OF JAPANESE LACQUER COATING (MEASURING DYNA-
MECHANICAL PROPERTIES AGAINST STORAGE TIME)

Despite historically long human experience of the use of the
lacquer wares, there was no scientific data evidencing the excelling
durability of Japanese lacquer when we began the study of Japanese
lacquer in our laboratory.

Let us first look at structural characteristics. Generally
speaking, for polymeric material such as this coating, measuring the
viscoelastic properties should give information relating to chemical and physical structures of the material.Therefore, as shown in
Fig. 2, viscoelastic properties of a specimen of Japanese lacquer
films were measured[12], which exhibited intermediate and in between
viscoelastic properties of the oil-based coating films made from
polymerized linseed oil [13]/oil-soluble phenolic resin varnish (oil/
p-tert-butylphenol formaldehyde resin (3:2))[14]/alkyd resin,60% oil-
length[13] and a rigid phenolic resin cured with 5% hexamethylenetetra-
amine[15]. As may be noted, the Japanese lacquer film, when compared
with the   other   coating films shown here, shows a higher rubbery modulus in the high temperature region and a broader λ peak
than the others, showing that the Japanese lacquer film is more
highly crosslinked and complicated than the others.

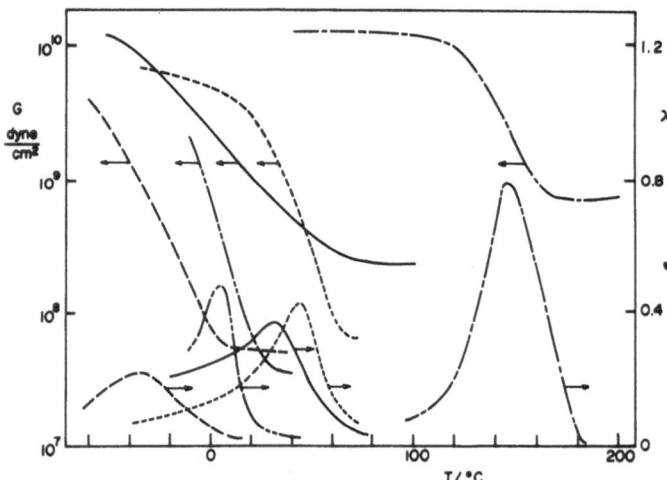

Fig. 2 Rigidity(G) and logarithmic decrement(λ) vs. temperature relation of films by torsional pendum method
made from:--- phenol novolac cured with 5% hexa;
--- oil/p-tert-butylphenol novolac (3:2) varnish;
— Japanese lacquer;--- alkyd, 60% oil-length and
--- polymerized linseed oil

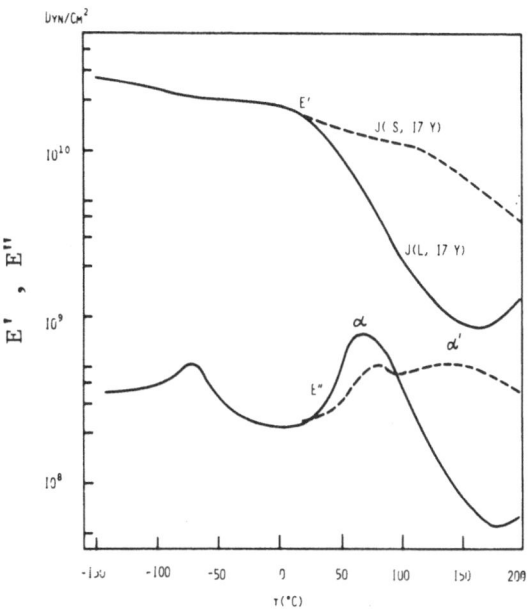

Fig. 3. Storage modulus E' and loss modulus E" vs.
temperature(T) relation at 11 Hz; a sap film
J(S, 17Y) and lacquer film J(L, 17Y).

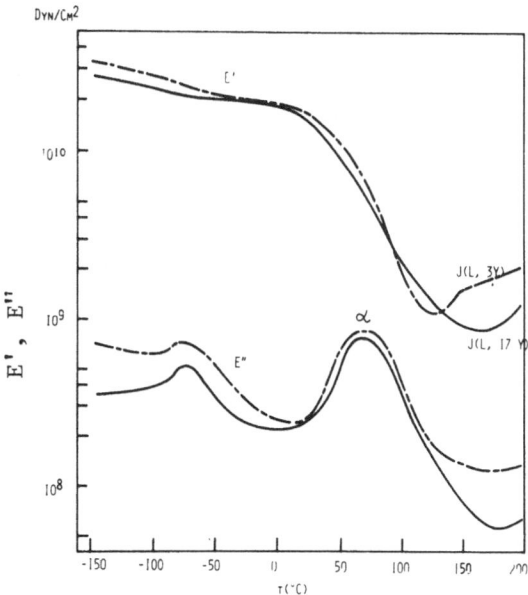

Fig. 4. Storage moudulus E' and loss modulus E" vs.
temperature (T) relation at 11 Hz; lacquer
films J(L, 17Y) and J(L, 3Y).

Now let us look at the change in durability over time. By measuring the viscoelastic properties we found that the sap film had changed and developed extensive chemical crosslinking and that the degree of crosslinking had remained almost unchanged for the Japanese lacquer films during its 3 years of storage[16]. The same was found to be true in regards to the lacquer and sap films which had aged over 20 years when dynamechanical property measurements were made recently on them[17].

Figs. 3 and 4 show the measured storage modulus E' and loss modulus E" as a function of temperature for the sap and lacquer films aged 17 years and 1 year, using a Vibron DDV II (Toyo Baldwin, Tokyo) at 11 Hz over a temperature range −150− +200°C. It is obvious that there is a distinctive difference between the sap and lacquer films in their dynamechanical properties, particularly, in the higher temperature region. As may be seen in Fig. 3, comparing sap and lacquer films, the modulus E' at 150°C is 7 x 10^9 dyne/cm^2 for the sap film J(S, 17Y)and 1 x 10^9 dyne/cm^2 for the lacquer film J(L, 17Y), indicating that the former is more highly crosslinked than the latter(J,S,L, Y and M are symbols for a dried Japanese lacquer film, sap, lacquer , year and month,respectively, and J(S, 17Y) means a sap film aged 17 years). For the sap films, the loss modulus E"$_{max}$ peaks appeared at 120°C, 48°C, −70°C and −100∼−150°C(designated as α, α', β andγ peaks,in this order, respectively), whereas, for the lacquer film J(L, 17Y), a rather sharp α peak was found at 67°C together with the β peak (−70°C), indicating that the sap film J(S, 17Y) has a more complicated structure than the lacquer film J(L, 17Y). Fig. 4 shows that the dynamechanical profile measured for the lacquer film J(L, 3Y) was found surprisingly to be identical to that of the older lacquer film J(L, 17Y). This indicates that the dynamechanical properties of the lacquer film remained unchanged for about 20 years.
This proved to be the first scientific data evidencing the high durability of the Japanese lacquer films, though obviously the 20 years span in this case is very short to make a comparison with an unchanging record of 1,000 years.

In Table 1, density(25°C),a parameter for oxidation, crosslinking and deterioration measured for the sap and lacquer films is shown. It is obvious that in agreement to the above variation of the dynamechanical properties for the sap and lacquer films, the lacquer films J(L, 19Y) and J(L, 3Y) showed almost the same density, however, density of the sap film J(S, 3M) which was smaller than that(1.180) of the lacquer film J(L, 1Y) increased after 19 years to 1.217.

All the above findings were endorsed and confirmed by IR spectral data for the films. Fig. 5 shows the IR spectra for the sap and lacquer films, J(S, 18Y), J(L, 18Y) and J(S, 1M) together with that of the plant gum separated from a sample of the sap, indicating

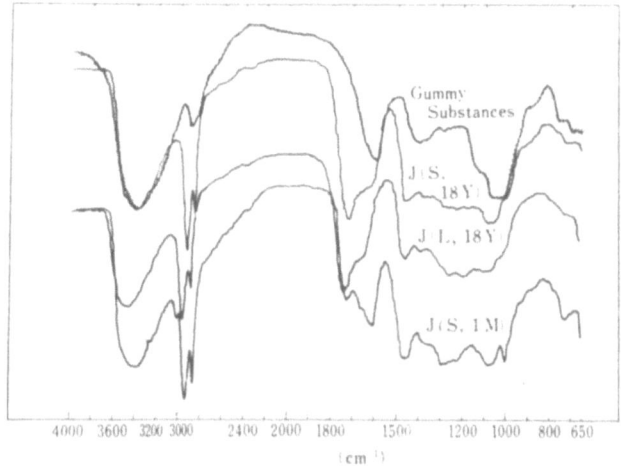

Fig. 5. IR Spectra for the plant gum separated from a sap
of lacquer trees (Rhus vernicifera), a Japanese
lacquer film J(L, 17Y) and sap films J(S, 1M) and
J(S, 17Y).

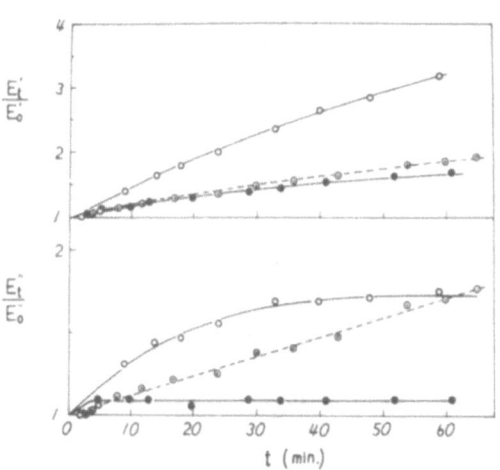

Fig.6. Variation of a ratio of (storage modulus E'/loss modulus E")
vs baking time at a temperature of a minimum E' value in
the rubbery plateau region: E'_0 and E''_t mean E''and E" at
baking time 0 and t(min),respectively: o J(S, 1M), ⊚ J(1,
17Y) and ● J(B, 17Y)(a baked J(S,1M) at 100°C-30 mins and
storaged for 17 years).

Table 1. Density at 25°C of Japanese lacquer
and Sap Films

Film	J(S, 3M)	J(S, 19Y)	J(L, 19Y)	J(L, 1Y)
Density	1.128	1.217	1.185	1.180

that the IR spectra of the sap film J(S, 18Y) in the finger print
region became gradually similar to that of the plant gum, demon-
straing occurence of considerable degradation of the catechol nucleus of the polymerized urushiol in the film during the storage of 18
years. However, the lacquer film J(L, 18Y) showed a spectrum similar
to that of the younger film J(S, 1M), though outside the finger
print region, an unsaturated ketone-absorption peak at 1730 cm^{-1}
could be observed for the both, J(L, 18Y) and J(S, 18Y).

Occurence of this oxidation of the unsaturated side chain of
urushiol is understandable, because the γ peak due to the local mode
of the side chain of urushiol in the sap and lacquer films,
though less distinctive, could be observed in the low temperature
region, allowing occurence of oxidation of the side chain even at
room temperature.

As may be seen in Fig. 6, when the sap and lacquer films which
had been aged 17 years were heated in the air at a certain temperature T where the lowest rubbery modulus E'_h was shown in the plateau
region of the E'-temperature curve, the E'_h value at T increases
along with baking time similarly for both films, but the E' value at
T increases only for the lacquer film being heated, indicating that
the remaining unsaturated side chain of urushiol in the lacquer
film underwent oxidative crosslinking during baking. On the other
hand, the sap film which already had degraded considerably seemed
to undergo oxidative crosslinking due to the reaction of the remaining catechol nucleus or phenolic nucleus of (3),(5) or (4) in
Scheme 2, judging from the observed changing of E' and E" (related
to viscosity of a system)values and IR spectra as a function of
storage time as shown in Figs. 6 and 5,respectvely.

ORIGIN OF α AND α' RELAXATION PEAK IN THE SAP AND LACQUER FILMS[16,17]

Having proven the basis of durability of the lacquer, we next
turn to the structural relation between sap and lacquer as represented by α and α' relaxation peaks and the origin of these peaks.

It may be assumed that the higher E' values and broadened E"
peak in the aged films result from oxidation of the sap films during
storage; in the drying process of the sap, the plant gum, dissolved
in water droplets that disperse in the oil phase,remains in the dried

sap film, suggesting the sap film is composed of some two components, polymerized urushiol and plant gum domains. Since urushiol is an oxidizable phenolic compound, the polymerized urushiol domain in a sap film is also susceptable to oxidation followed by crosslinking reaction, resulting in the growing of the α' peak in the sap film.

To elucidate as to the α' peak, therefore, the variation of the dynamechanical properties were measured for the sap films as a function of storage period. Fig. 7 shows that the sap film J(S, 1M) exhibited E"$_{max}$ peak at 65°C with a shoulder at 85°C at 11 Hz. This smaller shoulder peak grew larger, becomming prominent in the higher temperature region with lapse of time. In the sap film J(S, 18M),the shoulder peak at 85°C of J(S, 1M) shifted to 118°C and became more prominent. In the case of J(S, 17Y), the α' peak became smaller and flattened with its base becomming lengthened beyond 200°C. In addition, the rubbery modulus E', a parameter of degree of crosslinking, increased with lapse of time pararell to growing of the α' peak.During these variations of the α' peak, the α peak,on the other hand, remained unchanging in the temperature range 65-75°C.

The nature of the α' peak was acertained further by measuring the dyanmechanical property of the baked urushiol film at 160°C for 3-24 hrs in the air (see Fig. 8).For the baked urushiol film at 160-24 hrs, E"$_{max}$ peak appeared at 75°C, 100°C and 150°C. It is clear that with progress of baking, the E'vs. temperature curve shifted to higher temperature region, E" peak becomming larger and broader. These variations of E' and E" vs.temperature in the baked urushiol films are seen to correspond to the changes of the α and α' peaks in the sap films during storage. Therefore, it is concluded that the variation of the α and α' peaks which took place in the sap films during storage is stemed from the oxidative crosslinking of the urushiol domain.

Consequently, assuming the above, it is seen conclusive that the sap is inherently capable of making α and α' peaks when it is dried into a film; in other words, the dried sap film is composed of two domains of polymerized urushiol and plant gum, and the α' peak grew as a result of oxidation followed by crosslinking and/or degradation in storage. It should be noted that the α peak remained in a narrow, limited temperature range 65-75°C, during the deterioration of the sap films, which is coincident with the α peak temperature of the lacquer films.

On the other hand, the observed durable property of the lacquer films, in contrast to the deteriorable property of the sap films, would seem to conflict with logic, because the sap and lacquer films originate from the same material, but this may be interpreted as the difference of the morphological features of the sap compared with lacquer films, which will be discussed next.

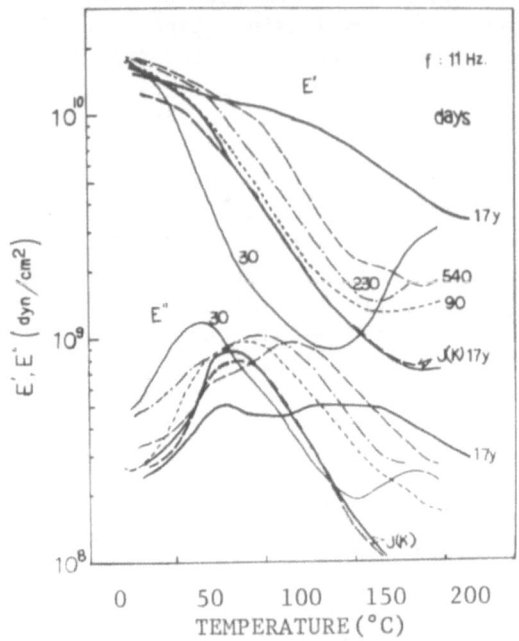

Fig. 7. Variation of a dynamechanical profile for a sap film
         vs. storage period (days) up to 17 years (Y) in compar--
         ison with a Japanese laquer film aged 17 years J (K, 17Y).

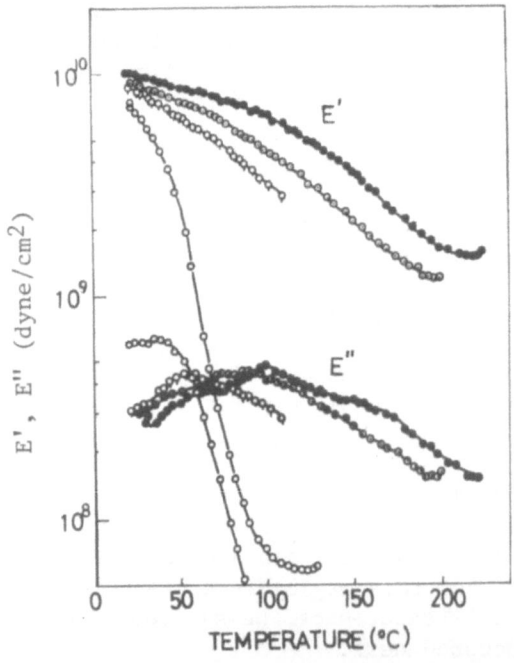

Fig. 8.
Dynamechanical Profile
for the baked urushiol
film at 160°C for 3hrs
(o), 4.5 hrs(ǫ), 6 hrs
(◉), and 24 hrs(●).

MORPHOLOGICAL FEATURES OF THE SAP AND LACQUER FILMS

Sap of lacquer trees, which appears dark under a polarized light optical microscope, showed a "Maltese cross" pattern with a number of circles of about 10 μm when it was dried into a film as shown in Fig. 9. This indicated the formation of considerable molecular orientation of the plant gum in the film[18], corresponding in pattern to the water droplets in the oil(urushiol) phase of sap.

Fig. 10 shows the presence of grains in the sap film in the process of drying. These support the previous assumption that the sap film is composed of polymerized urushiol and plant gum domains.

As may be seen in Fig. 11, the sectioning of the older sap film J(S, 18Y) shows a structure more complex than that of the younger film (compare with Fig. 9 and 10), this being presumably, due to the occurence of oxidative crosslinking and degradation in storage of the sap film with some grains still being detectable. On the other hand, as may be seen in Fig. 12, the section of the lacquer film J(L, 18Y),does not show any special characterizing feature. However, Fig. 13 shows that the lacquer film J(L,18Y), after being etched with ionized air in a Ion Coater IB-3 (Eiko Co., Tokyo) discloses aggregated grains densely packed in the film with a size of 0.1 μm.

A PROPOSED CELL STRUCTURE PROVIDING DURABILITY OF THE LACQUER FILMS

The basis for high durability in the cell structure is seen as;

1) High durability of the lacquer films has been recognized through long human experience and use of lacquer wares.
2) No variation in the dynamechanical profile of the lacquer films could be found after 20 years in contrast to that of the sap films.
3) The α peak in the lacquer film is a sharp, single peak, indicating that the dynamechanically observed chemical and physical structure of the lacquer film is simple and has a low susceptability to oxidation or deterioration.
4) The scanning electron microphotograph of the section of the lacquer film shows a very characteristic structure with aggregated grains of 0.1 μm size.
5) The higher barrier feature toward oxygen of polysaccharides is well known in literature[19].
6) Urushiol is a phenolic compound susceptable to oxidation, but the matrix made thereof in the lacquer films in contrast to the sap films, is low in susceptability to oxidation.

Considering these findings we here propose a cell structure for the durable Japanese lacquer film; the lacquer film is composed of cells or grains of 0.1 μm size packed densely in the film. The cells have plant gum wall with polymerized urushiol inside, and are firmly

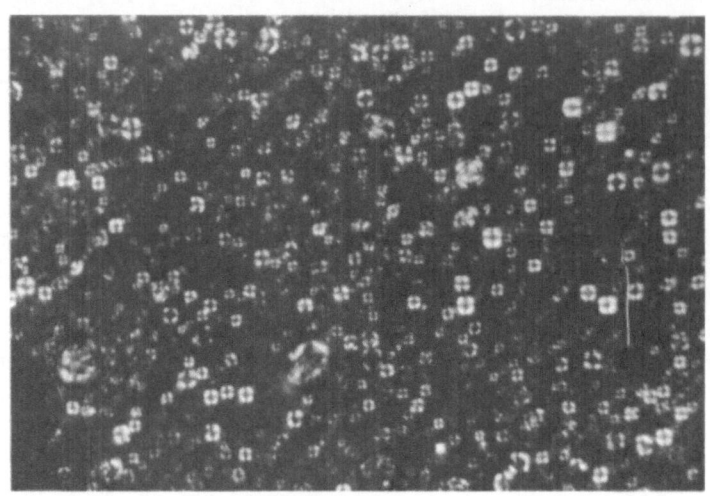

Fig. 9. An optical microphotograph under polarized light of a sap of
       lacquer tree (Rhus vernicifera); the major circle size
       is about 10 µm.

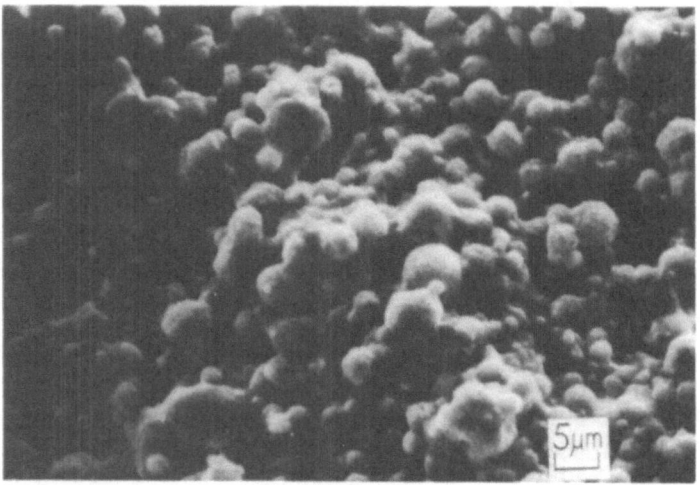

Fig. 10. A Scanning electron microphotograph for growing
        grains in a drying sap film.

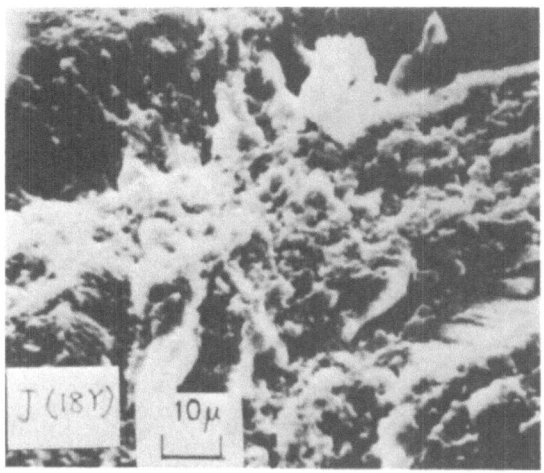

Fig. 11. A scanning electron microphotograph for the section of
a sap film aged 18 years.

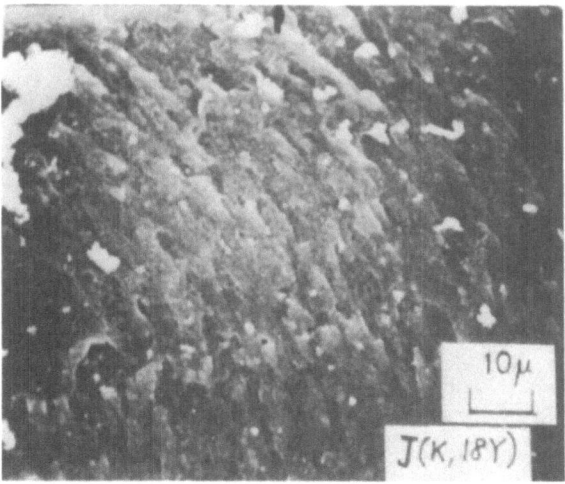

Fig. 12. A scanning electron microphotograph for a section of
a lacquer film aged for 18 years.

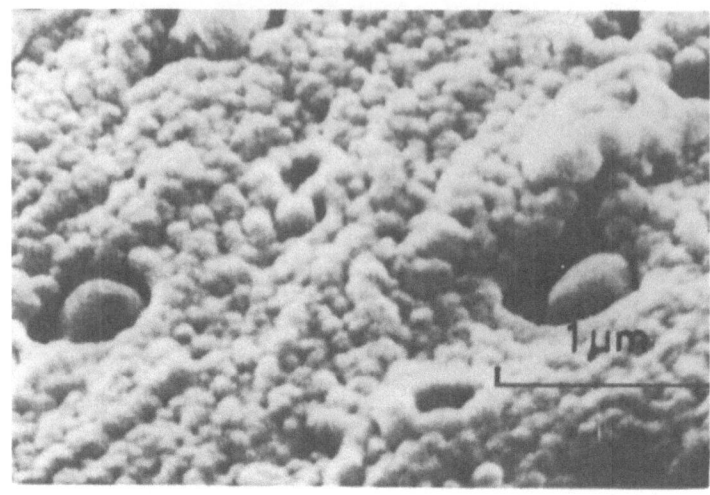

Fig. 13. A scanning electron microphotograph for an etched
Japanese lacquer film aged for 18 years.

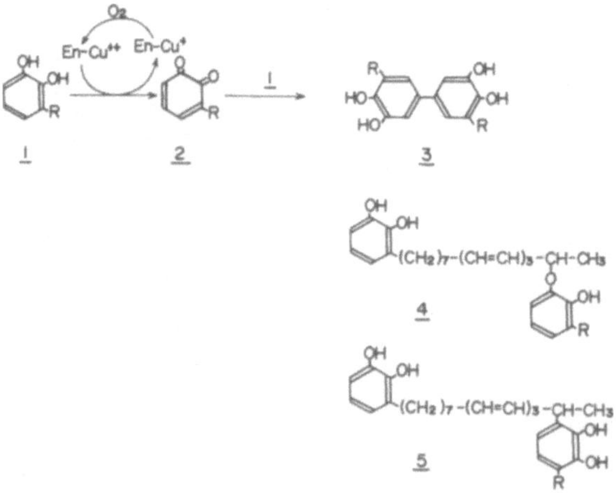

Scheme 2.   Mechanism of Laccase-catalyzed polymerization
of urushiol.

cemented by the polymerized urushiol and urushiol soluble or water insoluble plant gum of the sap. Repeated coating in making lacquer wares is seen to introduce a greater density in IPN structure[20] in the Japanese lacquer films.

In this cell structure model, the plant gum provides a protective barrier towards diffusion of oxygen into the polymerized urushiol inside, resulting in and explaining the high durability of Japanese lacquer films. Since the plant gum is hygroscopic, absorption of humidity would break down the barrier characteristics, and thereby causes degradation of the polymerized urushiol in the lacquer film, but a combination with hydrohobic polymerized urushiol results in a cell structure wall of low humidity absorption.

In order to elucidate the formation of aggregated cell structure made from plant gum and urushiol in Japanese lacquer films proposed above, it is most important to know the interaction between urushiol and plant gum in the sap being treated or in the lacquer in the film formation process.

EXPERIMENTS ON FORMATION PROCESS OF CELL STRUCTURE

The interaction between plant gum and urushiol was investigated by measuring the variation of viscoelastic properties and solubility in toluene of the sap under treatment and IR spectra of the plant gum separated from the sap and the lacquer made thereof.

1) Alteration of Viscoelastic Properties of the Sap in the Lacquer-making Process[21] Dynamic rigidity(G') and viscosity(η') were measured using a coaxial double cylinder rheometer(Rheometer All Mighty Iwamoto Seisakusho, Kyoto) and an audio rheometer[22]. The frequency dependence of G' and η' was obtained at angular frequency ranging from 0.1 to 10^3 sec^{-1}. In Fig. 14 in the low frequency range (0.1-1.0 sec^{-1}) the frequency dependence of η' is large, whereas, the frequency dependence of G' was smaller than that in the audio frequency range. These results may be explained as follows. Fine droplets of an aqueous gum solution dispersing in urushiol interact with each other and construct temporary network structure with long relaxation times, which vanish as making the lacquer from a sap progresses. The behavior of the lacquer is similar to that of the ordinary synthetic polymer solutions in a low frequency range, but the values of G' and η' in the audio frequency range are most independent of the frequency (see Fig. 15). These behaviors are likely to be intermediate between those for rigid rods and flexible chains in solutions[23]. It is thought that as the lacquer-making process preceeds, water molecules gradually evaporate, and molecular chains of the plant gum may take more shrunken conformation. It is presumable that the restricted segmental motion of the polysaccharides may contribute to the observed peculiar viscoleastic behaviors of Japanese lacquer at the higher frequencies.

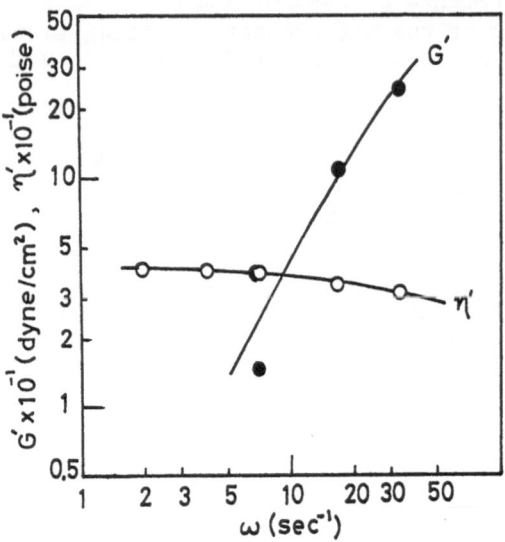

Fig. 14. Frequency dependence of dynamic rigidity (G') and
          viscosity (η') for the sap of lacquer trees(Rhus
          vernicifera).

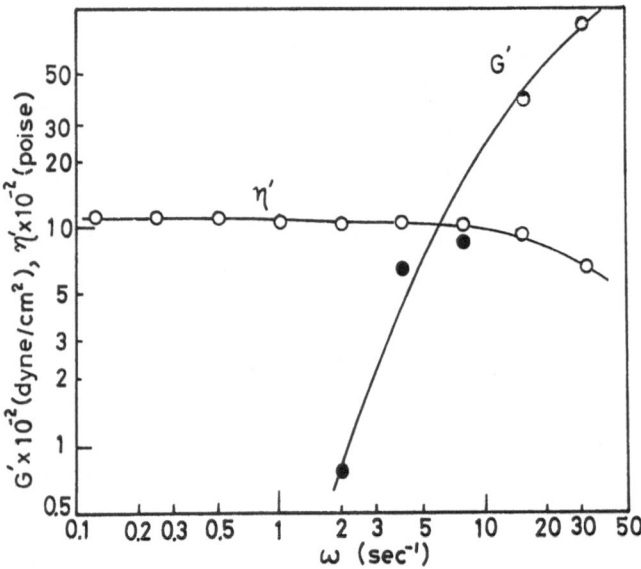

Fig. 15. Frequency dependence of dynamic modulus (G') and
         viscosity (η') for the lacquer.

Plant gum, separated from a sap as an acetone insoluble part, showed an IR spectrum characteristic of the polysaccharides in a salt form. However, the aceone insoluble part of the lacquer showed an IR spectrum having characteristic peaks of the plant gum and urushiol. Furthermore, though sap is partly soluble in toluene, it becomes completely soluble in toluene when made into lacquer. These results may suggest that the water soluble gum becomes soluble or dispersible stable in the oil(urushiol)phase as a result of grafting of urushiol on the plant gum(polysaccharides) in the lacquer-making process.

2) Polymerization of Urushiol[24,25]   Since urushiol is a major component of the sap or lacquer, participation of urushiol, in the lacquer-making process from the sap or in the film formation process from the sap or lacquer is significant.

By a series of studies of the enzymatic and non-enzymatic polymerization mechanism of urushiol, Scheme 2 has been established. Urushiol (1) is oxidized into the corresponding quinone (2)[26],and the formed quinone (2) undergoes C-C and C-O coupling reaction with catechol nucleuse or the triene side chain urushiol, a major component of urushiol, giving dimeric urushiol (3), (4) and (5) [27-30]. The catechol nucleus of these dimeric urushiol undergoes enzymatic oxidation into the corresponding quinones following the same type C-C and C-O coupling reaction with urushiol or with each other. Thus urushiol grows in its polymerization up to M_n of 20,000-30,000 in the lacquer-making process from sap.

The finally obtained lacquer usally contains 3% water. The content of water in the lacquer or the sap under treatment seems to correlate to the activity of laccase. Furthermore, from a study of a model reaction with 4-tert-butyl-o-quinone and methyl linoleate, it is found that the existence of water and its content may influence the coupling reaction rate of the urushiol quinone (2), and the ratio of the C-C and C-O coupling of the urushiol quinone(2)[31] .

Rotational energy barrier around the C-C bonding in the aliphatic carbon-phenyl or phenyl-phenyl groups in the dimeric urushiol of (5) or (3) is larger than the C-O bonding in (4). The former bonding may introduce larger rigidity in the polymerizing urushiol molecules or the lacquer film made therefrom than the latter bonding. On the other hand, toughness of the lacquer films is expected to be influenced by the proportion of the C-C and C-O coupling bonds in the film. Therefore, controlling of water content in the lacquer-making process seems to be one of the important factors to make a good quality lacquers,which may vary the C-C and C-O coupling proportion of urushiol quinone(2).

It is notworthy that the C-C coupling product (4) may make a

network structure in which the rigid catechol nucleus is bonded
between a rather long olefinic aliphatic chain, resulting in, probably, that these structures contribute to the formation of the lacquer films with toughness balancing the rigidity and flexibility of
the lacquer films.

3) <u>Factors Controlling Cell Structure Formation</u>   Admitting some
speculation, the following factors are seen to contribute to the
process of cell structure formation in the lacquer films.

1) phase separation due to the difference in polarity of the plant
gum (polysaccharides in a salt form) and urushiol in the urushiol
grafted plant gum[20]

2) Enzymatic film formation reaction, resulting in the durable film
structure.

The first phase separation concept is well accepted in polymer
chemistry; formation of the cell structure arises from the larger
difference in polarity or cohesive energy between the plant gum and
urushiol.

With reference to the 2nd reaction, dynamechanical properties
between three sap films were compared. One was made enzymatically
at room temperature. The second one was made by baking the sap directly at 160°C for 6 hrs. The last one was made by baking the previousely dried film enzymatically at 160°C for 30 mins.

Fig.16 shows that the enzymatically dried sap film shows larger E
values against temperature than the baked sap film at 160°C for 6
hrs directly. The E' values as a function of temperature increased
remarkably after baking of the previously enzymatically dried film
at 160°C for 30 mins. In parallel to these E' values changes, the
E"-temperature curve grew larger and broader with baking time in
the higher temperature region.

Furthermore, larger values for the activation energy, 77 and 79
kcal/mole, were obtained for the α peak relaxation of the lacquer
films than the value 57 kcal/mole for the α peak of the sap film
J(S, 1M) by the Arrhenius plots dynamechanically, probably, resulting from mutual interpenetration of the α and α' peak domains.

Considering these results, it is clear that the enzymatic drying process brings out a more dense network structure in the sap
film than the baking process, and that, in the presence of laccase,
the lacquer gives highly and tightly crosslinked structure in which
segmental motions are much more restricted than in the sap film.

These results indicate an important difference between the
enzymatic and thermal film formation processes.

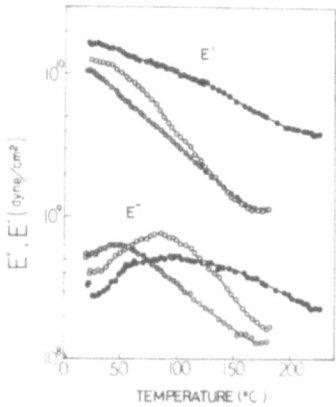

Fig. 16. Variation of a dynamechanical profile for sap films
made enzymatically(o), by baking at 160°C for 30 mins
of a sap directly(⊙) and by baking a previousely dried
sap film enzymatically at 160°C for 30 mins(●).

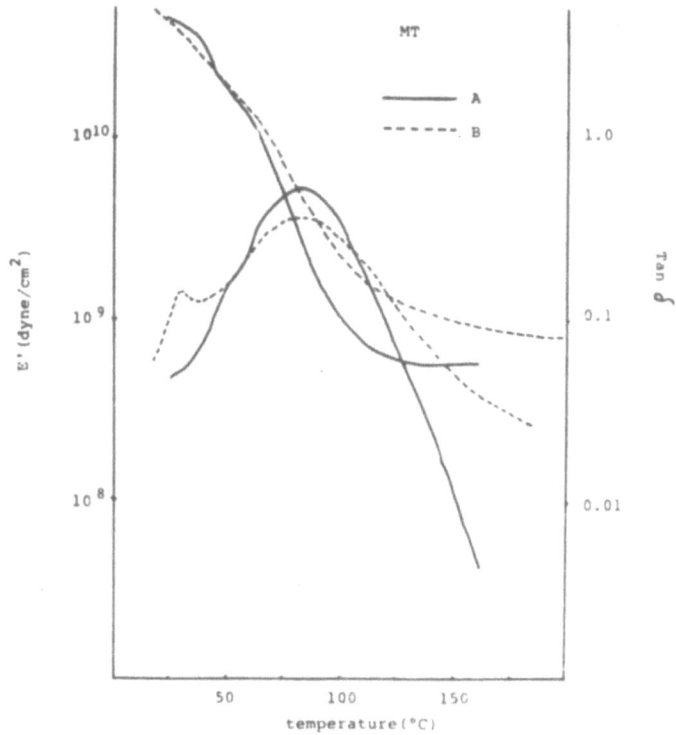

Fig. 17. A dynamechanical profile for alkyd resins cured with
butylated melamine resins: common alkyd(---) and
collodial alkyd (-) at 11 Hz.

It should also be noted that capability of gelation of the plant gum (polysaccharides) seems to take part significant role in the enzymatic film formation process[32] .

ATTEMPTS TO SYNTHESIZE JAPANESE LACQUER

Based on the aforementioned results, probable methods of synthesizing Japanese lacquer involve two routes: the morphological synthesis of Japanese lacquer independent of constituents of the sap of lacquer trees, and the synthesis of Japanese lacquer like materials using an enzyme.

1) Morphological Synthesis   Assuming the aforementioned cell structure model is reasonable for durable Japanese lacquer, a morphological synthesis is one possible way to synthesize the Japanese lacquer, independent of the constituents of the sap. We thus succeeded in the preparation of reactive colloidal particle-containing alkyd resins of about 0.1 μm size[33,34]. The alkyd resin was crosslinked with butylated melamine resins, resulting in high quality coatings. Fig. 17 shows the dynamechanical properties of the colloidal alkyd cured with butylated melamine resins in contrast to that of a common alkyd resin. The observed single sharp E''_{max} peak may indicate that crosslinking occured surrounding the alkyd resin colloidal particles by the melamine resin, probably, making aggregated cell structure in films.

In addition, MMA-grafted pullulan (a water soluble polysaccharide, M_n, 100,000) was prepared by using ceric ammonium nitrate as initiater in an aqueous medium[35]. A scanning electron microphotograph of the film made from a butyl acetate-acetone solution showed an aggregated cell structure in the film like in Japanese lacquer films as shown in Fig. 18.

2) Semi-synthesis of Natural Japanese Lacquer   The shortest way to approach synthesis of natural Japanese lacquer is to make substitutes for the constituents of the sap, which is easily done in practice.

On the basis of Scheme 2, it is clear that a polyphenol oxidase which can oxidize not only monophenol into diphenol, but also the formed diphenol into the corresponding quinone can substitute for laccase.

We attempted to attain drying properties by applying a polyphenol oxidase from alternaria tenuis[36] to non-drying sap which resulted from deactivation of the laccase involved in the sap, and found that the drying properties of the sap were improved satifactorily. From our studies, improvement of the thermal reisistance of the polyphenol oxidase would provide for wider application of the polyphenol oxidase and would contribute to expediting the drying of the various kinds of the sap of lacquer trees planted in Asian countries.

Furthermore, as substituents for urushiol, we found that catechol derivative, made by the Friedel-Craftes type alkylation of catechol with linseed oil, linseed oil alcohol made by the reduction of linseed oil[37] or oligomeric butadiene[38] can be used. Though there are number of papers relating to synthesis of urushiol components[39] synthesis of the triene side chain urushiol is faced with difficulties. On the other hand, recently, a report appeared about the formation of urushiol by tissue cell structures of Rhus vernicifera stock[40]. This method seems to be the most interesting to synthesize urushiol, suggesting a need for a further studying.

In south east Asian countries, the lacquer trees are easily cultivated and grow rapidly, and the yield of the sap there is large in comparison with that of the lacquer trees, Rhus vernicifera of China, Japan and Korea. Furthermore, the content of plant gum (20-30%) is larger than in the sap of lacquer trees (Rhuse vernicifera). Therefore, it appears positively promising, despite difficulties in quality as base for lacquer; the poor drying properties of the lacquer made from the sap of the south east Asian countries may be improved by using the urushiol-substituents and the polyphenol oxidase already mentioned.

Fig. 18. A scanning electron microphotograph for a film of a MMA grafted pullulan.

CONCLUSION

Since the petroleum crises, all renewable natural resources have been intensively sought and studied to develop urgently needed materials for the future. Since Japanese lacquer will dry at room temperatures in the presence of laccase, this Japanese lacquer is an energy conserving coating material, an excellent characteristic which may be added to its great durability.

It is believed that this lacquer holds great promise for the future as a coating material or durable polymeric material. Its development awaits work of not only polymer chemists but also plant- and biochemists.

In addition, considering that the super surable properties of Japanese lacquer is dependent on the cell structure made up of urushiol (a phenolic compound) and plant gum (polysaccharides), based on this model, this cell structure in theory, generally speaking, could be expanded to improve the durability of all synthetic polymeric materials in a combination with polysaccharides which are most aboundant, being made by cultivable plants utilizing the endless resources of sun rays, water and carbon dioxide. Much research is contemplated.

This study was supported partly by grants in aid for scientific research, Ministry of Education, Culture and Science, Japan. The writer is most obliged and wiashes to thank all my co-workers, particularly, Mr. S. Kazama, Dr. T. Kato, Mr. A. Hikosaka, Mr. H. Mitsui, Mrs. M. Ishiwatari, Dr. R. Oshima, Mr. M. Achiwa, Mr. K. Adachi, Mr. Y. Yamauchi, Mr. N. Iwatsuki and Mr. K. Nagase for their cooperation in conducting this work.

Literature Cited

1. Hu Jigao, Reports, 3rd International Symposium on the Conservation and Restoration of Cultural Property-Conservation of Far East Art Objects(1979, Tokyo),84
2. Y. Yamauchi and J. Kumanotani, 28th IUPAC (Vancouver, Canada) 1981, OR 30.
3. R. Majima, Ber., 55:172(1920); J. Kumanotani, Makromol. Chem., 179:172(1978) and literature 4.
4. W. F. Smyes, C. R. Dawson, J. Am. Chem. Soc.,76:2959, 5070 (1954)
5. J. Kumanotani, M. Achiwa, R. Oshima,and K. Adachi, Proc. 2nd ISCR-CP, Cultrual Property and Analytical Chemistry, 51(1979)
6. R. Oshima, A. Yoshikawa and J. Kumanotani, J. Chromatgr. 213:142 (1981).
7. R. Oshima and J.Kumanotani, Chemistry Letters(Japan), 943 (1981).
8. T. Nakamura, Biochim. Biophys. Acta, 30:40(1958)
9. R. A. Holwerda and H. B. Gay, J. Am. Chem. Soc., 96:6012(1974)
10. B. G. Malmstrom, B. Reinhammar and T. Vanngar, Biochim. Biophys.

Acta, 205:48(1970).

11. T. Kuwata, J. Kumanotani and S.Kazama, Bull. Chem. Soc. Japan, 34:1682 (1961).

12. L. W. Chen and J. Kumanotani, J. Appl. Polymer Sci., 9:3521 (1965).

13. L. W. Chen and J. Kumanotani, J. Appl. Polymer Sci., 9:2789 (1965).

14. J. Kumanotani, I. Akado, and T. Kuwata, Kogyo Kagaku Zasshi, 65: 678(1961).

15. T. Kuwata, J. Kumanotani and S. Kazama, Bull. Chem. Soc.Japan, 34:1678(1961)

16. J. Kumanotani and M. Achiwa, Report on Progr, Polym. Phys. Japan, 633 (1978).

17. J. Kumanotani, 7th International Conference on Organic Coatings, Science and Technology (1981), Athens, 265-282.

18.P. K. Stumpf and E. E. Conn " The Biochemistry of Plants" 3 Carbohydrates: Structure and Function, J. Preiss, ed., Academic Press, New York (1980), 329.

19. Polymer Handbook, J. Brandrup and E.H. Immergut, ed.,John Wiley & Sons, New York (1975),III-238.

20. L. H. Sperling " Interpenetrating Polymer Networks and Related Materials", Plenum Press, New York(1981).

21. T. Amari, J. Kumanotani, and M.Achiwa, Shikizai, 53:629 (1980)

22. S. Kuroiwa, M. Ogawa, and M. Nakamura, Kogyo Kagaku Zasshi, 70: 212 (1969).

23. T. C. Warren, J. L. Schrag, and J. D. Ferry, Biopolymer, 12:1905 (1973).

24. J. Kumanotani, Makromol.Chem., 179:47 (1978).

25. J. Kumanotani, Fatipec XIII (1976), 360-368.

26. J. Kumanotani, T. Kato, and A. Hikosaka, J. Polymer Sci.,Part C, 519 (1968).

27. T. Kato and J. Kumanotani, Bull. Chem. Soc., J. Polymer Sci., A-1, 7:1455(1969).

28. T. Kato and J. Kumanotani, Bull. Chem. Soc.Japan, 42:2375 (1969).

29. T. Kato, Y. Yokoo, M. Taniai,and J.Kumanotani, Can. J. Chem.,47: 2106 (1969).

30. J. Kumanotani and T. Kato, IUPAC International Symposium on Macromol. Chem., Butapest (1969), Preprint Vol. 5, 105.

31. M. Ishiwatari, Nippon Kagaku Zasshi, 5: 859 (1975).

32. D. A. Rees " Polysaccharides Shape ",Haltred Press, New York (1977).

33. H. Hata, J. Kumanotani and T. Tomita, Fatipec XIV:356-362 (1978)

34. H. Hata, T. Tomita, and J. Kumanotani, Fatipec XIV:485-362 (1980)

35. J. Kumannotani, T. Eto, and R. Oshima, Polymer Preprint, Japan, 29: 1077 (1980)

36. Unpublished data. See S. Motoda, J. Ferment. Technol.,Japan, 57:79 (1979) with respect to the polyphenol oxidase.

37. T.Kuwata, J. Kumanotani, and S. Kazama, Bull. Chem. Soc.,Japan, Kogyo Kagaku Zasshi, 64: 1629 (1961).

38. Mitsubishi Petrochemical Co., TOKUKAISHO 51- 54520.

39. J. S. Byck and C. R. Dawson, J. Am. Chem. Soc., 32: 1084 (1979);
    H. Halim and H. D. Locksley, J. C. S. Perkin I, 2331 (1981).
40. O. Hasimoto and K. Minami, Mokuzai Kagaku Zasshi, 26:49 (1980).

SIMULTANEOUS INTERPENETRATING NETWORKS FROM EPOXIDIZED TRIGLYCERIDE

OILS: MORPHOLOGY AND MECHANICAL BEHAVIOR

Shahid Qureshi, J.A. Manson, L.H. Sperling and
C.J. Murphy*

Materials Research Center
Coxe Laboratory #32
Lehigh University
Bethlehem, Pa. 18015

INTRODUCTION

Concern for the stability of future petroleum supplies has stimulated much attention in the past few years to feedstocks for polymers based on renewable resources. (1-7). Among these have been the botanical oils, which are usually triglycerides. Indeed, botanical oils have long been employed in applications such as coatings, plasticizers, and in the synthesis of monomers such as brassylic and sebacic acid. More recently, other monomers and polymers have been made, including simultaneous interpenetrating networks (SINs) based on elastomers from various botanical oils coupled with polystyrene networks (PSN) (8-13). In the present work, the goal has been to develop new polymers based on industrial-type oils derived from plants (8,11). To demonstrate feasibility, several unsaturated oils (linseed, crambe, lunaria and lesquerella) were epoxidized and converted to elastomeric components of SINs; thalictrum oil has also been partially epoxidized and converted to a flexible crosslinked product. The use of epoxy-containing vernonia oil in SINs is described separately (12), and the possible adaptation to reactive injection molding (RIM) processes is proposed elsewhere (13). This paper updates earlier reports (8,11) and reviews the synthesis, morphologies, and properties of SINs based in part on epoxidized botanical oils.

---

* Visiting scientist, on leave from East Stroudsburg (Pa.) State
  College.

Chemistry

The basic concept involves the essentially simultaneous polymerization by independent routes of epoxidized triglycerides (in this case linseed, crambe, and lesquerella oils) and styrene to make simultaneous interpenetrating networks. A typical commerical epoxidized linseed oil is considered to have the average structure described in structure (1). Crambe and lesquerella oils differ in that the corresponding fatty acids have the following compositions: crambe 57% C_{22} (monounsaturated), and 32% C_{18} (mono-, di-, and triunsaturated); and lesquerella, 70% C_{20} (-OH at C_{14}).

$$CH_2 -O-C-(CH_2)_7-CH-CH-CH_2-CH-CH-CH_2-CH-CH-CH_2-CH_3$$
$$CH-O-C-(CH_2)_7-CH-CH-CH_2-CH-CH-(CH_2)_4-CH_3 \qquad (1)$$
$$CH_2-O-C-(CH_2)_7-CH-CH-(CH_2)_7-CH_3$$

EXPERIMENTAL

Epoxidation and Polymerization

Two different methods were used for epoxidation. With method I, epoxidation was carried out using an ion-exchange (cationic) catalyst. The proper amount of dried Dowex 50W X-8 was charged with glacial acetic acid in a stirred three-necked flask. After equilibration to 50°C, 400 g oil and 171 g toluene were added, followed by the dropwise addition of the desired amount of H_2O_2. After stirring for 5-7 hours, the catalyst was filtered off, and the oil layer was washed with dilute aq. $NaHCO_3$ and hot water. The toluene and water were then removed from the neutralized oil layer using a rotary evaporator, and the epoxidized oil filtered and dried.

With method II, m-chloroperbenzoic acid was slowly added in small portions to a magnetically stirred mixture of dichloromethane and aq. $NaHCO_3$. The mixture was stirred at room temperature for 6-8 hours. Then the organic phase was washed successively with 1 N NaOH and dried over $CaSO_4$; the methylene chloride was removed using a rotary evaporator. In all cases, the acid-to-oil ratio was based on moles of acetic acid per mole of ethylenic unsatruation in the oil. The values of oxirane (epoxy) contents are listed in Table 1.

Preliminary experiments were also conducted with oil obtained from thalictrum polycarpum. However, in contrast to the other oils,

Table 1. Epoxidation of Natural Oils by Different Methods

Oil	Percent Oxirane Calcd.	Percent Oxirane Exptl.[a]	% Epox'n	Moles Epoxy/ Molecule(av.x)	Mol. Wt. Epox'd Oil
Linseed[b]	10.0	8.7-9.5[a,b]	84-92	5.5	980
Soybean[b]	8.0	6.9-7.2[a,b]	84-90	4.5	1,000
Crambe	5.9	5.3-5.5[a]	90-93	3.4	993
Lesquerella	---	4.9-5.0[a]	-----	3.0	1,008
Vernonia[c]	4.9	-------	-----	2.8	927
pauciflora					

[a] By method I; others by method II

[b] Commerically available

[c] Contains epoxy groups as received

[d] Calculated

characterization by infrared spectroscopy revealed a high concentration of free acid. Epoxidation by the usual method (H_2O_2) in acetic acid/Dowex 50W X-8 gave a product with an oxirane content of only 1.47% and a very high hydroxyl content, as shown by the infrared spectrum. Under the reaction conditions employed, hydroxylation was apparently favored over epoxidation. The iodine value dropped from 105.5 for thalictrum oil to zero for the epoxidized oil.

The synthetic sequence is shown in Fig. 1. First the step-growth polymerization of the epoxidized oil to form an elastomer was started at 140°C using dimer acid (Emery Industries) as the cross-linking agent. Polymerization was continued for various times, and the conversion to polymer determined by analysis of the acid content following the ASTM test D1639. The epoxidized oil was reacted to achieve an elastomeric prepolymer having an acid value in the range of 45-100 (see below). In some cases, the polymerization mixture was used directly in order to permit examination of the effect of prepolymerization time. The monomer or prepolymer mixture was then charged into a 500-ml three-necked flask, followed by a mixture of styrene with 1 vol-% divinyl benzene (DVB). The temperature was raised to 80°C, and polymerization was carried out with continued stirring. Before gelation occurred, the mixture was poured into molds and polymerization was completed at an elevated temperature, typically 140°C (for further details see reference 8).

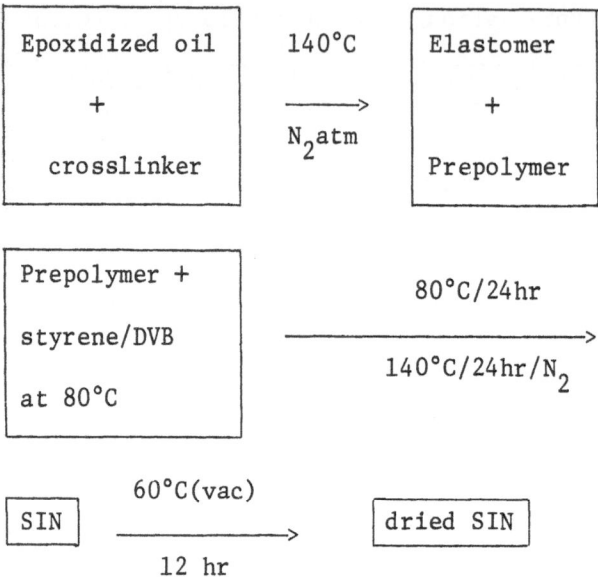

Fig. 1.   Process flow sheet for synthesis of SINs based on
          epoxidized oils.

     Attempts to polymerize thalictrum oil gave products less suit-
able for SINs than the other oils. Reaction of epoxidized thalictrum
oil with sebacic acid at 110° for two hours under a nitrogen atmosphere gave a semi-solid grease. However, reaction of epoxidized
thalictrum oil with tolylene-2,4-diisocyanate in a 3:1 to 2:1 weight
ratios at 80°C for 2 hours under vacuum gave very hard but flexible
light brown coatings.

Characterization

     Morphologies were characterized by thin-section electron micro-
scopy. Samples were first stained with osmium tetroxide (which reacts with residual double bonds); they were then embedded in an epoxy
matrix, trimmed to a truncated pyramid shape, and microtomed using a
Porter MT-2 ultra-microtome equipped with a diamond knife. Transmission electron micrographs were taken using a Phillips 300 electron
microscope.

     Shear modulus (G)-temperature plots were obtained using a Gehman
torsional stiffness tester following ASTM test D1053-61. Stress
relaxation experiments were performed to cover the entire range of
viscoelastic behavior from the glassy to the rubbery region. Modulus
readings were taken at various intervals of time between 10 sec and
1000 sec. Master curves were obtained by shifting data obtained at
various temperatures with respect to a reference temperature (15).
Values of the relaxation time τ were taken as the time to relax to

a modulus corresponding to log E(t) = (log E_g + log E_r)/2, where E and E_r are the glassy and rubbery Young's moduli, respectively; E was taken to be approximately equal to three times the shear modulus, 3G. Dynamic mechanical spectra (DMS) were obtained at 110 Hz using an Autovibron DDV IIIC over a temperature range from -100°C to 200°C at a programmed heating rate of 1°C/min. The instrument applies a sinusoidal tensile strain to one end of a sample, and measures the stress sensed at the other end. Transducers permit the reading of the absolute dynamic modulus, E_g (the ratio of maximum stress to maximum strain) and the phase angle between the strain and the stress. Data are taken automatically, and values of the loss and storage moduli (E" and E', respectively) are computed and plotted on a Hewlett-Packard plotter (model 9825A). Values of T_g were estimated from the temperatures corresponding to maximum slopes in the modulus-temperature plots, and to the peaks in the loss modulus E"-T curves.

Stress-strain behavior was examined at room temperature, according to ASTM test D638-68, with a crosshead speed of 0.085 mm/sec. Young's modulus (3), ultimate tensile strength (TS), and ultimate elongation to break (ε_B) were determined. Notched Izod impact tests were made following ASTM test D256-73; usually 4 to 5 specimens were tested.

Fatigue crack propagation (FCP) tests were conducted on notched wedge-opening-load specimens (0.5 cm x 4.9 cm x 6.3 cm) under ambient conditions at a sinusoidal frequency of 10 Hz, and a value of 0.1 for the ratio of minimum to maximum load, using an electrohydraulic closed-loop testing machine and procedures described previously (16). FCP rates were plotted in terms of the crack growth rate per cycle, da/dN, as a function of ΔK, following the Paris equation (da/dN = $A\Delta K^n$). Here ΔK is the range in stress intensity factor ($\Delta K = Y\Delta\sigma\sqrt{a}$), where Y is a geometrical factor, $\Delta\sigma$ the range in applied stress, and a the crack length.

RESULTS

Synthesis

As shown in Table I, method II gave values of conversion to epoxy somewhat lower than those typical of commercially epoxidized linseed and soybean oils (method I) but higher than those reported for method I by Dabhade et al. (14). (Of course, for the purpose of this work, complete conversion is not desirable, for the elastomer should have a fairly low crosslink density). With the crambe oil both methods gave high conversions. It should be noted that this study is the first to apply method II to the epoxidation of oils.

Prepolymers of the various oils were mixed with styrene and DVB, and polymerized at 80°C and 140°C following the scheme in Figure 1. All materials were vacuum dried before examination.

Slabs of 10cm x 10cm x 0.6cm of SIN material were prepared
in all cases. The products were opaque white in appearance, indicating a two-phased morphology. Some of the products, especially
those containing above about 15% oil were softer, indicating that
the oil phase was continuous.

In general, the characteristicsof SINs resembled those of
castor-oil-based SINs (9). For example all elastomer prepolymers
were mutually soluble with styrene monomer, but shortly after the
prepolymerization of styrene began, phase separation occurred, as
indicated by the onset of turbidity. The time required for the
phase separation was greater, the higher the acid value of the
elastomer prepolymer. As the polymerization of styrene proceeds, a
phase inversion may occur if stirring is continuous and vigorous;
in that case the elastomer phase breaks down into large droplets,
and PS becomes the more continuous phase. For 15/85 SINs, phase
inversion was observed to depend on the acid value of the elastomer
prepolymer (see the discussion).

Morphology

Clearly all the SINs exhibit a two-phase morphology (Fig. 2);
the dark and white areas correspond to the (stained) elastomer and
(unstained) polystyrene network (PSN). For 10/90 (elastomer/PSN)
SINs, phase inversion has taken place so that the elastomer phase
is discontinuous (A-E). While the elastomer domains do not differ
greatly in size (Table 2), their shape does vary depending on the
oil used. Also, the morphology of the elastomer phase depends
on the mode of polymerization. With a 10/90 SIN in which the styrene
was mixed with the oil prior to polymerization of the oil phase, no
cellular domains of PS were observed within the elastomer domain.
In contrast (Fig. 2E), when the oil was prepolymerized (but not to
the point of gelation) prior to addition of the styrene, such PS
domains were evident, as found previously for a castor-oil-based
SIN (9) and high-impact polystyrene (HIPS) (16). It is evident from
Figs. 2A and 2B that when the oil is not prepolymerized prior to
adding the styrene-DVB the resulting elastomer phase exhibits a
lower rubber-phase volume fraction (RPVF) than is the case when the
oil is prepolymerized.

Prepolymerization prior to styrene addition is shown in Figs.
3 and 4. For a typical 10/90 SIN (Fig. 3), the RPVF increases with
the extent of elastomer prepolymerization but the elastomer remains
the discontinuous phase. In contrast, in a 15/85 SIN (Fig. 4) not
only the RPVF increases but also the phase continuity changes. At
a higher extent of prepolymerization, the rubber becomes the continuous phase. Hence for 15/85 SIN compostions, the morphologies can
be controlled by the extent of elastomer prepolymerization (as measured
by the acid value). For a 15/85 composition, the phase continuity
depends on the conversion to elastomer and on the nature of the oil.

Table 2. Morphological Characteristics of SINs

Base Oil	A.V.[a]	Specimen	Cont. Phase	Elast. Domain Size, nm	Av. Elast. Size, nm	Cellular Domain Size, nm
linseed	47	10/90 ELODAN/PSN	plast.	500-2600	1200	64-194
linseed	79	10/90 (not prep)	plast.	400-2000	1300	-----
crambe	47	10/90 ECrODAN/PSN	plast.	300-1800	800	130-260
lunaria	47	10/90 ELuODAN/PSN	plast.	650-1950	1130	66-176
lesq.	52	10/90 ELqODAN/PSN	plast.	600-1760	1070	60-120
lunaria	47	15/85 ELuODAN/PSN	plast.	620-2600	1600	130-500
crambe	47	15/85 ECrODAN/PSN	plast.	450-1700	1050	-----
lesq.	52	15/85 ELqODAN/PSN	elast.	------	----	60-120
crambe	47	40/60 ECrODAN/PSN	elast.	------	----	-----
linseed	47	40/60 ELODAN/PSN	elast.	------	----	200-1800
lunaria	47	40/60 ELuODAN/PSN	elast.	------	----	100-800
lesq.	52	40/60 ELqODAN/PSN	elast.	------	----	60-1000

a Acid value is defined as the number of mg of potassium hydroxide required to reneutralize the alkali-reactive groups (in this case epoxy groups) in 1 g of material under standard test conditions. Thus the conversion of epoxy to ester groups is reflected in a decrease in acid value.

10/90 Lin. not Prep.   10/90 Linseed
ELO DAN/PSN   ELO DAN/PSN

10/90 Crambe   10/90 Lunaria
Ec$_r$o DAN/PSN   ELuO DAN/PSN

15/85 Lesq.   10/90 Lesq.
ElqO DAN/PSN   ElqO DAN/PSN

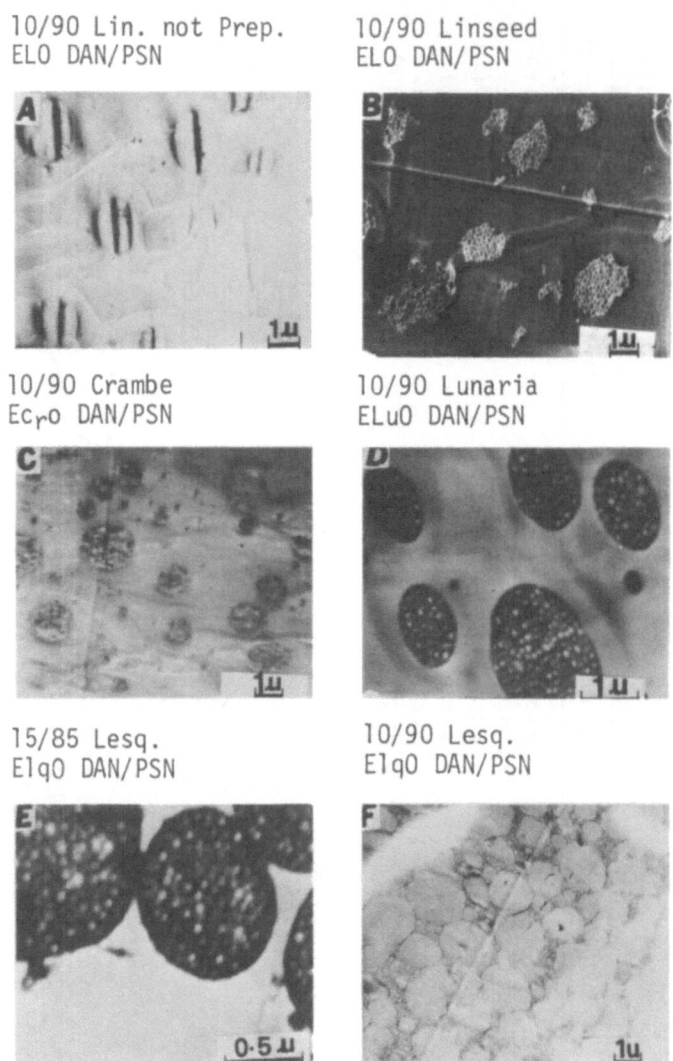

Fig. 2.   Transmission electron micrographs of SINs based on
          epoxidized oils and polystyrene networks (PSN).

     For similar acid values, epoxidized crambe and lunaria oils
yield discontinuous elastomer phases (at 15/85 elastomer/PSN ratios),
while epoxidized lesquerella and linseed oils yield a continuous
elastomer phase (Fig. 5). [It seems likely that the high viscosity
of the mixture (as is the case with lesquerella and linseed) prior
to casting may inhibit the phase inversion.]

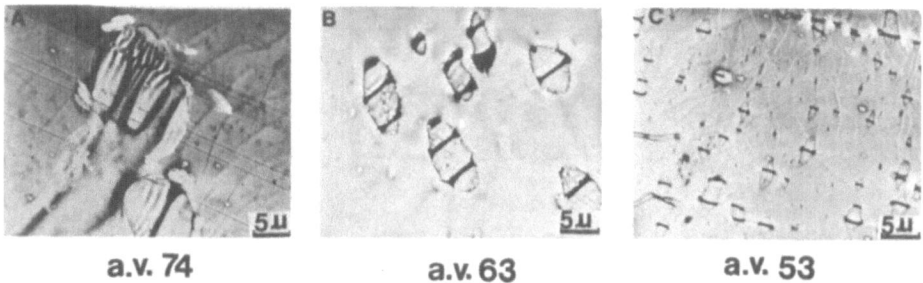

a.v. 74   a.v. 63   a.v. 53

Fig. 3.   Effect of elastomer prepolymerization on Rubber Phase
          Volume Fraction 'RPVF' for 10/90 Crambe. a.v.=acid value.

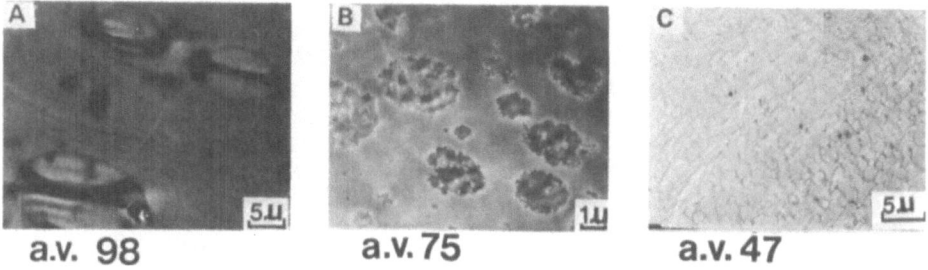

a.v. 98   a.v. 75   a.v. 47

Fig. 4.   Effect of elastomer prepolymerization on RPVF and phase
          continuity for 15/85 Lunaria

     When the ratio of elastomer to PSN is increased to 40/60, the PS
phase became the discontinous one (i.e. no phase inversion), through
the elastomer inclusions (phase-with-in-a phase) varied in size, depending on the oil. Thus, not surprisingly, the morphology depends
on the type of oil, the composition, and the extent of prepolymer
formation prior to mixing with the styrene. [In all these cases,
stirring was kept constant and slow.]

Effect of Composition on Tg Behavior

     Fig. 6 shows the dynamic mechanical response of SINs based on
epoxidized linseed oil and Table 3 gives T_g data obtained by both the
Gehman (Fig. 7) and dynamic tests (the latter giving the higher values
due to the higher frequency). The finding of well-defined peaks in
E" corresponding to the elastomeric and glassy phases is consistent
with the two-phase morphology.

Crambe   Lunaria

Lesquerella   Linseed

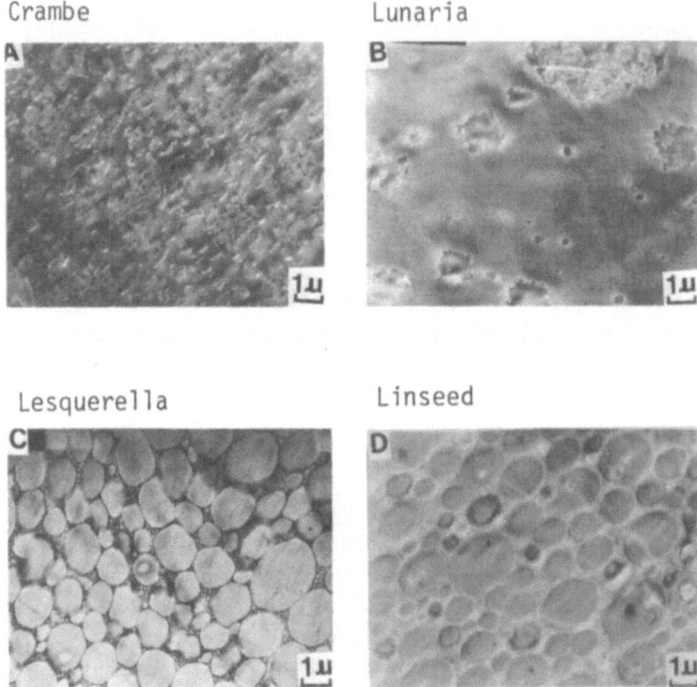

Fig. 5.   Transmission electron micrographs of 15/85 SIN's at acid
          value (a.v.=68). A. Crambe, B. Lunaria, C. Lesquerella,
          D, Linseed.

Table 3. Glass Transition Temperatures for Epoxidized-Linseed-
          Oil/PSN SINs.

Composition	T_g, °C[a]		Composition	T_g, °C[a]	
	Elastomer	Plastic		Elastomer	Plastic
0/100	−	92(116)	40/60	−13 (5)	100 (130)
5/95	−	92	75/25	−14 (−3)	103 (135)
15/85	−12	96	100/0	−18 (−3)	−
25/75	−13(9)	96(131)			

[a] First value by Gehman tester, second by DMS.

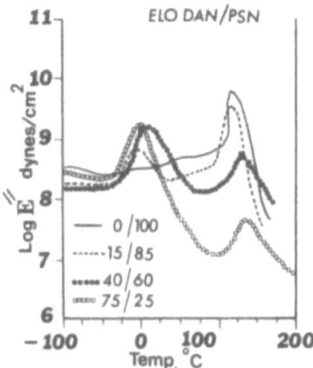

Fig. 6. Loss modulus (E'')-temp. plots from DMS for SINs based on
epoxidized linseed oil and PSN.

As mentioned in the section on morphology, the extent of
elastomer prepolymerization prior to styrene addition affects the
rubber-phase volume fraction (RPVF) and the phase continuity of the
SINs. As shown in Table 4 and 5, the T_g behavior of SINs also
depends on the acid value.

Effect of Elastomer Prepolymerization Conditions on T_g

For the 10/90 SINs, the values of T_{g1} (elastomer phase) and
T_{g2} (plastic phase) shown in Table 4 tend to be lower than those
of the homopolymers (9,10). The observed trend suggest that molecular mixing between the plastic and rubber phases is more extensive
for the case of no prior prepolymerization. Similarly for 15/85
SINs (see Table 5), the T_g behavior changes with the acid value (depending on the type of oil) until the rubber becomes the discontinuous phase, shown in Figs. 5 and 6. However, after exceeding this
critical value, the 15/85 compositions exhibit an anomalous upward

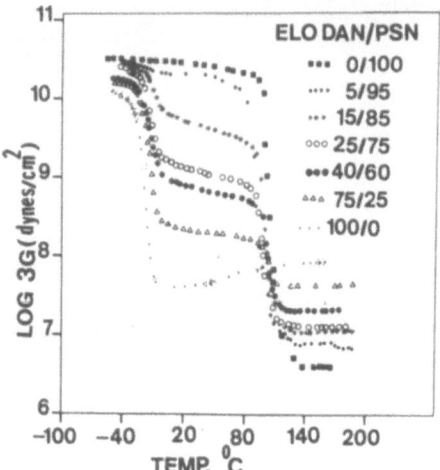

Fig. 7. 10-sec. modulus-tem. curves for SINs based on epoxidized
       linseed oil and PSN. (E∿3G).

shift in the T_gs as shown in Table 5. It can be concluded from the
morphology and T_g behavior that SINs exhibit an inward T_g shift for
a discontinuous rubber phase and an upward anomalous shift for a continuous rubber phase. Interestingly, precendents for such an upward
shift were noted previously (20,21) for polystyrene/polybutadiene
blends and poly (n-butyl acrylate)/epoxy SINs.

Table 4. Glass Transition Temperatures as a Function of Acid
         Value for 10/90 SINs

Base Oil	Acid Value	T_g, °C [a]	
		Elastomer	Plastic
Linseed	79	− 5	101
	74	3	106
Crambe	75	−22	101
	63	−22	104
	53	−21	105
Lesquerella	55	−17	105
Lunaria	47	−20	105
PSN	--	---	108

[a] By DMS, 110 Hz

Table 5. Glass Transition Temperatures as a Function of
          Acid Value for 15/85 SINs

Base Oil	Acid Value	T_g^a °C Elastomer	Plastic
Linseed	79	− 5	105
	47	12	108
Lesquerella	78	−16	106
	54	− 1	110
Lunaria	98	−27	103
	47	−12	109
PSN	--	---	108

[a] by DMS at 110 Hz

Stress-Strain and Impact Behavior

All specimens exhibited stress-whitening on failure (presumably
due to crazing), and higher values of toughness than the PS control,
as indicated by increased areas under the stress-strain curve and
by values of impact strength (Fig. 8, Table 6).

As noted before for castor-oil-based SINs (9,10), general
behavior ranged from that of toughened plastics to that of toughened
elastomers, depending on the concentration of elastomer.

All low-acid-value 10/90 SINs yielded before failure, and had
the highest ultimate elongation. Such behavior was observed to depend on the acid value of the elastomer prepolymer (as noted in the
cases of morphology and T_g behavior); Table 7 illustrates this response for 10/90 crambe/PS SINs. The higher acid value (low RPVF)
SINs exhibit less toughness (smaller area under the stress-strain
curves) than those with low acid values (high RPVF).

As mentioned in the morphology section, for 15/85 SINs, elastomer prepolymerization not only affects the RPVF but also the phase
continuity. The 15/85 lunaria system (a.v.=47) suggests that the
elastomer phase is continuous. In contrast, a 15/85 lunaria system
(a.v.=98) did not yield and exhibited low elongation, consistent
with the morphology shown in Fig. 4 (discontinuous elastomer phase
and low RPVF). Similarly a 15/85 lesquerella material (a.v.=54)
which gave a much higher elongation (and toughness), and which
behaves like a flexible plastic (similar in modulus to low density
polyethylene), is actually a reinforced elastomer having a continuous elastomer phase (Fig. 2E).

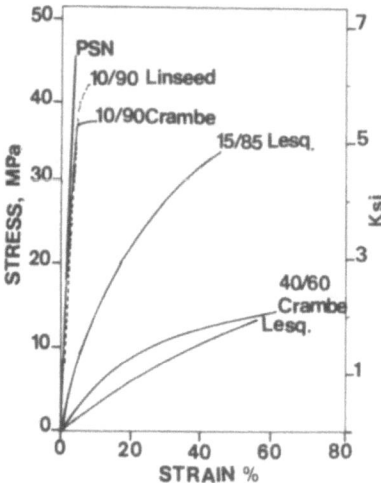

Fig. 8. Stress-strain curves for SINs based on epoxidized oils
and polystyrene networks (PSN)

        Impact behavior also generally parallels the stress-strain
toughness, and resembles that of castor-oil-based and other SINs (9).
As noted earlier, high impact strengths are associated with the use
of elastomers having low T_gs (8,11); particle and cellular domain
sizes cannot constitute a major factor, for no significant difference
in domain sizes are observed between systems. Thus the best 10/90
system is based on epoxidized Crambe oil (a.v. 53) whose T_g is -21°C
compared to 3°C for a linseed-oil-based elastomer (shown in Table 4).
Impact response is also affected by the polymerization procedure·
once again differences in 10/90 systems are seen depending on the
prepolymerization of the elastomer phase. In general, the higher
the volume fraction of occluded PS, the higher the effective volume of
of the rubbery particles (RPVF), and the higher the RPVF, the higher
the notched impact strength. This observation is consistent with
observation on HIPS (22). Of course, the higher the overall elastomer concentration, the higher the impact strength (Table 6); with
from 10 to 15% elastomer, impact strengths are increased by factors
of up to 4 or 5 relative to that of the control PSN.

Table 6. Stress-Strain, Relaxation, and Impact Characteristics of SINs

Base Oil	A.V.[a]	Specimen[c]	TS,MPa	ε_B,%	E,MPa	IS,J/m	τ,sec[b]
PSN Control	--	--	45	4	830	16	1.6
Linseed	47	10/90 ELODAN/PSN	42		689	34	5.4
Linseed	79	10/90 (not prep.)	40		746		1.3
Crambe	47	10/90 ECrODAN/PSN	37	10	758	59	12
Lunaria	47	10/90 ELuODAN/PSN	35	5	726	43	18
Lesquerella	52	10/90 ELqODAN/PSN	36	6	778	45	23
Linseed	47	40/60 ELODAN/PSN	18	37	51	--	--
Crambe	47	40/60 ECrODAN/PSN	15	63	76	--	--
Lesquerella	52	40/60 ELqODAN/PSN	13	56	21	--	--

a Acid Value.   b at 100°C

c Nomenclature:

ELO   Epoxidized Linseed Oil
ECrO   Epoxidized Crambe Oil
ELuO   Epoxidized Lunaria Oil
ELgO   Epoxidized Lesquerella Oil
DAN   Dimer Acid Network
PSN   Polystyrene Network

Table 7. Effect of Elastomer Prepolymerization on Stress-Strain
          and Impact Response.

SIN	Acid Value	TS, MPa	E, MPa	% Elong	IS,J/M
10/90 Crambe	74	34	600	10	20
	63	33	700	12	22
	53	33	750	15	57
15/85 Lunaria	98	31	1000	7	19
	75	30	850	8	45
	47	20	210	22	70
15/85 Lesquerella	78	34	1100	8	20
	54	33	200	45	80

Stress Relaxation

Stress relaxation reflects either molecular reactions such as
degradation, or molecular motion. In the present case, the onset
of coordinated chain motions associated with the glass transition
are of interest.

Typical stress-relaxation moduli are given in Fig. 9 as a
function of log time, for several SINs and a homopolymer PSN. In
each case, the data were correlated through the use of the time-temp-
erature superposition principle in the form of a master curve, with
a reference temperature of 100°C. This reference temperature is near
the glass transition temperature of homopolymer polystyrene.

As seen in Fig. 9, with one exception (ELO not prepolymerized),
the relaxation times are increased over that of the homopolymer PSN
control. While the exact reason is not known, the general phenomenon
probably involves a reduction of free volume, and analogous tendencies
have been noted for epoxies (20, 23, 23a). Often this is caused by
the visiting molecule (the oil) taking up free volume in the host
(the epoxy), thus raising T_g.

A plot of the experimental shift factors, a , vs temperature
is shown in Fig. 10 for all of the 10/90 SINs. The theoretical WLF
equation is shown for comparison (15), a T_g of 95°C was assumed.
Naturally, the data do not agree with the WLF equation below T_g as
this equation is valid only between T_g and T_g+50°C. In this range,
good agreement is indicated in Fig. 10.

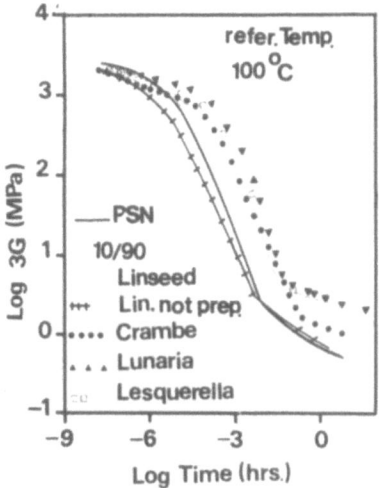

Fig. 9. Master curve relaxation modulus (converted from torsional
creep curves from Gehman experiments as a function of log
time for 10/90 SINs and PSN control. (E∿3G).

The apparent activation energies, E_a, are shown in Table 8.
The values are in the same range as those reported for bisphenol-A-
type epoxies, ca. 85 kcal/mole (23), and equal to the value
reported for creep in PS at its T_g (23b). However, Ferry (23b)
points out that this activation energy is actually that for an
elementary flow process. The values listed are for the onset
of segmental motion, at T_g.

It is of interest to return to the values of τ,(see Table 6).
The values shown are the times necessary to reach half the total
relaxation between the plastic and the rubbery states, shown in log
form. The differences in τ of a factor of 10 are also an indication
of the extent of molecular mixing free volume changes. Hence in-
directly, they bear on the trend of impact strengths noted.

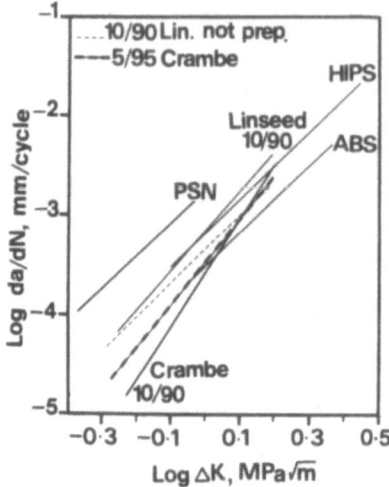

Fig. 10 Composite curve of experimental shift factors as a
         function of temp. for 10/90 SINs and PSN control.

Table 8. Apparent Energies of Activation*for Stress
          Relaxation of Typical SINs. (E_a determined
          at T_g for PS.)

Epoxidized Oil	Acid Value	E_a, kcal/mole	E_a, kJ/mole
PSN Control	–	95	400
10/90 Linseed (not prep)	79	94	390
10/90 Linseed	47	88	370
10/90 Lesquerella	52	81	340
10/90 Lunaria	47	82	340
10/90 Crambe	47	80	340

*Presumably corresponding to the activation of the cooperative
 flow of many chain segments at T_g.

Fatigue Response

Fig. 11 presents FCP results for several epoxidized-oil-based SINs; curves are also presented for the PSN control and typical HIPS and ABS resins (24). All the SINs exhibit lower FCP rates at all values of ΔK (and hence $\Delta\sigma$) and fail at values of ΔK nearly double the maximum value of ΔK for the PSN. While the maxi-

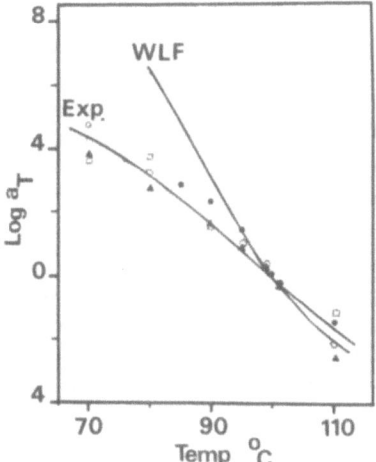

Fig. 11   Relationship between crack growth rate per cycle and stress intensity factor range ΔK, for SINs, PSN control, HIPS and ABS.

mum values of ΔK are less than for HIPS and ABS, the FCP rates are in the same range as for HIPS and ABS.

Indeed, at low values of ΔK, the 10/90 crambe-oil SIN grows a crack up to 40% slower than ABS. Thus the improvements in impact strength noted above (Table 6) are also reflected in increased fatigue resistance. Interestingly, as is the case with rubber-modified PVC (25) and poly(ethylene terephthalate) (26), FCP resistance may be increased significantly even though impact re-

sistance may be low in an absolute sense; in fact, good FCP resistance was obtained with the elastomer having a low T_g (i.e. 10/90
Crambe as compared to 10/90 Linseed). This suggests that a low T_g for
the elastomer phase is important in resistance to fatigue crack propagation, in agreement with the well known case of impact resistance.
While studies of the micro-mechanisms of failure are not yet complete
(18), clearly the fatigue response of the SINs is favorable.

DISCUSSION

On mixing the oil prepolymer and styrene-DVB, the system is
homogeneous. On polymerization at 80°C, the styrene-DVB component
(only) reacts. At a certain point, the materials phase separate.
Immediately after phase separation, the oil is the continuous phase
for some compositions, especially those containing 90% PS, there is;
a phase inversion. It is interesting to examine the morphological
path.

It is well-known that phase continuity in polymer blends depends
on the volume fraction, $W\phi$, of each phase, the viscosity η of each
phase, and the presence or absence of stirring (27) see Figure 12
In general, that phase with the larger volume fraction or lower
viscosity will tend to be continuous, if there is a reasonable
rate of stirring.

Immediately after phase separation, the majority phase is the
oil prepolymer. Droplets of PS form inside the oil, while the un-
polymerized styrene is partitioned between the two polymer phases.
As polymerization proceeds, the polystyrene phase grows in volume
at the expense of the styrene-swollen oil phase. As styrene mi-
grates out of the oil phase, the mixture becomes more viscous.

The polystyrene phase also becomes more viscous as styrene is
consumed. While changes in the ratio of the viscosities are not
known experimentally, it may be safely assumed that the polystyrene
has a much higher molecular weight than the oil prepolymer, and
eventually gels.

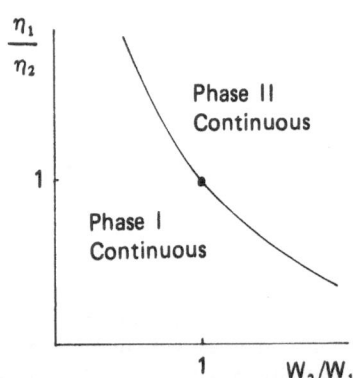

Fig. 12 The phase continuity
depends on the relative melt
viscosities and volume fractions
of the two phases.

For those materials experiencing phase inversions, the inversion must take place before gelation. It can be assumed that the reaction path increases the PS volume fraction during this time faster than the ratio of the PS viscosity to the oil viscosity. When the line is crossed in Fig. 12, moving in a left-to-right direction, (this assumes that phase I is the oil and phase II is the PS) then a phase inversion occurs and PS becomes the continuous phase. If the line in Fig. 12 is not crossed, of course, the oil remains the continuous phase. In any case, the stirring is stopped and the material put in the molds before the PS gels.

After the PS has fully reacted at 80°C, the temperature is raised to 140°C, allowing the oil phase, now reasonably pure and free of styrene monomer, to polymerize and gel. Of course, some of each polymer remains soluble in its visiting phase, as dictated by thermodynamics.

Those compositions having polystyrene as the continuous phase form, as a class, impact-resistant plastics. Those having the rubbery oil as the continous phase are softer, and behave as reinforced elastomers. In the latter case, the PS takes the place of the carbon black as the reinforcing component.

The T_g of the oil-network phase, near -40°C is most cases, is close to the upper temperature limit for rubber-toughening of the glassy PS phase. This is because the rubber must remain elastomeric under impact loading condition. (See Ref. 17, Chapter 3). Under fatigue loading conditions, however, a higher T_g may be acceptable because the rate of loading is lower than in an impact test.

CONCLUSIONS

The use of epoxidized oils to prepare the SINs discussed in this paper illustrates the broad range of chemistry possible with triglyceride oils. In the past, most oils chemistry was limited to the polymerization of the double bonds. However, research on vernonia oil (12), which also contains epoxy groups and castor oil (9), which contins hydroxyl groups, shows that many chemical derivatives of the oils can and should be used to prepare polymers from renewable resources.

As part of an investigation of the use of renewable feedstocks to synthesize polymers, a series of simultaneous interpenetrating networks (SINs) were prepared in which an elastomer network based on epoxidized botantical oils was combined with a glassy polymer network of crosslinked polystyrene (PSN). Thus SINs were successfully prepared containing from 5% to 40% elastomer based on epoxidized linseed, crambe, lunaria, and lesquerella oils. Epoxidized thalictrum oil was also crosslinked using a diisocyanate.

Electron microscopy showed that two-phased systems were obtained in all cases, the morphology depending on the oil and detailed synthetic procedure. Products ranged from toughened plastics to reinforced elastomers, with typical impact strengths of the plastics (e.g., a 10/90 crambe oil/PSN system) being several-fold higher than that of the PSN control. The resistance to fatigue crack growth was also significantly improved, even at low levels of elastomer. Mechanical properties depended on the size, continuity, and composition of the phases, and hence on the oil and synthetic details. It is concluded that the formation of potentially useful SINs based in part on a wide variety of botanical oils is indeed feasible.

ACKNOWLEDGMENTS

We gratefully acknowledge partial support from the National Science Foundation, Program on Alternate Biological Sources of Materials (Grant No. PFR 7827336) and from the Materials Research Center. We also appreciate advice and assistance from Ms. Sally M. Webler and Messrs. K. Earhart and J. Michel (Lehigh), and provision of the samples and advice from Drs. L.H. Princen and R. Kleiman (USDA, Peoria, IL).

REFERENCES

1.   "Future Sources of Organic Raw Materials", L.E. St. Peirre and G.R. Brown, eds., Pergamon, New York, 1980.
2.   L.H. Princen, J. Coat. Technol., 49(635), 88 (1977).
3.   I.S. Goldstein, Science 817, 189 (1975).
4.   "Renewable Resources for Industrial Materials", Nat. Acad. Sci., 1976.
5.   K. Gidanian and G.J. Howard, J. Macromol. Sci-Chem, A10, 1391 (1976).
6.   D. Swern, Ed., "Bailey's Industrial Oil and Fat Products", Wiley, NY, 1979.
7.   W.J. Schneider, L.E. Gast, V.E. Sohns and J.C. Conan, J. Paint Technol. 44(575), 58 (1972)
8.   S. Qureshi, J.A. Manson and L.H. Sperling, "Org. Coat. Plast. Chem., 43, 7 (1980).
9.   L.H. Sperling, N. Devia, J.A. Manson and A. Conde in "Modification of Polymers", (C. Carraher and M. Tsuda, Eds.), American Chemical Society, Washington, DC, 1980.
10. G.M. Yenwo, L.H. Sperling, J. Pullido and J.A. Manson, Polym. Eng. Sci., 17, 251 (1977).
11. L.H. Sperling, J.A. Manson, S. Qureshi, and A.M. Fernandez, Ind. Eng. Chem. Prod. Res. Dev. 20, 163 (1981).
12. A.M. Fernandez, L.H. Sperling and J.A. Manson, this symposium.
13. Shahid Qureshi, J.A. Manson, and L.H. Sperling, "Technology Transfer", National Technical Conference Series, Vol. 13, SAMPE, October 1981.

14.   S.D. Dabhade, P.K. Mataai and G.C. Patil, Paint India, 25, 16
       (1975).
15.   J.D. Ferry, "Viscoelastic Properties of POlymers", 2nd ed.,
       Wiley, New York, 1970, Ch. 11.
16.   S.L. Kim. et al., Polym. Eng. Sci., 17, 194 (1977); ibid. 19,
       145 (1979).
17.   J.A. Manson and L.H. Sperling, "Polymer Blends and Composites",
       Plenum Press, New York, 1976, Ch. 3.
18.   S. Qureshi, Ph.D. Thesis in progress, Lehigh University.
19.   E. Chang and A. Takahashi, Polym. Eng. & Sci. 18, 350 (1978).
20.   S. Manabe, R. Murakami, M. Takayanagi, and S. Vemura, Int.
       J. Polym. Mater. 1, 47 (1971).
21.   R.E. Touhsaent, D.A. Thomas, and L.H. Sperling, J. Polym.
       Sci. 46, 175 (1974).
22.   S.G. Turley and H. Keskkula, Polymer, 21, 466 (1980).
23   S.L. Kim, J.A. Manson, and S.C. Misra, ACS Symp. Ser. 114,
       183 (1979).
23a.   R.E. Touhsaent, D.A. Thomas, and L.H. Sperling, Adv. Chem,
       Ser. No. 154, R.D. Deanin and A.M. Crugnola, Eds., ACS, (1976).
       Wiley, N.Y., 1980. pp. 289-290.
24.   R.W. Hertzberg and J.A. Manson, "Fatigue of Engineering
       Plastics", Academic New York, 1980, Ch. 3.
25.   M.D. Skibo, J.A. Manson, S.M. Webler, R.W. Hertzberg, and
       E.A. Collins, ACS Symp. Ser. No. 95, 311 (1979).
26.   A. Ramirez, J.A. Manson and R.W. Hertzberg, Org. Coatings
       Plast. Chem. (1981); in press, Polym. Eng. Sci., 1981.
27.   L.H. Sperling, "Interpenetrating Polymer Networks and Related
       Materials," Plenum, 1981, Ch. 2.

VERNONIA OIL CHARACTERIZATION AND POLYMERICATION, AND SIMULTANEOUS INTERPENETRATING POLYMER NETWORKS BASED ON VERNONIA OIL-SEBACIC ACID/POLYSTYRENE-DVB COMPOSITIONS

A.M. Fernandez, C.J. Murphy*, M.T. DeCosta,
J.A. Manson and L.H. Sperling
Materials Research Center #32
Lehigh University
Bethlehem, Pennsylvania 18015

INTRODUCTION

The term "renewable resources" has come to mean sources of energy or products that can be used, grown, or replenished naturally, time after time, as opposed to mineral and petroleum products, which once depleted, are gone forever. Among the renewable resources available in the world, plant products rank very high. Examples include wood, cellulose, starch, rubber, and triglyceride oils. These oils are usually obtained by pressing or extracting various seeds. For industrial purposes, the oils may be classified according to the principal types of chemical reactivity.

Classically, the presence of multiple double bonds has allowed for ready polymerization, providing the basis for paints, adhesives, and other industrial uses. Many of these oils are also edible.

Most commercial triglyceride oils contain only double bond reactive sites. While a few oils contain another reactive group contains another type of reactive site, an hydroxyl group, has become commercial. Oils containing hydroxyl groups or other reactive become commercial. Oils containing hydroxyl groups or other reactive

---

*Visiting scientist, on leave from East Stroudsburg (Pa.),
State College.

The authors wish to thank the National Science Foundation for support through Grant No. PFR 7827336. The authors also wish to thank Dr. R. Kleiman of the USDA, Peoria, Ill., for supplying the vernonia oil.

groups, as discussed below, are nonedible. However, because of
their high reactivity, they offer special industrial advantages.

Recently, the USDA has pointed out that new oilseed crops,
originating from wild plants, bear oils containing various interesting chemical groups (1). Besides other hydroxy bearing oils,
keto, epoxy, and long chain fatty acids are being researched (1).

Lehigh's Oils Research Program

Beginning at the time of the 1974 petroleum crisis, the
Polymer Laboratory at Lehigh University, in cooperation with the
Universidad Industrial de Santander, UIS, in Colombia, South America,
undertook a study of the preparation of tough plastic and reinforced elastomers based on castor oil (2-13). The synthesis
route involved making either the polyurethane or polyester of
castor oil, both of which are soft elastomers. These polymers were
combined with cross-linked polystyrene to form an interpenetrating
polymer network, IPN (14-16).

An IPN may be defined as a combination of two polymers in network form, at least one of which was polymerized and/or cross-linked
in the immediate presence of the other. Both sequential IPN
syntheses (2-8) and simultaneous interpenetrating network, SIN,
syntheses were undertaken (9-13). The latter involving simultaneous
but independent polymerizations via step and chain-growth mechanisms
yielded the more practical of the two routes.

Beginning in 1979, research on different oils was undertaken.
Experimental oils from potential new oilseed crops, suggested by
the USDA, were investigated. In particular, epoxy-bearing oils
from vernonia (1) as well as chemically epoxidized linseed, crambe,
lunaria, and lesquerella oils were used (1,17).

This paper will highlight new work on vernonia oil. The related
work on epoxidized linseed oil will be presented in another paper
(17).

EXPERIMENTAL

Materials

By way of review, castor oil is the triglyceride of ricinoleic
acid (1).

$$
\begin{array}{l}
CH_2 - O - \overset{\overset{\displaystyle O}{\|}}{C} - (CH_2)_7 - CH = CH - CH_2 - \overset{\overset{\displaystyle OH}{|}}{CH} - (CH_2)_5 - CH_3 \quad (1)\\[2mm]
CH - O - \overset{\overset{\displaystyle O}{\|}}{C} - (CH_2)_7 - CH = CH - CH_2 - \overset{\overset{\displaystyle OH}{|}}{CH} - (CH_2)_5 - CH_3 \\[2mm]
CH_2 - O - \overset{\overset{\displaystyle O}{\|}}{C} - (CH_2)_7 - CH = CH - CH_2 - \overset{\overset{\displaystyle OH}{|}}{CH} - (CH_2)_5 - CH_3
\end{array}
$$

As discussed above, the interesting reactive sites are the three hydroxyl groups. This compound is 90% pure in ordinary pressed castor oil.

By contrast, vernonia oil is a natural triglyceride that contains a bountiful supply of epoxy groups. It consists mainly of the triglyceride of 12, 13-epoxyoleic acid (Vernonia galamensis, native of Kenya, Africa, collected by Dr. R.E. Perdue, USDA) (18).

$$
\begin{array}{l}
CH_2 - O - \overset{\overset{\displaystyle O}{\|}}{C} - (CH_2)_7 CH = CH - CH_2 - CH \overset{O}{\diagup\diagdown} CH(CH_2)_4 CH_3 \quad (2)\\[2mm]
CH - O - \overset{\overset{\displaystyle O}{\|}}{C} - (CH_2)_7 CH = CH - CH_2 - CH \overset{O}{\diagup\diagdown} CH(CH_2)_4 CH_3 \\[2mm]
CH_2 - O - \overset{\overset{\displaystyle O}{\|}}{C} - (CH_2)_7 CH = CH - CH_2 - CH \overset{O}{\diagup\diagdown} CH(CH_2)_4 CH_3
\end{array}
$$

Nuclear magnetic resonance studies, Figure 1, show that this oil contains 80% epoxy groups. The epoxy group is indicated by the absorption in the 2.50 - 2.80 PPM region. This corresponds to an estimate by the USDA (on other samples) of 72-78% epoxy groups (18). While 80% epoxy groups suggest an average functionality of 2.4 epoxy groups per molecule, kinetic studies, below, show it behaves as if it were 80% of structure (2) and 20% diluent oils.

Infrared studies were also made on several vernonia oil samples, see Figure 2. The epoxy absorption is at 825 cm^{-1} (18, a,b). Attention is drawn to the hydroxyl absorption band at 3800 cm^{-1}. The data obtained to date indicate that the oxirane band is essentially constant with time, while the hydroxyl absorption (of the order of 5% hydroxyls) increases somewhat with the age of the samples and/or exposure to atmospheric oxygen or water. Thus, the change in the hydroxyl absorption is probably not due to oxirane hydrolysis, but may be due to some oxidation reaction involving the unsaturation of the fatty acid. Perhaps the type of oxidation

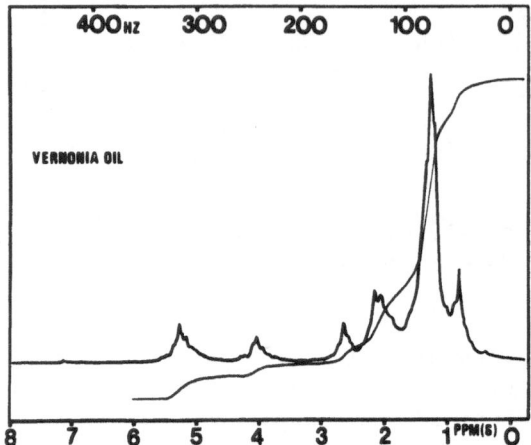

Fig. 1.   NMR Studies of vernonia oil. From Structure (1), the
          following identifications were made; based on the hydrogen
          nuclei:

$$\underset{a}{\underline{H_3C-}}\ \underset{b}{\underline{(CH_2)_3}}\underset{c}{\underline{-CH_2}}\overset{O}{\underset{d}{-\overset{\triangle}{CH}-\underline{CH}}}\underset{c}{\underline{-CH_2}}\underset{f}{\underline{-CH=CH}}\underset{c}{\underline{-CH_2}}\underset{b}{\underline{-(CH_2)_5}}\underset{c}{\underline{-CH_2}}\ -\ \overset{\overset{O}{\|}}{C}-O-\underset{e}{CH_2}$$

$$-CH \quad f$$

$$-CH_2 \quad e$$

Signal Identification:

a - 0.70 - 0.90 ppm
b - 1.00 - 1.70 ppm
c - 1.80 - 2.30 ppm
d - 2.50 - 2.80 ppm
e - 3.90 - 4.20 ppm
f - 5.00 - 5.40 ppm

commonly occuring at the α-site, through the breakdown of hydroperoxides, is responsible. In any case, the kinetic results shown
below are influenced to an unknown extent by the small concentration
of free hydroxyl present in the oil.

When vernonia oil reacts with difunctional sebacic acid, a
three dimensional esterification reaction occurs. Chain growth is
not restricted to one direction and the polymer produced forms a
network. This kind of system has the peculiarity that at some

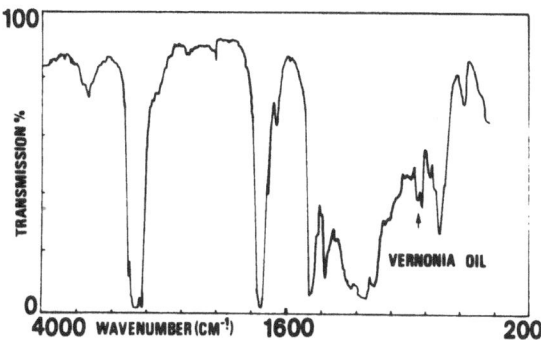

Fig. 2.   Infrared spectrum of vernonia oil.   Oxirane absorption
          doublet is indicated by the arrow.

state of the reaction, the network becomes infinite in dimension
and gelation occurs. At this stage, the reaction mixture consists
of two components: the gel, which is a swollen three dimensional
insoluble network, and the sol, which is a branched, still soluble
polymer.

Hydroxyl and epoxy groups react to form cross-linked polyesters with dibasic acids, but by different mechanisms. Sebacic
acid, itself commerically obtained from castor oil (and hence
constituting a renewable resource), serves as an excellent crosslinker:

$$H — O — \overset{O}{\overset{\|}{C}}(CH_2)_8\overset{O}{\overset{\|}{C}} — O\text{-}H \tag{3}$$

The reaction between a dibasic acid, such as sebacic acid, and
an epoxy bearing oil can be written as eq. 4. The reader should
note that a free hydroxy group forms in eq. 4, a possible locus of
further reactions.

The eight methylene units in sebacic acid in eq. 3 contribute
to the low glass transition temperature, T_g, of the final product.
This aspect attains importance in the development of novel
elastomers for impact-resistant plastics.

Instrumental

Kato's osmium tetroxide staining technique was employed, which
attacks the double bonds of the triglyceride oils. Specimens with
cross sections of 0.2 x 0.2 mm and lengths ranging from two to ten

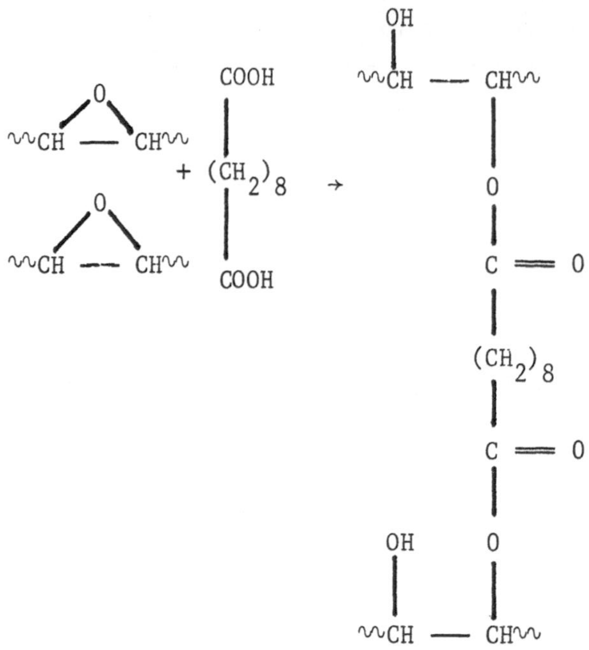

$$\underset{\text{CH}}{\text{CH}} \underset{O}{\diagup\!\!\diagdown} \text{CH} + \begin{matrix} \text{COOH} \\ | \\ (\text{CH}_2)_8 \\ | \\ \text{COOH} \end{matrix} \rightarrow \begin{matrix} \text{OH} \\ | \\ \text{CH} - \text{CH} \\ | \\ O \\ | \\ C = O \\ | \\ (\text{CH}_2)_8 \\ | \\ C = O \\ | \\ \text{OH} \quad O \\ | \quad | \\ \text{CH} - \text{CH} \end{matrix} \qquad (4)$$

mm were exposed to osmium tetroxide vapor at room temperature for one week.

Portions of the stained specimens were embedded in an epoxy resin, trimmed to a truncated pyramid shape, and microtomed on a Porter Blum MT-2 ultramicrotome using a diamond knife. Ultrathin sectioning at room temperature to a thickness of 60–80 nm (600–800 Å) yielded satisfactory results. Transmission electron micrographs were taken employing a Philips 300 electron microscope.

Dynamic mechanical spectroscopy (DMS) studies employed a Rheovibron direct reading viscoelastometer model DDV-II (manufactured by Toyo Measuring Instruments Co., Ltd. Tokyo, Japan). Samples were cut from the molded sheets to dimensions of about 0.03 x 0.15 x 2 cms. The measurements were taken over a temperature range from –80 to 160°C, using a frequency of 110 Hz. and a heating rate of about 1°C/min.

RESULTS

Kinetics of Gelation

The reaction between epoxy and carboxylic acid groups is a step-growth reaction. The presence of three or more epoxy groups leads to network formation when reacted with a diacid. When the reaction has consumed more than two of the epoxy groups, the

material undergoes gelation as described by Flory (19). The time
to gelation at 140°C for typical compositions of vernonia oil and
sebacic acid, is about 420 min. (20).

This experiment bears comparision to castor oil polymerizations.
At 190°C, castor oil and sebacic acid reached the gel point in 8-10
hrs. depending on the catalyst (10). Since water removal is difficult
beyond the gel stage, final network formation was improved by finish-
ing the cross-linking with the formation of urethane bonds.

Homopolymer vernonia oil-sebacic acid, HVOSA, components
were mixed in various molar ratios by weighing the reaction components separately, mixing quickly, stirring, and placing the
reaction vessel in a silicone oil bath at 140°C. Equivalent amounts
of the two substrates were reacted in order to obtain a ratio of
unity between their respective functional groups. In addition, 2:1
and 1:2 equivalent amounts were used. The reaction between vernonia
oil and sebacic acid was carried out with continuous magnetic
stirring under a nitrogen atmosphere.

The reaction progress was followed by titrating the unreacted
carboxyl groups in samples removed from the reaction mixture at
different times of reaction, with a standard methanolic potassium
hydroxide solution. The oxirane groups were also titrated in some
of the samples, using the hydrogen bromide in glacial acetic acid
titration technique (21).

The extent of reaction, as well as the acid value, were determined from the carboxyl group titer and plotted against time, see
Figure 3. Gelation usually took place after about seven hours at
140°C., and after about four hours at 160°C. For the 2:1 and 1:2
vernonia oil: sebacic acid compositions, the gelation took between
6 and 24 hours at 140°C. The studies below used 140°C as the
temperature of interest.

The progress of the reaction can also be followed by studying
the disappearance of the oxirane absorption at 825 cm^{-1}, see
Figure 2. A plot of oxirane absorption _vs_ time at 140°C. is shown
in Figure 4. The result is in qualitative agreement with that
shown in Figure 3.

The extent of reaction at the gel point is of special interest.
Two theories were compared. According to Carothers (22), the critical extent of reaction, P_c, at the gel point is given by

$$P_c = 2/f_{avg} \qquad (5)$$

where f_{avg} represents the average functionality of the system.

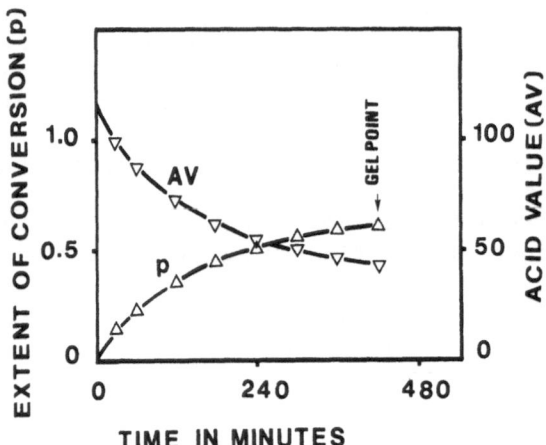

Fig. 3.   Reaction kinetics of vernonia oil with sebacic acid.

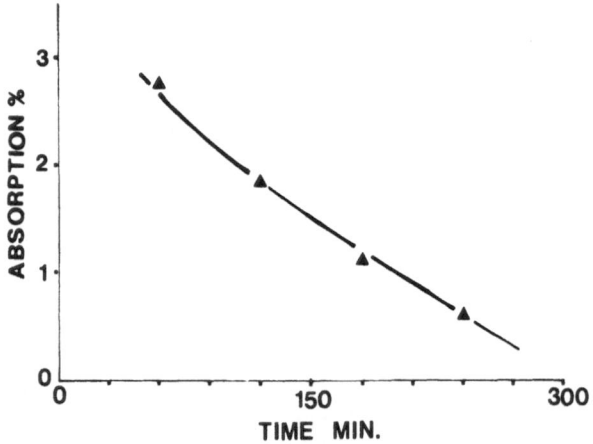

Fig. 4.   Infrared absorption at 825 cm^{-1} vs time, at 140°C.,
          after appropriate subtraction of background.

The quantity f_{avg} is given by

$$f_{avg} = \Sigma N_i f_i / \Sigma N_i \qquad (6)$$

where N_i is the number of molecules of monomer i with functionality f_i. Equation (6) yields $f_{avg} = 2.57$ for an equimolar vernonia oil-sebacic acid mix, assuming that vernonia oil is trifunctional. Equation (5) then yields $P_c = 0.788$.

Flory and Stockmayer (19,23) arrived at a more general relationship,

$$P_c = 1/(f-1)^{1/2} \qquad (7)$$

where f represents the functionality of the branch units, ie., of the monomer with functionality greater than two. Taking f = 3 from structure (2) yields $P_c = 0.707$. These values are compared in Table 1. The experimental value of P_c at 140°C was determined to be 0.62, Table II.

Table I. Comparison of P_c Values: Vernonia Oil-Sebacic Acid Network

	Assumed Vernonia Oil Functionality			
	2.4	3.0	4.8	6.0
Carothers Theory	0.917	0.777	---	---
Flory-Stockmayer Theory	0.845	0.707	0.513	0.447

Table II. Experimental and Theoretical Values for P_c for the System Vernonia Oil-Sebacic Acid at Gelation

Equivalent Composition Ratio, oxirane:acid	Values of P_c		
	Carboxyl, Experimental	Oxirane Experimental	Theoretical
1:1	(a) 0.62	(b) 0.73	(c) 0.707
2:1	0.93	0.63	0.500
1:2	0.57	∿ 1.00	1.00

(a) From titration of carboxyl groups.
(b) From titration of oxirane groups.
(c) Flory-Stockmayer theory.

Values were also calculated taking the average functionality of vernonia oil as 2.4, see Table I. Better agreement with experiment is noted if the oil is assumed to have a functionality of 3.0 with a diluent. (Castor oil is also thought to be trifunctional, with a diluent.)

One further point should be emphasized. When an epoxy reacts with a carboxyl group, an hydroxyl group is formed, see reaction (4). Thus it could be argued that in this esterification reaction, the vernonia oil has a functionality of six (or 4.8 if it is assumed that the epoxy groups are randomly distributed). These values are also shown in Table I. Of these calculations, the Flory-Stockmayer trifunctional value fits best. Use of equation (7) suggests an actual functionality of 3.61.

The reaction is somewhat more complicated when stoichiometry on an equivalent basis is different from 1:1. In this case, the more general Flory-Stockmayer equation reads,

$$P_c = \frac{1}{[r + r(f-2)]} \tag{8}$$

where P_c is the critical conversion at the gel point, referring to the oxirane groups in this case, and the quantity r is the ratio of oxirane to carboxyl groups. These values are shown in Table II for three composition ratios. While the experimental values of P_c are slightly higher than theory for the 1:1 and 2:1 cases, agreement is satisfactory.

It was of interest to analyze the kinetic order of the reactions, see Figures 5 and 6. In Figure 5, a plot of inverse free acid remaining vs time is linear, showing that the reaction is second-order in carboxyl groups. A plot of log (epoxy) vs time, Figure 6, indicate that the reaction is first order in epoxy, or oxirane groups. Apparently, the hydroxyl groups generated in reaction (4) do not react to any great extent.

The final kinetic equation can be written,

$$\frac{-d[-COOH]}{dt} = k\,[-COOH]^2\,[-\overset{O}{\overset{\diagdown\!\!\!/}{C}-C-}] \tag{9}$$

where $k = 5.61$/moles-min. at 140°C. Since one carboxyl group is consumed for every oxirane,

$$\frac{-d[-COOH]}{dt} = \frac{-d[-\overset{O}{\overset{\diagup\diagdown}{C}-C-}]}{dt} \tag{10}$$

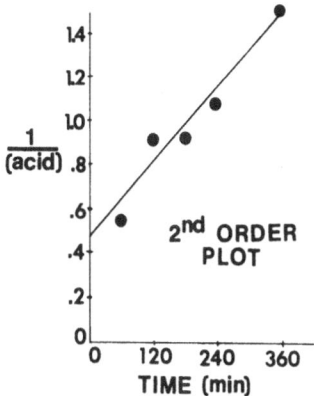

Fig. 5. A plot of remaining acid vs time, 1:1 equivalent concentrations. The linear results suggest that the reaction is
second order in carboxyl groups.

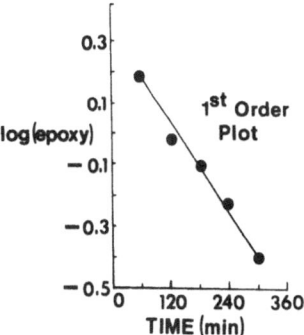

Fig. 6. The linear character of this 1:1 concentraction plot indicates that reaction is first order in epoxy (oxirane) groups.

That the reaction is second order in acid and first order in oxirane is typical of uncatalyzed reactions of this type (24). These results are important because they verify the fact that vernonia oil epoxy groups are ordinary epoxies, and can be treated as such for chemical reaction purposes. That these epoxy groups originate from a biological source imparts no special or unusual reactions to them, and impurities and hydroxyl groups do not interfere substantially.

Elastomer Characterization

The vernonia oil-sebacic acid elastomer is yellow in color, transparent, soft and of low tensile strength. Via Gehman torsion instrumentation, 3G(10)-temperature curves revealed a glass transition temperature of -50°C, where 3G(10) refers to three times the shear modulus (equaling Young's modulus), at 10 seconds.

Vernonia Oil Sebacic Acid/Polystyrene (VOSAN/PSN) Simultaneous Interpenetrating Networks (SIN'S)

An SIN is a combination of two monomers or prepolymers, plus their respective crosslinkers, which are polymerized simultaneoulsy by independent, non-interfering routes (25). In this case, styrene (S) was mixed at room temperature with divinylbenzene (DVB) at a 1% concentration level, and 0.4% of benzoyl peroxide was added as an initiator.

Vernonia oil-sebacic acid prepolymer was charged into a three necked flask, followed by the styrene mixture. The temperature was then raised to 80°C to initiate the styrene polymerization. During this stage, the reaction vessel was stirred under a nitrogen atmosphere. Just before the styrene-DVB component gelled, the mixture was poured into preheated molds and the reaction continued at 80°C for 24 hours. After this time, the temperature was raised to 140°C for 30 hrs. to complete the VOSA network formation.

Characterization of the VOSAN/PSN SIN

For initial characterization, a 50/50 VOSAN/PSN SIN composition was selected. It was opaque white, and had tough elastomeric properties. Similar to other epoxy-based SIN's (17).

Dynamic mechanical spectroscopy, Figure 7, on a Rheovibron DDV III revealed a T_g for the vernonia oil phase of -45°C., while the polystyrene phase exhibited a T_g of +120°C., at 110 Hz. These results suggest little mixing between the components. Morphological studies via transmission electron microscopy were done with the aid of osmium tetroxide staining of the vernonia oil component. Figure 8 shows that the vernonia oil phase, stained, is continuous. The PSN domains averaged 0.25 - 0.30μ in diameter, with a bimodal size distribution. The larger domains are 0.4 - 0.6μ, while the smaller

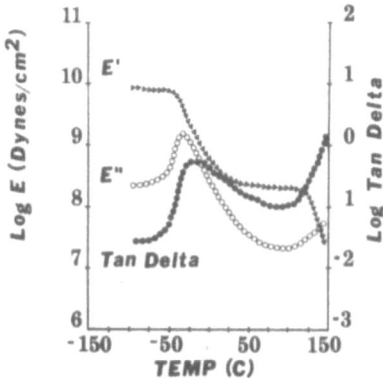

Fig. 7. Dynamic mechanical behavior of a 50/50 VOSAN/PSN at 110 HZ. This data indicates extensive phase separation of the two polymers.

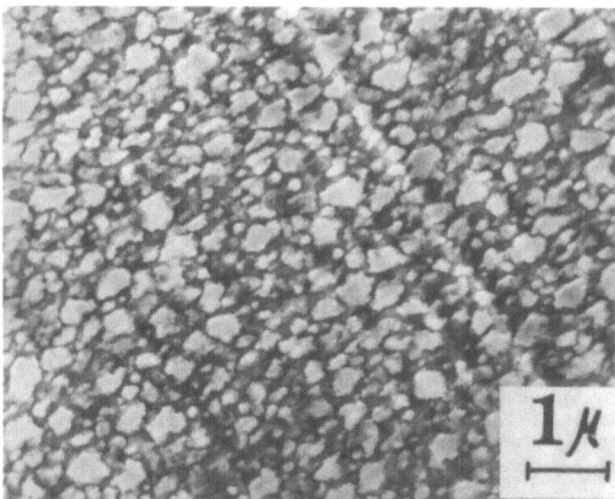

Fig. 8. VOSAN/PSN 50/50 morphology via transmission electron microscopy. The dark phase is the vernonia oil-based polymer.

domains are in the range of 0.10 - 0.15μ, and were the more numerous
of the two.

Calculation of Domain Sizes

The phase domain diameters can also be estimated theoretically.
It is important to note that phase separation takes place during
the stirring part of the reaction (phase separation before gelation).
In that case, the equation for semi-II IPN's applies (26). The
following equation holds for the diameter of the polystyrene phase,
D_2:

$$D_2 = \frac{4\gamma^0}{RT \left\{ B\nu_2 - \dfrac{\rho_1}{M_1} \dfrac{\phi_1}{\phi_2} \ln \phi_1 \right\}} \tag{11}$$

where γ^0 is the interfacial tension, ν_2 is the crosslink density of
polymer II, ρ_1 and M_1 are the density and molecular weight of polymer
I, respectively, ϕ_1 and ϕ_2 represent volume fractions, RT is the gas
constant times the absolute temperature, and $B = 1/2 (\ln \phi_2 - 3\phi_2^{2/3} + 3)$.

Assuming the following values: γ^0 = 3.65 dynes/cm (from the
poly(n-butyl acrylate)/polystyrene system, polymer I of which is also
a type of polyester), M_1 = 5 x 10^4 gms/mole (estimate), ρ = 1
gm/cm^3, $\phi_1 = \phi_2 = 0.50$, ν_2 = 2 x 10^{-4} (from 1% crosslinker), R =
8.31 x 10^7 dyne-cm/mole °K, and T = 413°K, a value for D_2 of 0.076
is obtained.

An examination of the theory reveals this to be a minimum value,
since it assumes that the polystyrene was fully crosslinked at the
time of phase separation. Separate studies (7) reveal that polystyrene polymerized linearly under similar conditions has an M_w =
3 x 10^5 gms/mole. This suggests a lower limit of 3 x 10^{-6} for
ν_2, or D_2 = 1.3μ. These two theoretical numbers span the experimental range, at least. A better determination will have to wait a
molecular characterization at the incipient point of phase separation.

CONCLUSIONS

Vernonia oil, a natural bearer of epoxy (oxirane) groups, can
be polymerized with sebacic acid to form a polyester-type network.
The products are soft elastomers, which bear significant resemblence
to the polyesters formed from castor oil and sebacic acid. A kinetic
analysis of the reaction shows it to be second-order with respect to
acid, and first-order with respect to oxirane.

Simultaneously polymerized with styrene and DVB, an SIN can be made with promising properties. Midrange compositions are tough elastomers.

Vernonia oil is relatively rare among the products of nature in bearing epoxy groups naturally, in high concentration. Commercially grown, this oil could significantly enhance our library of important chemicals, and provide for a new source of soft elastomers. It should be pointed out that most commercial epoxies now on the market yield either glassy plastics or leathery products on polymerization. As a renewable resource, vernonia oil appears to have an important future.

REFERENCES

(1) (a) L.H. Princen, J. Coat. Technol., 49 (12), 88, 1977;
    (b) J. Am. Oil Chem. Soc., 56(9), 845, 1979.

(2) G.M. Yenwo, J.A. Manson, J. Pulido, L.H. Sperling, A. Conde, N. Devia, J. Appl. Polym. Sci., 12, 1531, 1977.

(3) G.M. Yenwo, L.H. Sperling, J. Pulido, J.A. Manson, Polym. Eng. Sci., 17(4), 251, 1977.

(4) J.E. Pulido, G.M. Yenwo, L.H. Sperling, J.A. Manson, Rev. UIS (Colombia), 7(7), 35, 1977.

(5) G.M. Yenwo, L.H. Sperling, J.A. Manson, A. Conde, "Chemistry and Properties of Crosslinked Polymers", Labana, S.S., Ed.; Academic Press; New York, p. 257, 1977.

(6) L.H. Sperling, J.A. Manson, G.M. Yenwo, N. Devia, J.E. Pulido, A. Conde, "Polymer Alloys", D. Klempner, K.C. Frisch, Ed.; Plenum: New York, 1977.

(7) N. Devia, A. Conde, L.H. Sperling, J.A. Manson, Rev. UIS (Colombia), 7(7), 19, 1977.

(8) N. Devia-Manjarres, A. Conde, G. Yenwo, J. Pulido, J.A. Manson, L.H. Sperling, Polym. Eng. Sci.; 17(5), 294, 1977.

(9) N. Devia, J.A. Manson, L.H. Sperling, A. Conde, Polym. Eng. Sci., 18(3), 200, 1978.

(10) N. Devia, J.A. Manson, L.H. Sperling, A. Conde, Macromolecules, 12(3), 360, 1979.

(11) N. Devia, L.H. Sperling, J.A. Manson, A. Conde, Polym. Eng. Sci., 19(12), 870, 878, 1979.

(12) N. Devia, L.H. Sperling, J.A. Manson, A. Conde, J. Appl. Polym. Sci., 24, 567, 1979.

(13) L.H. Sperling, N. Devia, J.A. Manson, and A. Conde. In "Modification of Polymers", C.E. Carraher, M. Tsuda, Ed.; ACS Symposium Series No. 121, American Chemical Society: Washington, D.C., 1980.

(14) S.C. Kim, D. Klempner, K.C. Frisch, N. Radigan, H.L. Frisch; Macromolecules, 9, 258, 1976.

(15) Y.S. Lipatov, L.M. Sergeeva, Russ. Chem. Rev., 45(1), 63, 1976.

(16) D.A. Thomas, L.H. Sperling in "Polymer Blends", Vol. 2, R. Paul, S. Newman, Ed.; Academic Press: New York, 1978.

(17) Shahid Qureshi, J.A. Manson, L.H. Sperling and C.J. Murphy, published elsewhere in this volume.

(18) K.D. Carlson, W.J. Schneider, S.P. Chang and L.H. Princen, accepted for publication in "New Sources of Fats and Oils", E.H. Pryde and L.H. Princen, Eds., American Oil Chemists' Society, publication expected 1981.

    (a) R. Kleiman, F. Spencer, L. W. Tjorks, and F. R. Earle, Lipids, 6, 617 (1971).
    (b) C. F. Krewson, J. S. Ard, and R. W. Riemenschneider, JAOCS, 39, 334 (1962).

(19) P.J. Flory, "Principles of Polymer Chemistry", Cornell University Press, Ithaca, New York Ch. 9, 1953.

(20) L.H. Sperling, J.A. Manson, Shahid Qureshi and A.M. Fernandez, I&EC Products Research and Development, 20: 163 (1981).

(21) ASTM D 1652 - 73.

(22) W.H. Carothers, Trans. Faraday Soc., 32, 29 (1936).

(23) W.H. Stockmayer, J. Chem. Phys. 11, 45 (1943).

(24) G. Odian, "Principles of Polymerization", McGraw-Hill, New York, 1970.

(25) L.H. Sperling, "Interpenetrating Polymer Networks and Related Materials", Plenum Press, New York, 1981.

(26) J.K. Yeo, L.H. Sperling, and D.A. Thomas in preparation.

MONOMER REACTIVITY RATIOS OF TUNG OIL AND

STYRENE IN COPOLYMERIZATION

A.M. Fernandez
Materials Research Center No. 32
Lehigh University, Bethlehem, PA 18015

A. Conde
Ingenieria Quimica
Universidad Industrial se Santander
Bucaramanga, Columbia

INTRODUCTION

Triglyceride oils such as linseed oil, oiticica oil, tall oil, and tung oil have long been used in varnishes, alkyds, and other coating formulations. When copolymerized with vinyl monomers, a wide range of commercial products are obtained. The overall objective of the present research program is directed toward the modification of the mechanical properties of glassy polystyrene, PS, by blending it with random copolymers of tung oil and styrene, TO-S. The specific objective of this paper is the understanding of the structural characteristics of the TO-S copolymer from studies of the monomer reactivity ratios during bulk copolymerization. The values of the reactivity ratios provide knowledge about the general composition of the copolymer; control over the fabrication process, molecular design; and criteria for optimizing the modifier effects of the rubber compound on the PS. New polymers based on industrial oils from agricultrual sources represent an important field of research, presenting an alternative way of reducing the actual dependence on petrochemical derivatives (1,2).

TUNG OIL, OCCURRENCE AND PROPERTIES

Tung oil, also known as Chinese wood oil, is native to central and Western China. Aleurites Fordii, the commercially most important species of tung oil apparently has very specific soil preferences, due to its being a large consumer of

nitrogen and water. Therefore, favorable soil and climatic
conditions not only accelerate the growth of the tung oil trees,
allowing longer productive life and increasing the quality and
volume of see yield per tree, but also largely determine the
regions of successful cultivable area.

The rainfall regime is particularly important during the
periods of growing and fruiting activity in the spring, summer and
early fall seasons. Rather heavy and fairly evenly distributed
precipitation is required. Generally a minimum rainfall of 30 inches
per year is considered essential to satisfactory growth, while an
average rainfall of 40 inches gives the best results (3,4). The
tung oil tree is deciduous and therefore develops vigorously where
there is sufficient cool weather to allow for a dormant period, but
only in regions where frost will not injure the fruit-bud growth,
such as temperatures of 26° and 28°F, which have caused damage to
trees of all ages when not thoroughly dormant (3,4).

Figures 1 and 2 show a tung oil tree in full flower in the
early spring season, and cluster-type tung oil fruits, respectively.
The oil, of course, is pressed from the seeds.

In 1905, the United States Department of Agriculture introduced tung oil trees to the United States in an attempt to

Fig. 1.   Tung oil tree in full flower in the early spring
          season (4).

Fig. 2. Cluster-type tung oil fruit

develop an independent domestic production. The plantations were
mainly located in Mississippi, Louisiana, Florida, Alabama, Georgia
and Texas, but in recent years repeated freezes and hurricanes have
limited production. No domestic tung mills operated between 1973
and 1976, and today there no longer exists any expectation of
significant tung oil production. From 1905, the industrial demands were continuously increasing and consequently the introduction
of tung oil trees was expanded to different countries around the
world. In 1925, the plantings in Latin America began. The north-
east part of Argentina, and the southeast regions in Paraguay and
Brazil, which have a subtropical climate, appear to be the most
suitable areas for tung oil tree growth in South America. Figure
3 shows this region.

During the 19th and 20th Century, drying oils, mainly linseed
and tung oils, became the principal components in paints, varnishes
and enamel formulations, but since World War II have been displaced
in many of their traditional uses by petroleum-derived products.
It is clear that tung oil, and drying oils in general, are declining
in use. The reasons are not only the displacement by technologi-
cally new and improved materials, but also for land competition
by food crops.

Fig. 3.   Suitable areas for tung oil tree growth in
          South America.

The tung oil used in this research came from Argentina. The first
plantings of tung oil trees in that country were made in 1928 by
the North Eastern Argentine Railway from seed obtained from Florida.
During many years Argentina was one of the larger producing countries, but production is dropping continuously. However, together
with China, they remain the principal producers (5,6,7).

Figure 4 shows the fluctuations in tung oil production in
Argentina in recent years. The gradual decline in production was
caused by low internal prices, except for the sharp decrease
during 1973-4, which was caused by a late freeze (8).

The international situation can be illustrated briefly as
follows for a similar time period: From 1951 to 1978, tung oil
production decreased by a factor of five, which from 1970 to 1978

the global price increased eight times.   In Argentina, the price
decreased about 50% during this decade.

The gradual reduction in price, the low yield per acre, and
old plantations with freeze-susceptible tung oil species has led
to a general decline of the industry.   The combination of all of
these factors reduce the expectation of any significant tung oil
production increases during the present decade (5-8).   Because of
the length of time between the planting of new trees and the first
harvest, however, plans must now be made if production is to be
increased in the period of 1990-2000.

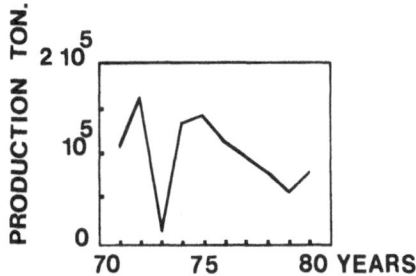

Fig. 4.   Tung oil production in Argentina.

For many centuries tung oil has been extensively used in paints
and water proofing materials, also in making insulating compounds
for electrical uses.   However, tung oil applications may be extended
to many other newer and more promising products than those tradition-
ally used.   References 9 and 10 give numerous examples, such as
lubricating greases, metallic dryers for paints and other protective
coatings, preparation of vinyl resin stabilizers, etc.

Table I. Chemical and Physical Properties of Tung Oil

Specific gravity 25°C	0.935
Viscosity, mP	165-250
Iodine value	158-166
Acid Value	0.3-0.4
n_D 25°C	1.515

According to its iodine value, tung oil is a typical drying oil. Table I shows some of its chemical and physical properties.

Tung oil is a mixture of triglycerides of saturated and unsaturated fatty acids. Table II shows its chemical composition.

For reaction purposes, its composition can be considered as the triglyceride of the oleosteric acid (9,11,13-octadecentrienoic acid) which has three conjugated vinyl bonds for each chain, equation (1) which make it capable of polymerization by chain mechanism reactions.

$$CH_2-O-CO-(CH_2)_7-CH=CH-CH=CH-CH=CH-(CH_2)_3-CH_3$$
$$CH-O-CO-(CH_2)_7CH=CH-CH=CH-CH=CH-(CH_2)_3-CH3 \qquad (1)$$
$$CH_2-O-C)-(CH_2)_7-CH=CH-CH=CH-CH=CH-(CH_2)_3-CH_3$$

BACKGROUND

Styrene, cyclopentadiene, α-methylstyrene, and other vinyl monomers react with tung oil by bulk, solution, or emulsion polymerization techniques to form copolymers with excellent film properties. There is a good deal of work related to the paint and coating industries concerning these materials, however there is little research concerning the chemical reactivity of tung and other dry-

Table II. Fatty Acids Composition of Tung Oil, %.

Name	% Fatty Acids
oleic	4
linoleic	8
α-oleostearic	82
palmitic	4
stearic	2

drying oils involving copolymer formation, and as modifying agents
in polymer blends.

The macroscopic properties of copolymers depends on their
structure, which in turn depends on the method of synthesis, relative amounts of the reactant monomers, final molecular weight
distribution, and especially on the composition and distribution of
the monomer sequences in the polymer chains. The process of copolymerization allows the synthesis of different products by varying
the nature and relative amounts of the monomers.

Early work in blending was directed toward the modification of
polystyrene to obtain tougher, improved materials. Polymer modification by the polymer blend approach has been expanded to include
other polymers besides polystyrene, and also new rubber modifying
materials.

Recently, new materials termed interpenetrating polymer networks, IPN's, based on castor oil and styrene have been synthesized
in the polymer laboratories at Lehigh Univeristy and at Universidad
Industrial de Santander, UIS, in Colombia, S.A. Continuing the oil
research program at Lehigh University, new promising polymers are
being made from naturally occuring epoxy oils, like vernonia oil,
and epoxidized linseed oil, lesquerella oil, and crambe oil, all of
actual or potential commercial significance (14-15). The present
work on tung oil was done at UIS in conjuction with these programs.

Kinetic Considerations

The composition of the copolymer is determined by assuming a
steady state condition, and that the reactivity of the propagating
chain is dependent only on the monomer unit at the end of the
chain, independent of the composition which preceeds it (16, 17).
Monomers A and B lead to two types of propagating species $A^{\bullet}$ and
$B^{\bullet}$.

The copolymer composition equation was used,

$$F_a = \frac{r_a f_a^2 + f_a f_b}{r_a f_a^2 + 2f_a f_b + r_b f_b^2} \tag{2}$$

where f_a and f_b are the mole fractions of the monomers in the feed
and F_a is the mole fraction of the monomer A in the copolymer.
The parameters r_a and r_b are defined by:

$$r_a = K_{aa}/K_{ab} \quad \text{and} \quad r_b = K_{bb}/K_{ba} \tag{3}$$

Nomenclature

The following nomenclature will be used:

S = styrene monomer
PS = polystyrene
TO = tung oil monomer
copolymer TO-S = copolymer of tung oil and styrene
f_s = mole fraction of styrene in the feed
f_{to} = mole fraction of the tung oil in the feed
F_s and F_{to} = mole fraction of the monomers in the copolymer
r_s and r_{to} = monomer reactvity ratios of the styrene and tung oil
         respectively
[S] = molar concentration of the styrene in the feed
d[S] and d[TO] = change of molar concentration of the monomers in
         the copolymer.

EXPERIMENTAL

Materials

Copolymers TO-S were synthesized by using a bulk copolymerization technique. Styrene monomer was washed and dried before it was used, TO monomer of commercial grade was used directly. Benzoyl peroxide was used as a free radical initiator for the copolymerization reaction. The material sources are shown in Table III.

Reaction and Measurements

The copolymerization was carried out at constant temperature of 60 $\pm$ 2°C. Starting with a mixture of 50/50 of the monomers by weight, ten samples were analyzed at the following reaction times (in hours): 5, 6.5, 21, 31, 33.5, 45.5, 55.5, 69.5, 73.5. A maximum conversion of 10 percent was obtained. After those times, each reaction was stopped by using hydroquinone dissolved in methanol.

Table III. Description of the Materials

Reactant	Observations	Source
styrene	commercial grade	Dow Quimica de Colombia Colombia, S.A.
tung oil	commercial grade	Fabrica de aceites (Sto. Pipo), Misiones. Argentina, S.A.
benzoyl peroxide	reagent grade	Merck
hydroquinone	reagent grade	Merck
Methanol	reagent grade	Merck

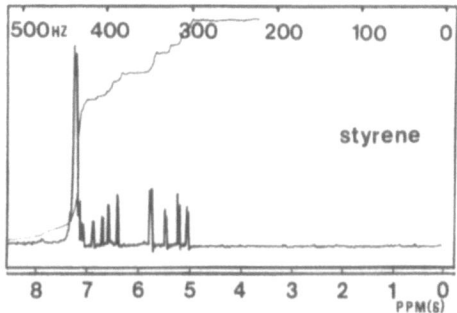

Fig. 5.   NMR spectra of S monomer.

The copolymers were quantitatively precipitated by pouring into
methanol. The copolymerized fractions were separated from the
unreacted monomers, and the unreacted monomer mix was washed and
dried. The reactions were followed by analyzing the mole fractions
of the unreacted monomers. For reference purposes, NMR analysis
of the pure monomers and of mixtures of different compositions
were made. An excellent agreement was obtained between the calculated and observed mole fractions by NMR.

Figure 5 shows the NMR spectra for S monomer. The signal at
7.40 to 7.00 ppm corresponds to the protons in the benzene ring.
The signals at 6.59 ppm (c), 5.59 ppm (b) and 5.11 ppm (a) corresponds to the non-equivalent vinyl protons (18,19) equation (4).

$$\text{(4)}$$

Figure 6 represents the NMR spectra for TO monomer. The signals at
6.20 to 5.00 ppm correspond to the vinyl protons and to the central
proton of the glycerol fraction. The signal at 4.10 ppm corresponds
to the methylene protons of the glycerol fraction. The signal at
2.40 to 2.10 ppm corresponds to the methylene protons next to the
vinyl bonds and next to the ester bonds. The signal at 1.28 ppm
corresponds to the rest of the methylene protons, and the signal
at 0.89 ppm corresponds to the methyl protons.

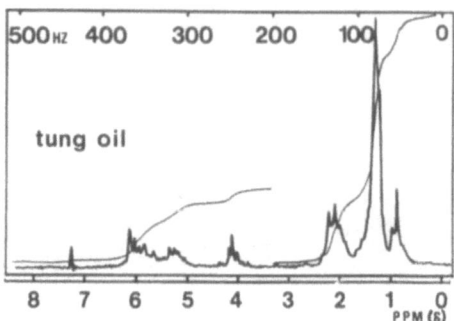

Fig. 6.   NMR spectra of TO monomer.

Figure 7 shows the NMR spectra for a styrene-tung oil monomer mixture. Because no interference exists, the signals at 7.40 to 7.00 ppm (A) (styrene), and at 2.40 to 0.80 ppm (C) (tung oil), were selected for the quantitative analysis. Thus, the equations for the evaluation of the mole fractions of the monomer components in the mixtures are given by:

$$f_s = \frac{A/5}{A/5 + C/69} \qquad f_{to} = \frac{C/69}{A/5 + C/69} \qquad (5)$$

RESULTS

Polymerization Kinetics

The monomer reactivity ratios of the TO-S copolymer were evaluated by a graphic solution of the copolymer equation (20). The equation was rearranged, giving:

$$H(1-h)/h = r_{to} - r_s(H^2/h)$$

where

$$H = [S]/[TO] \text{ and } h = d[S]/d[TO] \qquad (6)$$

Figure 8 shows a plot of $H(1-h)/h$ as a function of (H^2/h), which give a straight line. Its intercept value is $r_{to} = 3.20$, and its slope gives $r_s = 0.75$.

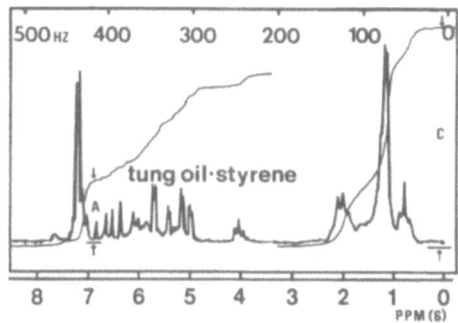

Fig. 7.   NMR spectra of the mixture of TO and S monomers.

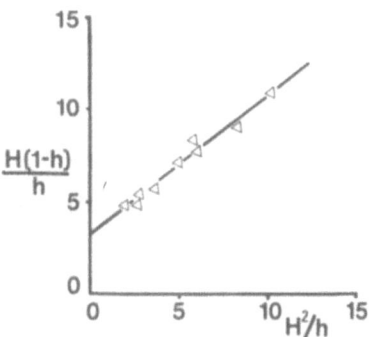

Fig. 8.   Evaluation of the monomer reactivity ratios by a graphic
solution of the copolymer equation.

Figure 9 shows the variation in the copolymer composition as
a function of the comonomer feed composition. This curve was made
using equation (2) and the experimental values of the monomer reactivity ratios. The importance of this kind of representation
arises in two ways; on the one hand it gives information about the
composition of the copolymer as a function of the concentration of
the monomers during polymerization, and on the other hand it permits
the control of the overall process. It is usually desirable to
produce a copolymer which is homogeneous in composition, which can
be done by using Figure 9 and data from reaction conversion.

The variation of the mole fraction of the monomers as a
function of the reaction time is shown in Figure 10. The first

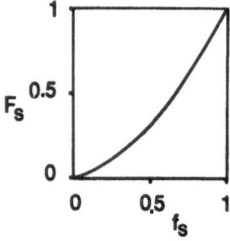

Fig. 9.   Instantaneous composition of TO-S copolymer (mole fraction
          of S, F_s) as a function of the monomer composition in the
          feed (mole fraction of S, f_s).

part of the curve, which represents the behavior of the polymerization at lower conversions (until approximately 1.46 percent), is
probably due to experimental error or to impurities. The tendency
observed is that the greater reactivity to the TO causes it to be
consumed faster than S. The system formed by TO and S shows a
compositional drift. Both radical growing chains show a preference
for adding TO monomer, according to this result.

From the values of the monomer reactivity ratios it is appropriate to assume that the first polymer formed will be richer in
the more reactive monomer, TO. Later, during the polymerization
process, this monomer is largely used up, and the last polymer
formed will be richer in the less reactive monomer, S.

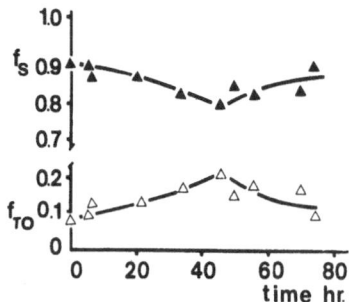

Fig. 10. Mole fraction of S monomer in the feed (f_s) as a function of the time reaction. Mole fraction of the TO monomer in the feed (f_{to}) as a function of the time reaction.

CONCLUSIONS

While there exists other useful instrumental methods of analysis, such as infrared spectroscopy and gas-liquid chromatography, NMR can be applied to the study of monomer reactivity ratios based on rates of monomer disappearance, giving good results. The straight line shown in Figure 4 agrees with the copolymer equation based on average values, and calculated by a least-square method. A correlation coefficient of 0.98 was obtained for this line. The reactivity ratio values show that TO has a reactivity four times greater than S, which agrees with the large amount of conjugated vinyl groups present to TO. Finally, the system formed by TO and S behaves according to the copolymer equation and steady-state assumptions.

ACKNOWLEDGMENT

The authors are pleased to acknowledge the support of the Colciencias in Colombia, and would also like to thank Dr. L.H. Sperling for his advice on the preparation of this paper and Prof. Louis E. Delfederico from Universidad Nacional de Misiones, Argentina, for his communication concerning domestic tung oil production. We also wish to thank Dr. R. Kleiman from the North Central Region USDA, for sending to us a sample of tung oil fruits.

REFERENCES

1. Yenwo, G.M., Manson, J.A., Pulido, J., Sperling, L.H., Conde, A., Devia, N., J. Appl. Poly. Sci., 12, 1531 1977.
2. Pulido, J.E., Yenwo, G.M., Sperling, L.H., Manson, J.A., Rev. UIS (Colombia), 7(7), 35 (1977).
3. U.S. Department of Commerce. Bureau of Foreign and Domestic Commerce. Trade Promotion Series, No. 129, 1932.
4. Blackmon, G.H. Tung Oil. A Gift of China. Economic Botany, Vol. 1, 1947.
5. U.S. Fats and Oil Statistics 1961-1976, Statistical Bulletin No. 574 Economic Research Service, U.S. Department of Agriculture, 1977.
6. Fats and Oils:Production, Consumption and Stocks, Bureau of Census, U.S. Department of Commerce, Jan. 1978.
7. U.S.D.A. Foreign Agriculture Circular FOP25-77, USDA, Washington, D.C., Dec. 1977.
8. Secretaria de Planification and Instituto Nacional de Techologia Agricola, Argentina.
9. Austin, R.O. and Long, J.S., "Tung Oil, Your Most Versatile Raw Material," Bulletin TO 67-1, Pan American Tung Research and Development League, August 1976.
10. Bailey's Industrial Oil and Fats Products. Wiley Interscience, 4th edition, 1979.
11. Thames, S.F., Long, J.S., Smith, O.D., Jen, S.J. and Evans, J.M., J. Am. Oil Chem. Soc., 45, 277 (1968).
12. Thames, S.F., Bufkin, B.J., Jen, S.J., Evans, J.M., and Long, J.S., J. Paint Tech., 48, 46 (1976).
13. Bufkin, B.J., Thames, S.F., Jen, S.J., Evans, J.M., Long, J.S., Smith, O.D., J. Am. Oil Chem. Soc. 53, 11 (1976).
14. Qureshi, S., Manson, J.A., Sperling, L.H., "New Interpenetrating Polymer Networks Based on Industrial Type Natural Oils," American Chemical Society Organic Coatings and Plastics Chemistry, 43(2), 7(1980).
15. Sperling, L.H., Manson, J.A., Qureshi, S., Fernandez, A.M., Ind. Eng. Chem. Prod. Res. Dev. 20, 163-166 (1981).
16. Mayo, F.R. and Lewis, F.H., J. Am. Chem. Soc., 67, 1774 (1945)
17. Lewis, F.M., Walling, C., Cummings, W., Briggs, E.R. and Mayo, F.R., J. Am. Chem. Soc., 70, 1521 (1948).
18. The Sadther Standard Spectra, Vol. 10 (1974).
19. Nathan, J., Diaz, "Introduccion a la Resonancia Magnetica Nuclear," Limuusa Wiley, S.A. (1970).
20. Fineman, M. and Ross, S.D., J. Poly. Sci., 5, 269 (1950).

PROCESSING EMULSIFIED OILS AND ALKYDS TO GENERATE POLYMERS

Jan W. Gooch, Bechtel Group, Inc.
George Bufkin, DAP, Inc.
Gary C. Wildman, University of Southern Mississippi

Research & Engineering, San Francisco, California
Research & Development, Dayton, Ohio; and Polymer
Science Department, Hattiesburg, Mississippi

INTRODUCTION

During the past one hundred years, the coatings industry has
employed large quantities of fats and oils. Table 1 demonstrates
that the coatings industry used larger quantities each decade
until the 1950's when their use began to decline.   The decline
in fats and oils has largely been a result of the substitution of
petroleum products for the vegetable oil. The alkyd resin contains a large percentage of oil or fatty acids and has been the
predominate binder used in the trade-sales and industrial paint
industry for over thirty-five years. Petroleum based product
usage increased dramatically over the past decade, but it is
evident from Table 2 that alkyd resins are still employed in
large quantities. A large percentage of the trade sales market
has switched to water-borne latex coatings based on petroleum
derived materials. The major reason for this is consumer con-
venience in clean-up, short dry times and low odor. As paints
based on these products further penetrate the existing market, the
usage of vegetable oil-based coatings will continue to decline.

Environmental factors have contributed to the reduction of
the utilization of vegetable oil-based products. The need has
been to employ coatings which do not pollute the environment,
i.e., water-borne, higher solds and powder types. Petroleum
based products have been acceptable as the solution to the
environmental problem and consequently, vegetable oil-based
materials have lagged behind, still being employed largely in
solvent-borne paints. The major method of employing vegetable
oil based polymers in non-polluting coatings has been in "exempt"

TABLE 1. USE OF FATS AND OILS IN SURFACE COATINGS[1]

Year	Million Pounds
1931–34	474
1940	652
1950	873
1960	716
1972	575

solvent systems as specified by Los Angeles Rule 66. However, the Environmental Protection Agency has stated that all organic solvents pollute the environment even though some are less reactive than are others and thus this use will also begin declining in the near future.

Table 3 lists the supply and demand of petroleum for 1965–1975 and predicts the supply through 1985. It is obvious from these data that the U.S.A. is becoming more dependent on imported petroleum. Table 4 represents U.S. land use for annual crops and primary sources of vegetable oils. The total land use of 336.3 million acres with a total of 470.0 million acres available if needed literally "paints" a more optimistic picture of the future than does continued and accelerated dependence on foreign petroleum. Referenced to total land use 334.3 million acres there are an additional 133.40 million acres available if needed. This indicates an opportunity for expansion of vegetable oil utilization compared to one of conversion and restriction with imported petroleum expected in the future.

TABLE 2.   CONSUMPTION OF SYNTHETIC RESINS
IN PAINTS AND COATINGS [2]

Type of Resin	Millions of Pounds
Alkyd	700
Vinyl	325
Acrylic	425
Epoxies	125
Urethane	70
Aminos	80
Cellulosic	55
Styrene-butadiene	25
Total resins	1,825

TABLE 3.   U.S. PETROLEUM BALANCES, IN
MILLION BARRELS PER DAY [4]

Year	Domestic		Imported	
	Supply	Demand	To Balance	% Total
1965	9.0	11.5	2.5	21.7
1970	11.3	14.7	3.4	23.1
1975	11.1	18.4	7.3	39.7
1980	11.8	22.5	10.7	47.5
1985	11.2	26.0	14.8	56.9

TABLE 4.   U.S. LAND USE FOR ANNUAL CROPS[4]

Crop	Planted Acres, millions
Wheat	70.0
Corn	66.9
Hay	61.9
Soybean	53.6
Sorghum	15.5
Oat	13.6
Corn silage	9.7
Cotton	9.1
All other	36.3
Total	336.6
Available if needed	470.0

As solvents become more expensive due to petroleum price increases and shortages, the need for water-borne and/or higher solids polymeric products will become more acute. This research is directed toward the generation of water-borne latex coatings from vegetable oils or vegetable oil derived materials as opposed to the more commonly used water reducible, water dispersible or water soluble products which normally require significant quantities of petroleum derived solvents and raw materials. A successful and competitive generation of new water-borne polymeric coatings based on vegetable oils could enhance the usage of vegetable oil products in place of petroleum based coating materials in the future.[3]

Alkyds, oils and oil/alkyd mixtures have been emulsified and subsequently autoxidatively crosslinked in the emulsion form to a near-gel or gelled state within the polymer particles. During the emulsification, the emulsifier type was carefully

selected such that a stable emulsion was generated. The
average particle size of the emulsion droplets was then reduced
to less than 1.0 micron and maintained at this size during the
autoxidative process. The autoxidative process is continued
until the maximum crosslink density that will allow proper
flow, which is a function of the crosslink density, particle
size and polymer type, is achieved. During coalescence, a
small amount of further crosslinking, as well as flow, generates
a dense uniform film.

This technology produces vegetable oil based emulsions
which will dry to touch rapidly and allow water clean-up
equivalent to acrylic and vinyl latex coatings.

THEORY

Emulsification of Alkyds and Oils

The emulsification of vegetable oils and vegetable oil
derived alkyd polymers requires suitable emulsifying agents. The
selection of an emulsifier for stabilization of vegetable oil
material will not necessarily be the most favorable emulsifier
for a vegetable oil derived alkyd. Emulsificatin experiments
must be conducted after determining solubility parameters of
the emulsifier in the aqueous and polymer phases such that the
hydrophile-lipophile balance, which is related to the solubility
parameter[5], may be obtained. Emulsifiers were selected on the
basis of the characteristics that appeared to be necessary for
emulsification of vegetable oils and vegetable oil derived
alkyds. McCutcheon's Detergents and Emulsifiers[6] was reviewed
for this general purpose. Figure 1 illustrates the geometrical
configuration of the particle in relation to the emulsifier
layer which presents a barrier to oxygen diffusion. It can be
seen that the lipophile segment of the emulsifier is soluble
in the alkyd particle phase and the hydrophile segment is
soluble in the aqueous phase. As indicated in Figure 1, the
oxygen must diffuse[7] from the aqueous phase through the
emulsifier region and into the alkyd particle in order to
initiate autoxidation. A discussion of mechanisms may be
found in references Lenz[8] and Odian[9].

Autoxidation of Emulsified Alkyds and Oils

Vegetable oils are obtained and refined from several plant
sources. They are largely comprised of the triglyceride esters
of fatty acids. Three particularly important oils are soya,
linseed and tung oil since these oils contain fatty acids with
wide variation in olefinic groups and conjugation. For example,

soya oil contains oleic, linoleic, linolenic, stearic, and
palmitic fatty acids, linseed oil contains the same fatty acids
but in different proportions, and tung oil contains a major
proportion of eleostearic fatty acid.

Stearic fatty acid is saturated, oleic contains one olefin
group at the carbn (9) position, linoleic fatty acid contains
two olefin groups at the carbons (9 and 12) positions. Linolenic fatty acid contains three olefin groups at the carbons
(9, 12 and 15) positions. Eleosteric acid contains olefin
groups at carbons (9, 11 and 13) and the double bonds are conjugated. In addition, a mixture of isomers exists within the
structure of these materials. Also, erythro- and threo-
conformations are present in these same structures, but of
undetermined proportions.

Historically, two fundamental theories of autoxidation
are dominant. First, there is allylic abstraction of a hydrogen

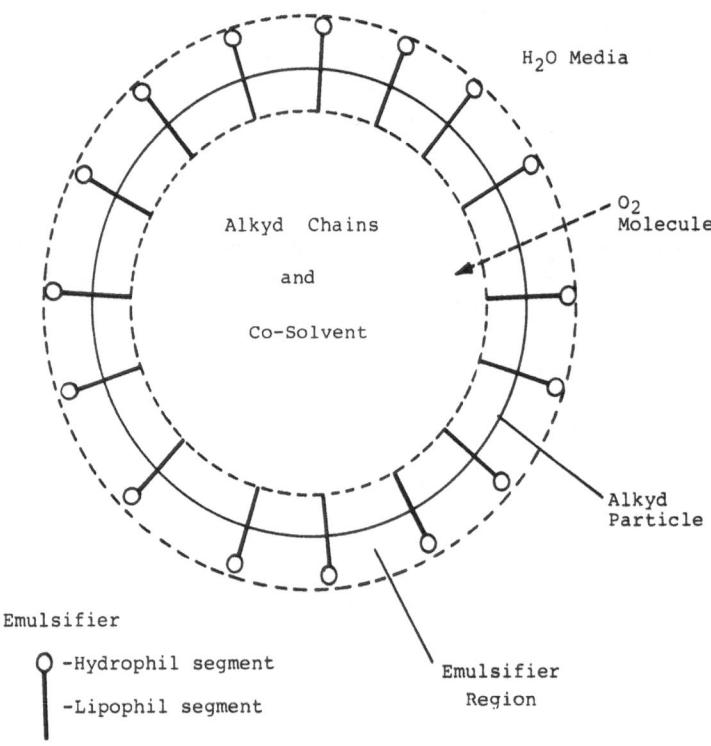

Figure 1.   Emulsified Alkyd Particle and
Oxygen Diffusion

atom and second, direct addition of oxygen to the double bond.
One of the most important consequences of this free radical
theory of olefin autoxidation is that the unpaired electron in
the initially formed allylic type hydrocarbon radical is
delocalized, being more or less equally distributed between the
"1" and "3" carbon atoms in alkylated ethylenes. This leads to
two consequences. The first is that more than one hydroperoxide
may be formed, the ratios of the various hydroperoxides depending on the electron densities on the carbon atoms involved. The
second is that the greater the rate of formation of hydroperoxide, the more delocalized an unpaired electron is. The
investigations of Farmer et al[10]   on polyenes showed the
existence of both of these phenomena.

    Autoxidative crosslinking of the above materials results
in the formation of three-dimensional polymers containing
carbon-carbon and ether linkages. The degree of crosslinking is
controlled by factors which enhance autoxidation including
oxygen concentration, concentration of unsaturated and conjugated groups and temperature. The oxygen concentration can be
controlled by the oxygen partial pressure as described by
Henry's law[11] , which states that "oxygen partial pressure is
directly proportional to oxygen concentration in dilute
aqueous solutions."

Control of Autoxidative Crosslinking with the Swelling Ratio

    There are numerous techniques for studying the network
structure in crosslinked polymers. An excellent technique
involves the swelling of polymer in solvent. If a crosslinked
polymer is added to a solvent for the uncrosslinked polymer,
then the crosslinked polymer will swell, but it will not
dissolve, the total volume will increase. The soluble fraction
of the polymer material will dissolve and diffuse out of the
swollen polymer. The uncrosslinked matter is the sol fraction
of the total polymer matter. The crosslinked material will
swell in the three-dimensional polymer gel phase until the
solution osmotic forces are balanced by the forces due to the
stretched segments of polymer chains. These elastic retractive
forces are inversely proportional to the molecular weight of
polymer between crosslinks. A highly crosslinked polymer will
not swell as much as a lightly crosslinked polymer[12] . A
method for measuring the degree of crosslinking is measurement
of the "swelling ratio[13]." The swelling ratio is the volume
of the solvent swollen polymer to the volume of the dry polymer.
As the crosslink density increases, the swelling ratio
decreases.

EXPERIMENTAL

Materials

Vegetable oil derived materials used for experimentation
are listed in Table 5. The alkyds were prepared by esterification of vegetable oils with phthalic anhydride and multi-
functional alcohols to obtain the correct oil length within the
structure of the alkyd molecule[14] . The raw soya oil was
refined with sodium hydroxide to remove free fatty acids and
purify the oil.

Conjugated oils were used for varying the percent conjugation of alkyds for autoxidizable emulsions which were
synthesized by the formulation described in Table 6.

Emulsification Procedure

The oils and alkyd materials were emulsified by the
following procedure:

1. The emulsifying agent was dissolved in distilled water
    at $25.0^{\circ}C$.
2. The oil (or alkyd) was mixed with the emulsifier
    solution while dispersed with the Braunsonic Ultra-
    sonic Vibrator which formed a pre-emulsion.

TABLE 5.   MATERIALS

Number	Material Description
I	Soya Oil, Alkali Refined
II	Short Oil Soya Alkyd
III	Medium Soya Oil Alkyd
IV	Long Oil Soya Alkyd
V	Extra-Long Oil Soya Alkyd
VI	Urethane Soya Alkyd

3. The pre-emulsion was stable enough to remain intact
    during transfer to the Gaulin Sub-micron Disperser
    and Homogenizer. The emulsion was recycled until the
    desired particle size reduction was achieved. The
    Gaulin Homogenizer was operated at 3500 pounds per
    square inch of shear force.

Table 8 lists the components of the formulation that were
found to be suitable for the emulsification of alkali refined
soya oil emulsion. Table 7 lists the components for the
formulation of the alkyd emulsions.

Particle Size Determination

A Coulter-Counter model TAII was used with a M3 Data Converter[15]
and 15 micron aperature to measure particle size diameter. The
limiting particle diameter for this instrument is 0.20 microns.
The emulsion samples were diluted with 2.0 percent NaCl electrolyte
before measurement. These data were reported in particle diameter
distribution by volume and population.

Processing of Alkyd and Oil Emulsions

The autoxidation of oil and alkyd emulsions were conducted
in the Bench Scale pressure vessel illustrated in Figure 2.
The vessel was air/oxygen pressure, temperature and agitation
controlled. The turbine agitator, controlled by a tachometer,
is capable of thoroughly circulating the emulsion within the
vessel. The temperature sensor rests within the liquid adjacent
to the turbine agitator which produces a continuous signal to
the proportional temperature controller. The proportional
temperature controller pulses heat to the vessel proportional to
the difference between the desired temperature setting in the
controller and the actual vessel temperature. The oxygen (or
air) pressure is controlled by a pressure regulator connected
to a delivery tube beneath the surface of the emulsion. The
autoxidation is thus controlled while periodic samples are
taken from the interior of the vessel beneath the surface of
the liquid through a sample valve.

Determination of the Swelling Ratio

The "swelling ratio" of the emulsified particles was
determined at frequent intervals to monitor the autoxidative
reaction. In each determination, a 0.2 gram sample of
emulsified solids was added to a deciliter of reagent grade
acetone to swell the crosslinked particles. Density, viscosity

TABLE 6. FORMULATION FOR CONJUGATION STUDY

Component	Percent (w/w)
Trimethylol Propane	31.90
Fatty Acid Oil, Vegetable	41.12
Phthalic Anhydride	33.10
Less Water	6.10

and percent solids measurements were made for each sample. Then, "swollen and unswollen volume fractions" were determined as by Gooch[16].

RESULTS AND CONCLUSIONS

The emulsification parameters and conditions for autoxidative crosslinking have been studied in detail for optimization of emulsion stability after processing together with film properties from the emulsions. The studies referred to in this paper merit further research in order to broaden usage of materials, but certain conclusions may be discussed with confidence.

Autoxidative Crosslinking of Soya Oil

Stable emulsions may be prepared from soya oil. The autoxidative crosslinking of the emulsified soya oil was slow and produced films of poor tensile strength which were slow to dry. These results were encouraging from the viewpoint that vegetable oil derived materials of higher molecular weight such as alkyds autoxidized faster and films formed from them dried more completely.

Observing the intrinsic viscosity[17] of the resins produced in the autoxidation reactions, it became obvious that an intrinsic viscosity of 0.046 of soya oil would necessarily be surpassed in order to achieve improved film integrity. Neither catalysts nor increasing heat of reaction for processing of the emulsion improveed film integrity from soya oil emulsions.

TABLE 7. FORMULATION OF ALKYD EMULSION

Component	Percent (W/W)
Alkyd, Vegetable Oil	50.0
Distilled Water	48.0
Emulsifier, 75% Nonylphenoxypoly (ethyleneoxy) alcohol, 2)% Dodecyl Sodium Sulfate	2.0

TABLE 8. FORMULATION OF SOYA OIL EMULSION

Component	Percent (W/W)
Soya Oil	50.0
Distilled Water	49.0
Dodecyl Sodium Sulfate	1.0

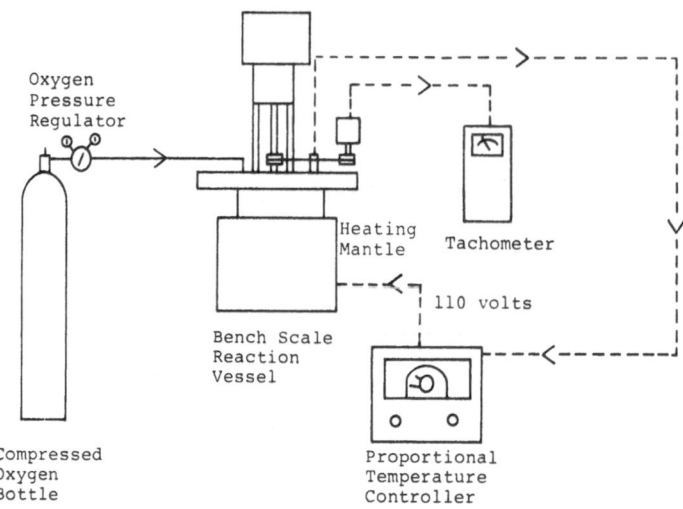

Figure 2. Emulsion Autoxidation Reaction Apparatus

Autoxidative Crosslinking of Soya Alkyds

After emulsifying an alkyd resin, processing conditions
were studied for autoxidation at atmospheric pressure utilizing
air and pure oxygen. Air autoxidation produced significant
gas voids within the film which reduced film tensile strength,
but did not impair drying of the films. Measurement of intrinsic viscosity demonstrated that the autoxidation reaction cross-
linked the alkyd beyond the maximum intrinsic viscosity obtained
with soya oil. The crosslinked alkyd produced an alkyd particle
that possessed a small swelling ratio, less than 10.0, which
indicates that the material is densely crosslinked. Oxygen
autoxidation of the alkyd emulsion developed crosslinks faster
than air and the emulsion formed a film of medium tensile
strength and elongation. Increased oxygen pressure further
increased the reaction rate by reducing the inductive period as
shown in Table 9. By observing the swelling ratio, the degree
of crosslinking could be monitored and a combination of drying
time and crosslink density could be compromised for maximum
tensile strength, (i.e., flow and drying time of the films).
Table 10 compares two leading acrylic emulsions with the vegetable oil alkyd films prepared in these laboratories. It can
be seen that the commericial emulsions possess higher tensile
strength than the medium soya alkyd, but the alkyd emulsion had
a comparable dry time. The soya urethane emulsion possesses a
tensile strength within the range of the commercial products,
but its elongation is lower, though adequate in our opinion.

The reaction rate increases with percent conjugation to about twenty percent conjugation and then, other parameters appear to be autoxidation rate determining as shown in Figure 3. From Table 11, increasing oil length of alkyds decrease the tensile strength of dried alkyd films, except for the urethane alkyd film tensile strength with sixty-two percent soya oil length. The urethane alkyd structure is responsible for the tensile strength property of the dried film.

Swelling ratios with dry time, tensile strength and percent elongation of films formed from alkyd III emulsions are listed in Table 12. It can be seen that the optimal films properties occur at a welling ratio of 3.7. However, the dry time continues to decrease. A swelling ratio of 3.7 was selected as the optimal degree of autoxidative processing, and it is utilized in following studies.

The effect of catalysis was studied at an oxygen pressure of 80.0 psi (6.13 x 10^{-3} moles/liter H_2O). The catalysts selected for this study were catalysts that were successful for catalyzing the alkyd in a solvent system. Cobalt naphenate catalysis in concentrations of about 0.04 percent (weight/weight alkyd) produced instability within the alkyd III emulsion and therefore

TABLE 9. OXYGEN CONCENTRATION, INDUCTION TIME,
CROSSLINKING PERIOD, TOTAL REACTION TIME
AND LOG REACTION TIME

Oxygen Concentration, psi.	Induction Period, hours	Crosslinking Period, hours	Total Reaction Time hours
80	4	5	9
70	7	5	12
60	12	4	16
50	16	5	21
40	19	5	24
30	24	7	31
20	29	16	42
10	36	22	58

TABLE 10. COMPARISON OF FILMS FROM COMMERCIAL
AND AUTOXIDIZED EMULSIONS

Emulsion	Tensile Strength, KG/CM^2	Percent Elongation	Dry Time, Hours
Rhoplex AC-64	35.5	475.0	0.5
Amsco 3077	24.6	707.0	0.7
Autoxidized:			
Alkyd III	10.7	95.0	0.5
Alkyd VI	21.5	51.2	1.0

TABLE 11. CHARACTERIZATION OF FILMS FROM ALKYD
EMULSIONS WITH VARYING PERCENT OIL

Alkyd	Percent Oil	Tensile Strength, KG/CM^2	Percent Elongation	Dry Time, Hours
II	30	5.2	43.4	1.0
III	52	10.7	95.0	0.5
IV	60	3.6	117.5	1.1
V	80	-	-	-
VI	62	21.5	51.2	1.0

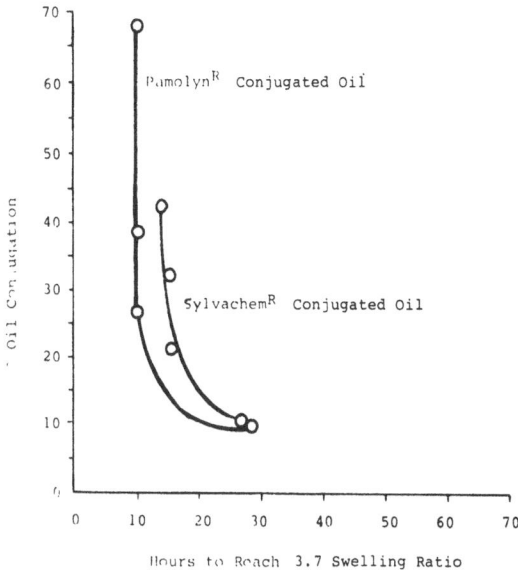

Figure 3.   Oil Conjugation vs. Reaction Time

this catalyst was not used singularly, but in combination with
zirconium naphthenate, which does not degrade the stability of
the emulsion. Catalysis of the emulsion by 0.02 percent cobalt
naphthenate and 0.02 percent zirconium naphthenate (weight/
weight alkyd) reduced the reaction time required to crosslink the
emulsion particles to 3.7 swelling ratio by about two hours less
than non-catalyzed autoxidation.

It would be obvious to increase the catalyst concentration in
order to further reduce autoxidation time required to densely cross-
link the alkyd polymer particles. However, formidable effects arise
from increasing the concentration of zirconium maphthenate to 0.05%
and cobalt naphthenate to 0.05%. When the emulsion was subjected
to 80.0 psi, the autoxidation reaction was uncontrollable. Therefore, the concentration of catalysts is critical.

Zirconium naphthenate does not impart instability to the
emulsion as cobalt naphthenate does. Considering reduction of
autoxidation time and emulsion stability, zirconium naphthenate
was selected to study catalyst concentration with autoxidation
reaction time. Figure 4 represents the effect on reaction time
with increasing zirconium naphthenate concentration. After 0.05
grams/1000 grams alkyd III, instability within the emulsion
occurs. Therefore, concentrations below 0.05 grams zirconium
naphthenate/1000 grams alkyd should be employed to catalyze the
reaction.

As indicated in Figure 1, the oxygen must diffuse from the aqueous phase through the emulsifier region and into the alkyd particle phase in order to initiate autoxidation. If the emulsifier is capable of retarding the diffusion of oxygen by radical stabilization or a formidable intermolecular reaction, then the crosslinking reaction by autoxidation is slowed. A total explanation of how emulsifiers retard the reaction is not within the scope of this research, but demonstration that the factor exists is clearly made here.

The effect of temperature on the autoxidation reaction was studied at 80.0 psi of oxygen pressure and uncatalyzed. The study was designed to include an alkyd III emulsion autoxidized at 20, 30, 40, 50 and 60 C at 80.0 psi oxygen pressure. Each reaction would be considered complete at 3.75 swelling ratio. Figure 6 represents the decreasing reaction times to reach 3.7 swelling ratio with increasing temperature. The effect of temperature is to exponentially increase the autoxidation rate. However, emulsion instability occurs after 55°C.

TABLE 12. REACTION TIME AND PHYSICAL PROPERTIES

Swelling Ratio	Dry Time, hours	Tensile Strength	% Elongation
4.9	2.0	--	--
4.0	1.0	4.6	67
3.7	0.5	10.7	95
3.0	0.4	8.4	55
2.3	0.3	8.5	50
2.0	0.3	8.4	57

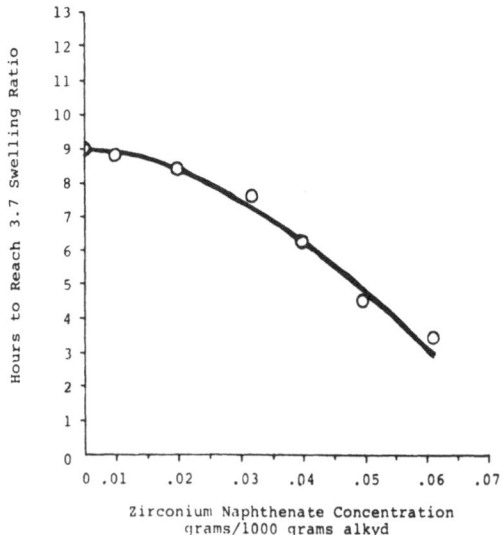

Figure 4. Zirconium Naphthenate Concentration
vs. Reaction Rate

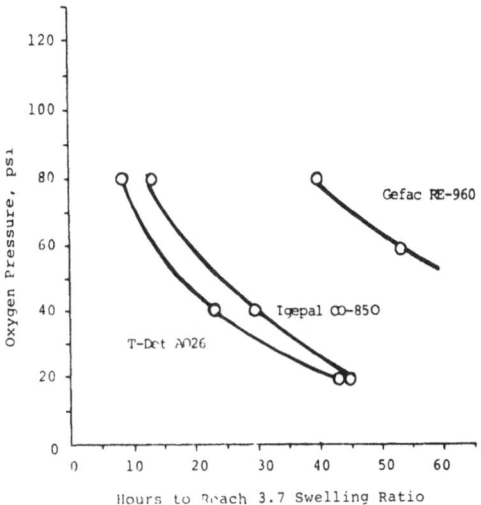

Figure 5. Effect of Emulsifier Structure on Reaction Time

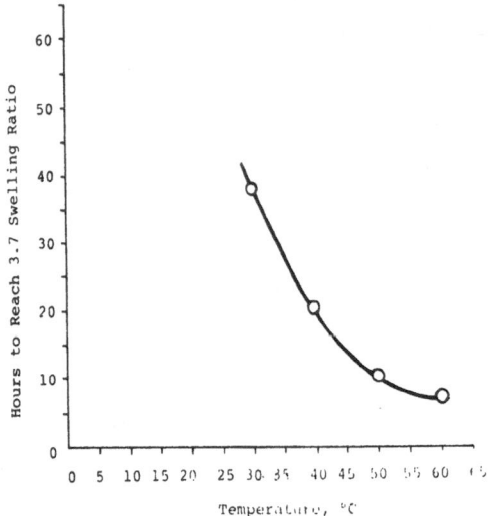

Figure 6.   Reaction Temperature vs. Reaction Time

        Many emulsifier variations have been used during the course
of this research effort. The effect of emulsifiers has been
observed on numerous properties of the emulsions generated. A
study was initiated to determine the effect of emulsifier structure on the autoxidation rate of reaction. Figure 5 represents
the oxygen pressure variation during successive reactions as a
function of the time required to reach a densely crosslinked
state. The Gafac RE-960 system is dramatically slower relative
to the autoxidation rate when T-Det and dodecyl sodium sulfate
or Igepal CO-850 and dodecyl sodium sulfate surfactants are
employed.

        Gafac RE-960 contains a phosphoric acid group together with
an aryl group. Both of these groups are capable of stabilizing
a radical through resonance. The Igepal Co-850 structure contains
an aryl group. The T-Det A026 does not contain any substituent or
groups capable of stabilizing a radical. Since the diffusion of
oxygen through the emulsifier region surrounding the polymer
particle is necessary if autoxidation within the polymer particles
is to occur, the emulsifying agent should not characteristically
possess an ability to retard oxygen diffusion. If it does, the
crosslinking reaction will proceed at a slower rate than expected
as evidenced in Figure 5. Figure 1 illustrates the geometrical
configuration of the particle in relation to the emulsifier layer
which presents a barrier to oxygen diffusion. It can be seen that
the lipophile segment of the emulsifier is soluble in the alkyd
particle phase and the hydrophile segment is soluble in the
aqueous phase.

Emulsification

Vegetable oil alkyds have been emulsified successfully and have remained stable after processing. The screening of emulsifying agents has eliminated slow emulsion reaction rates, low tensile strength and slow drying of films. Emulsifying agents preferable are aliphatic alcohols with ethylene oxide adducts in combination with dodecyl sodium sulfate.

These studies give evidence that the average particle size (0.9 microns) changes insignificantly during the crosslinking reaction. One year of aging of a soya urethane emulsion did not affect the average particle size significantly.

SUMMARY

Emulsified vegetable oil alkyds were autoxidatively crosslinked while in the emulsion form. The resultant emulsion produced films with acceptable tensile strength and drying rate comparable to some commercial emulsion products. The majority of petroleum solvents were eliminated from these emulsions, thus demonstrating the potential for significantly reducing petroleum products in these coatings. Further research and development are required to produce marketable products from these emulsions.

REFERENCES

1. J. C. Cowan, "Applied Polymer Science," edited by J. K. Craver and T. W. Tess, Organic Coatings & Plastics Chemistry, Division of the American Chemical Society, Washington, D.C., 1975.

2. H. J. Lanson, "Applied Polymer Science," edited by J. K. Craver and R. W. Tess, Organic Coatings & Plastics Chemistry, Division of the American Chemical Society, Washington, D.C., 1975.

3. J. A. McCarthy, "The Future of Hydrocarbon Emission Regulation," presented at the Water-Borne and Higher Solids Coating Symposium, February 1977.

4. L. H. Princen, J. Coatings Technology, 49, 88 (1977).

5. A. Beerbower and M. W. Hill, McCutcheon's Detergents and Emulsifiers, McCutcheon Division, MC Publishing Co., Glen Rock, NJ, 1971.

6. McCutcheon's Detergents and Emulsifiers, McCutcheon Division, MC Publishing Co., Glen Rock, NJ, 1977.

7. G. M. Barrow, "Physical Chemistry," McGraw-Hill Book Company, New York, 1973.

8. R. W. Lenz, "Organic Chemistry of Synthetic High Polymers," Interscience Publishers, New York, 1967.

9. G. Odian, "Principles of Polymerization," McGraw-Hill Book Company, New York, 1970.

10. F. J. Farmer, "Paint and Varnish Technology," Reinhold Publishing Co., New York, 1968.

11. G. M. Barrow, "Physical Chemistry," McGraw-Hill Book Company, New York, 1973.

12. P. J. Flory, "Principles of Polymer Chemistry," Cornell University Press, Ithaca, 1953.

13. G. M. Crews, G. D. Wildman, and B. G. Bufkin, paper presented at the Federation of Societies for Coatings Technology meeting, Houston, November, 1977.

14. T. C. Patton, "Alkyd Resin Technology," Interscience Publishers, New York, NY, 1962.

15. Coulter Electronics, Inc., Technical Bulletin-Coulter Counter Model TA II, Hialeah, FL, March (1978).

16. J. W. Gooch, "Autoxidative Crosslinking of Emulsions Prepared from Vegetable Oil Materials," Ph.D. Dissertation, University of Southern Mississippi, May 1980.

17. F. W. Billmeyer et al., "Experiments in Polymer Science," John Wiley and Sons, New York, 1973.

NOVEL UV AND RADIATION POLYMERISATION METHODS FOR MODIFYING

POLYOLEFINS, CELLULOSE AND LEATHER

Neil P. Davis, John L. Garnett, Chye H. Ang
and Lloyd Geldard

School of Chemistry
University of New South Wales
Kensington, N.S.W. Australia 2033

INTRODUCTION

The possibility of being able to modify existing trunk polymers with monomers by one-step grafting methods is a potentially very useful technique for increasing the range of renewable resource materials available for polymer applications. Both ionising radiation[1-7] and UV[6-11] are convenient initiatiors for these reactions. For high radiation grafting yields with certain materials, relatively large radiation doses or inert atmospheres are required during reaction. Methods for lowering the total radiation dose are thus useful, both economically and also for minimising competing degradation of the trunk polymer, especially where such materials are radiation sensitive.

An alternate approach to the above problem is to improve the grafting efficiency by lowering the homopolymer yield through the use of inorganic ions[12] such as Cu^{2+} and Zn^{+2}. The grafting efficiency can also be improved by the use of a styrene comonomer technique[13].

Recently a method was reported for enhancing the radiation grafting of monomers to polymers by the simple addition of mineral acid to the grafting solution[6,14]. Under some experimental conditions organic acids act in the same manner[15,16]. Extensive work on this acid effect has been published for the radiation copolymerisation of styrene to cellulose[6,14], wool[17], the polyolefins[16,18,19], PVC[20] and polyesters[21]. The acid effect has also been extended to the grafting of monomers other then styrene[22]. Preliminary studies of analogous acid enhancement in

323

UV grafting have also been reported, predominantly with the styrene in methanol to cellulose system[15,22]. The acid effect thus appears to be generally applicable to a wide range of grafting systems.

In the present paper further novel aspects of radiation copolymerisation are treated. In particular, a detailed mechanistic study of the role of alcohol solvents in the acid enhancement of the radiation grafting of styrene to the polyolefins, especially polypropylene, is reported. New additives for increasing the copolymerisation yield in this reaction will be discussed. The relative merits of ionising radiation will be compared with UV as initiators for grafting to polypropylene. Unique sensitisers for the UV copolymerisation processes will be considered especially for the grafting of monomers other than styrene. Finally the results of the above grafting work with polypropylene will be very briefly compared with analogous data obtained from copolymerisation to the natural materials, cellulose and leather. The importance of grafting reactions in radiation and UV sensitised solvent free, rapid cure processes will also be discussed.

EXPERIMENTAL

Styrene and methyl methacrylate were supplied by Monsanto (Aust.) Ltd. whilst divinylbenzene was purchased from Polysciences Inc. Monomers were purified by column chromatography on alumina. The polypropylene used was isotactic, doubly oriented film (thickness 0.06mm) whilst the cellulose was Whatman 41 filter paper and the leather was blue, full grain,side material.

Grafting Procedures

The following modifications of previously reported grafting methods[6,15,18] were used. For the gamma irradiation work all grafting experiments were performed in pyrex tubes (15 x 2.5 cm) with styrene/solvent solutions (20ml) at 20° ± 1°C. The trunk polymer films (4 x 2.5cm) were fully immersed in the monomer solutions and the pyrex tubes stoppered for irradiation which was performed immediately after preparation in a 1200 Ci cobalt-60 source. The pyrex tubes were uniformly positioned on a circular rack around the source such that the polymer films were perpendicular to the plane of the radiation. At the completion of the grafting, the films were immediately removed from solution, washed in an appropriate solvent to remove trapped homopolymer and extracted in a soxhlet apparatus for 72 hours. If acid had been used as additive, the films were washed with methanol/dioxan (1:1) prior to soxhlet treatment. With the polyolefin films at high graft of polystyrene (>50%), the films were soaked in chlorofom at 50°C for a further 3 hours after soxhlet extraction

to ensure complete removal of homopolymer. The films were dried
in air, then at 45°C to constant weight.

For the UV studies, sample solutions (monomer/solvent, 20 ml)
were prepared in pyrex tubes as before with the gamma irradiation
system and the tubes positioned on a motor driven ventilated
circulating drum at a distance of 30 cm unless otherwise stated
from the UV source (90W, high presure, Hg type 93110E E$_2$, Philips)
for 6-24 hours at 24 ± 1°C. The polymer films were so positioned
that, during irradiation, the surfaces of the films were
perpendicular to the incident radiation. After irradiation,
films were treated as for the gamma work.

In the rapid cure experiments, prepolymer mixtures, prepared
as previously described[23], were padded on to cellulose and leather
substrates and cured under a 6" Hanovia 200 W/inch high pressure
UV lamp. After exposure, the samples were extracted to constant
weight as described above.

RESULTS AND DISCUSSION

Radiation Grafting of Styrene in Alcohols and Miscellaneous
Solvents to Polypropylene

The grafting of styrene to polypropylene film is appreciable
in all of the alcohols reported in Figures 1 and 2, the high
yields being consistent with previous alcohol data[18] which were
obtained at lower dose-rates and with thinner polypropylene films.
Under the current radiation conditions used, a Trommsdorff effect
in all alcohols is observed at approximately 30% styrene. In
other polar solvents, the grafting properties of the styrene are
altered significantly. Thus from Figures 2 and 3, using dioxan,
acetone, carbon tetrachloride, chloroform, benzene and pyridine
as solvents,low grafting yields with styrene were obtained with
all solvents except dioxan. Even with dioxan, the yields were
still significantly lower than with the low molecular weight
alcohols. Pyridine gave moderate copolymerisation yields at 1.0
megarads whereas at 0.2 megarads, little grafting was observed.
Thus, overall, the alcohols remain the most efficient of the
current solvents studied for achieving relatively high yields in
polypropylene grafting.

Mechanistic Role of Alcohols in Radiation Grafting. An important mechanistic feature of these grafting data is the
accelerating effect of the low molecular weight alcohols.
Previously in the radiation grafting of styrene in methanol to
polyethylene[16] this accelerated reaction was attributed to a
Trommsdorff effect similar to that found in cellulose. According
to Odian and co-workers[24], in their original publications, the

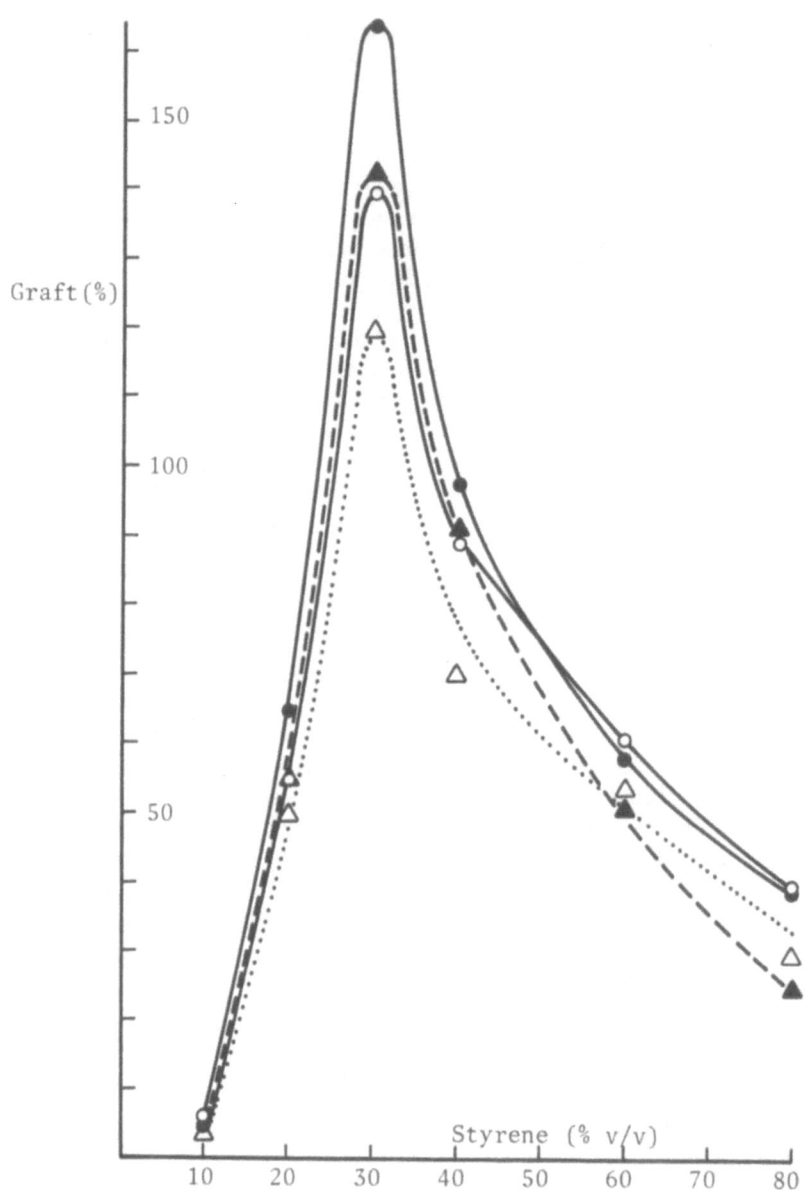

Fig. 1. Acid effect in radiation grafting of styrene to poly-
         propylene film (isotactic, doubly oriented, 0.06 mm) in
         methanol and ethanol at dose rate of 4.5 x 10⁴ rad/hr to
         total dose of 3.0 x 10⁵ rad.
         —●—methanol + H₂SO₄ (0.2M);--▲--ethanol + H₂SO₄ (0.2M)
         —○—methanol; ·····△·····ethanol.

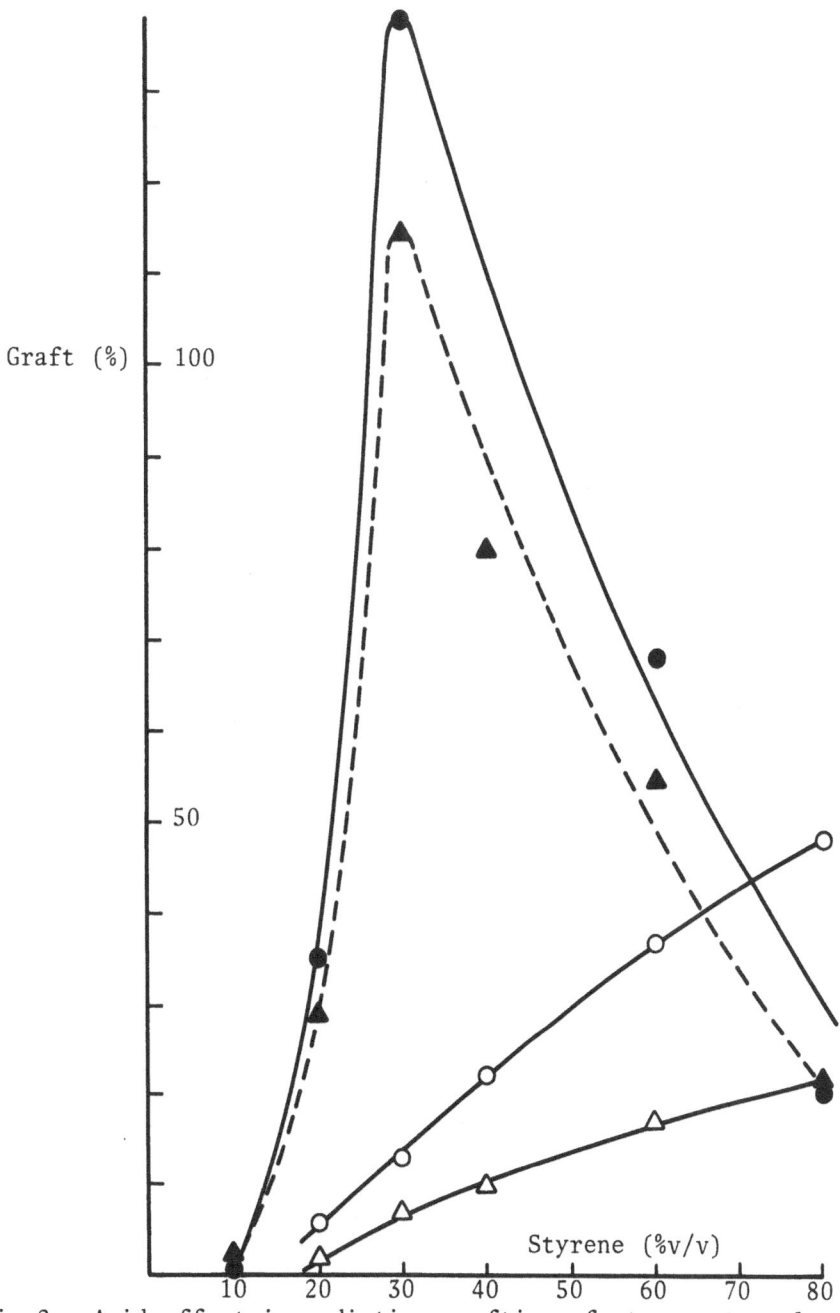

Fig 2. Acid effect in radiation grafting of styrene to polypropylene film in n-butanol and dioxan at dose rate of
4.5×10^4 rad/hr, total dose of 3.0×10^5 rad.
—●—n-butanol + H_2SO_4 (0.1M); ---▲--- n-butanol;
—○—dioxan + H_2SO_4 (0.2M); —△— dioxan

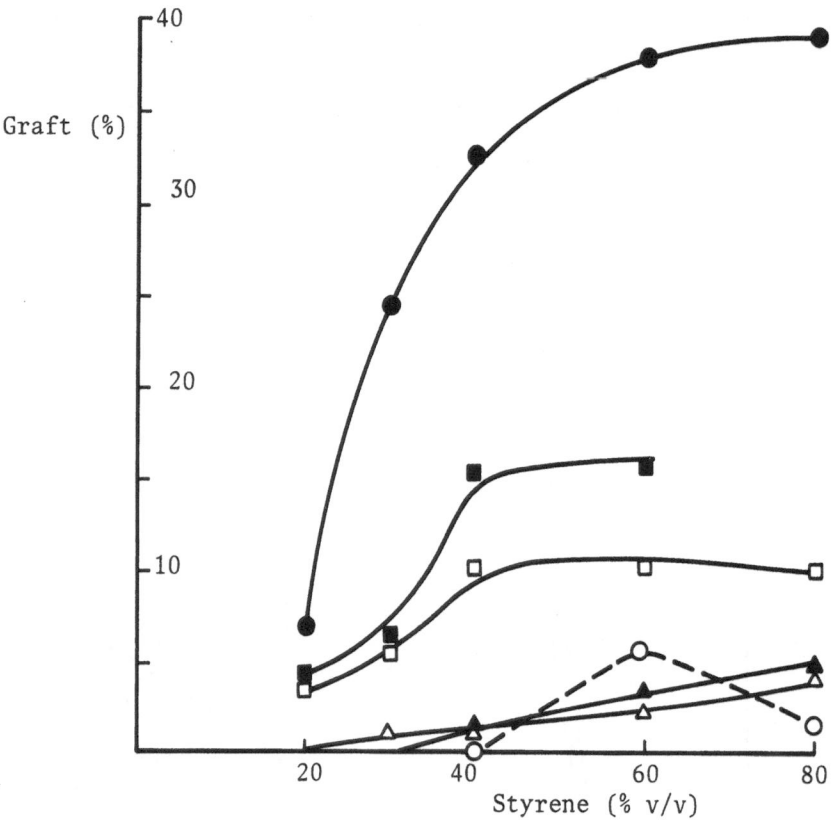

Fig 3. Acid effect in radiation grafting of styrene to polypropylene film in acetone, chloroform, carbon tetrachloride, benzene and pyridine at dose rate of 4.5×10^4
rad/hr to total dose of 2.0×10^5 rad except pyridine
(4.1×10^4, 1.0×10^6).

—●— pyridine; —■— acetone + H_2SO_4 (0.2M); —□— acetone;
--○-- benzene; —▲— chloroform; —△— carbon tetrachloride;

effect occurs because methanol is a precipitant of polystyrene and as the grafted polystyrene chains are precipitated they become immobilised. Further collisions with the precipitant polystyrene are inhibited, the termination rate is reduced with essentially no reduction in initiation thus grafting is increased. These authors emphasised that the termination rate in the grafting of undiluted styrene was already low due to the high viscosity of the amorphous polyethylene. In the presence of methanol, termination is further reduced because of the precipitation effect of the solvent. Thus the Trommsdorff effect which already exists in the grafting medium is enhanced by methanol addition. The mechanism developed for methanol is also applicable to all low molecular weight alcohols due to their similar precipitating properties with polystyrene. The conclusion is consistent with the alcohol data in Figures 1 and 2.

Later work on the system by Silverman and co-workers[25] led to further modifications of the Odian model. Thus, the Silverman group analysed mixtures of styrene and methanol which were absorbed into polyethylene films prior to grafting. They differentiated between the compositions of such "inside" solutions and those solutions in which the sample was placed ("outside" solutions)[24]. They found that the methanol fraction in the styrene-methanol mixture absorbed in polyethylene was too low to precipitate polystyrene chains as suggested by Odian, thus swelling of the polyolefin is entirely due to styrene. They proposed a mechanism for the Trommsdorff effect based on the viscosity of the amorphous region of the polyethylene swollen by styrene. Thus the presence of methanol reduces the concentration of occluded styrene in polyethylene and since it does not swell the polyethylene, the viscosity of the grafting medium increases. At low styrene concentrations, initiation and propagation are low and increase with increasing monomer concentration whilst termination is also low because of the high viscosity of the medium. At high styrene concentrations, propagation decreases while termination increases due to the decreased viscosity of the swollen polymer. Between these two extremes there is an optimum styrene concentration at which the grafting is a maximum.

Thus according to the proposed theories[24,25], solvents that are precipitants for polystyrene and are also insoluble in the polyolefin will produce Trommsdorff effects. Experimental data obtained from the grafting of styrene in benzene, dioxan, carbon tetrachloride, chloroform, pyridine and acetone (Figures 2 and 3) show that the grafting yields are low in these solvents and the corresponding grafting patterns are different to those in methanol and related alcohols. Thus the copolymerisation behaviour in all solvents studied in this work, except acetone, is in accordance with the proposed theory. Acetone, being a weak solvent for polystyrene and a non-solvent for polyethylene, should influence grafting in a manner similar to methanol, a conclusion which is in conflict with the actual experimental data. Further refinements to the Odian [24] and Silverman [25] models are thus indicated to explain such anomalous behaviour. The effects of radiation on the components in the grafting system obviously also needs to be considered for a more complete mechanistic treatment of the process.

Radiation Effects in the Grafting Mechanism

The acetone result is particularly important since it indicates that further parameters, in addition to those already proposed,[24,25] are needed to adequately define the conditions required to explain gel effects in these reactions. It is therefore suggested that the radiation chemistry of the system, particularly that of the monomer and solvent as proposed originally for the cellulose work[14] also needs to be considered in any mechanistic discussion of the radiation grafting to polyolefins. In this respect, it is significant that all of the above solvents have one common property, namely, under radiolysis conditions, they produce finite yields of H atoms. An analysis[26] of such G(H) data from these solvents (Table 1) suggests that there is a relationship between these hydrogen yields and both the yield and Trommsdorff effect in radiation grafting.

Table 1.   Literature Values of Hydrogen Yields from the Radiolysis of Various Solvents Used in Grafting Reactions[26]

Solvent	G(H)	G(H$_2$)
Methanol	2.4-2.5	4.0-5.4
Dioxan	-	2.1
Benzene	-	0.038
Chloroform	-	0
Pyridine	-	0.018
Acetone	0.6	-

With respect to these two grafting properties the order of reactivity in the solvents used was methanol > dioxan > acetone > remaining solvents and this is the order expected if a G(H) relationship were important. The acetone result indicates that, even with some solvents, the radiation chemistry parameter may be more significant in explaining grafting behaviour than either of the two physical models already proposed for the system[24,25].

Hydrogen atoms may participate mechanistically in grafting processes via abstraction reactions (Equation 1) yielding copolymerisation sites.

$$PH + H^{\cdot} \longrightarrow P^{\cdot} + H_2 \tag{1}$$

Such reactions have been proposed to explain the general effect of solvents in grafting to cellulose[27] and also, in preliminary studies, to the polyolefins[16][18]. As a consequence of irradiation, G(H) yields from hydrogen donor atom solvents in a radiation context such as methanol are high (Table 1). Styrene scavenges these radicals by either addition or substitution reactions to give styrene-methanol intermediates (MR·)[18]. Even after these scavenging reactions have occurred, at 30% styrene in methanol, data[18] indicate that the number of hydrogen atoms in solution remains sufficiently high to contribute mechanistically to the accelerated effects of the type observed by processes analogous to Equation 1. Species such as MR· may also participate in the copolymerisation process by hydrogen abstraction from the trunk polymer to give new grafting sites (Equation 2).

$$PH \ + \ MR^{·} \longrightarrow P^{°} \ + \ MRH \qquad (2)$$

grafting then being induced by a cage mechanism the complex depicted in Equation (3) being the intermediate in the process.

$$P^{°} \ + \ MRH \longrightarrow (P^{°} \longrightarrow MRH) \qquad (3)$$

The mechanism involving MR· species is supported by data from the radiolysis of benzene-methanol[18] and pyridine-methanol[18] mixtures where yields of scavenging products reach a maximum at 20-30% concentration of aromatic in methanol. This same concentration region is where styrene-methanol reaches a grafting maximum in the current work.

Effect of Acid in Enhancing Radiation Grafting

Inclusion of sulfuric acid in the styrene grafting solution (Figures 1 and 2), even with acetone (Figure 3), leads to an enhancement in copolymerisation to polypropylene, consistent with earlier studies[18] carried out under different radiation conditions. With the alcohols, the presence of acid increases the yield markedly at the Trommsdorff peak whereas with dioxan (Figure 2) uniform acid enhancement in graft occurs at virtually all monomer concentrations studied. The presence of the Trommsdorff peak gradually disappears for the styrene, methanol, polyethylene system as the dose rate increases (Figure 4). Addition of acid does not markedly affect the shapes of the grafting curves for this same system at different dose rates (Figure 4). Even at 112,000 rad/hr where the gel effect is not evident in Figure 4, acid still gives a significant improvement in grafting especially at 50% monomer concentration. However inclusion of acid leads to only a marginal increase in grafting efficiencies (Figures 5 and 6) at all dose rates except for the 21,000 rad/hr system, thus indicating that acid also influences the yields of homopolymer in these reactions.

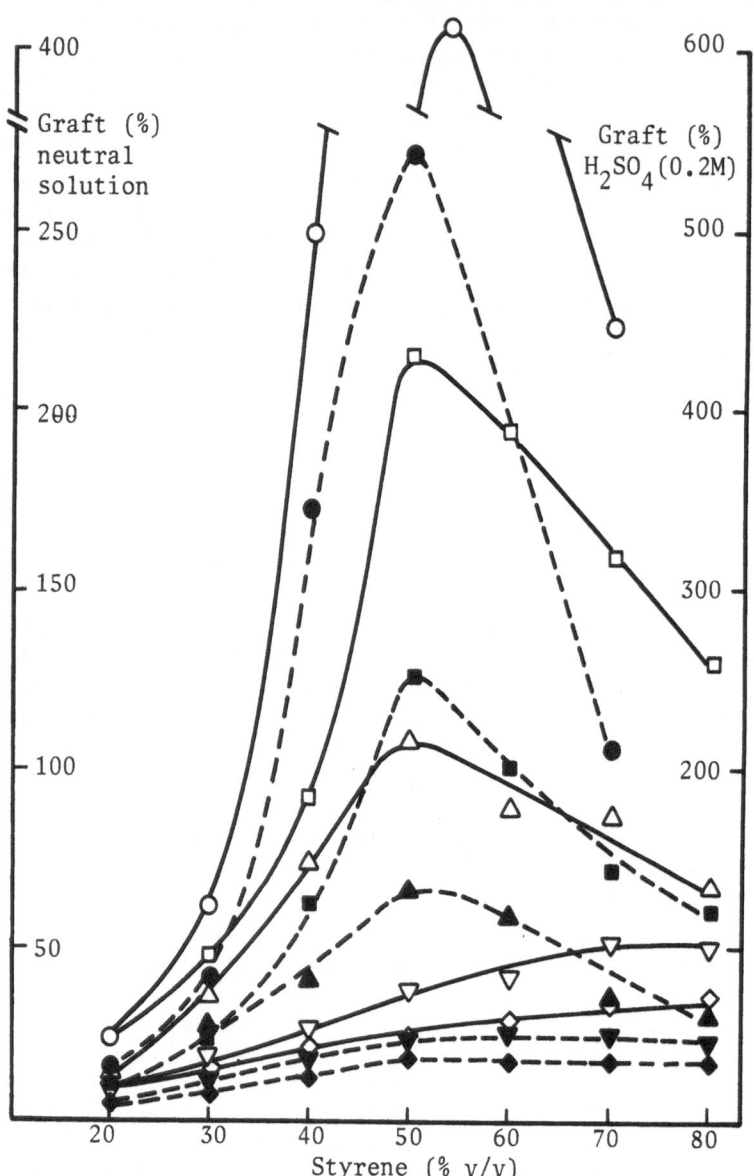

Fig 4. Effect of dose rate (rad/hr) on grafting of styrene in
        methanol in neutral and acid solution to polyethylene
        films at total dose of 2.3 x 10^5 rad.
        Neutral solution —O—10,000; —□—21,000; —△—41,000;
                         —▽—75,000; —◇—112,000.
        H_2SO_4 (0.2M) --●--10,000; --■--21,000; --▲--41,000;
                         --▼--75,000; --◆--112,000.

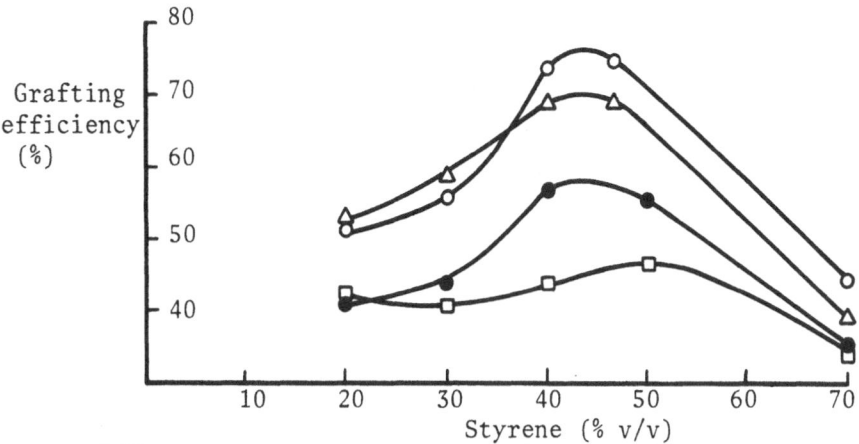

Fig 5.   Effect of dose rate on the efficiency of grafting styrene
to polyethylene film in methanol. Grafting efficiency =
graft / graft + homopolymer. Total dose 2.3 x 10^5rad.
—○—10,000 rad/hr; —△—21,000 rad/hr; —●—41,000 rad/hr;
—□—75,000 rad/hr.

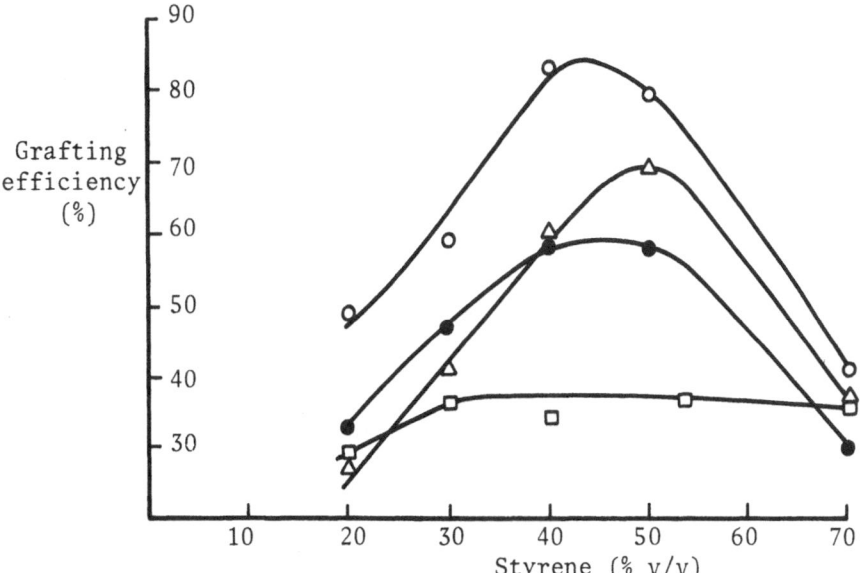

Fig 6.   Effect of dose rate on the efficiency of grafting styrene
to polyethylene film in methanol in the presence of acid
(H_2SO_4, 0.2M). Total dose, 2.3 x 10^5rad.
—○—10,000 rad/hr; —△—21,000 rad/hr; —●—41,000 rad/hr;
—□—75,000 rad/hr.

Mechanistically, the present acid results are consistent with the preceding radiation chemistry pathway proposed to explain the grafting reaction. Thus, inclusion of mineral acid, at the level used, should not markedly affect the precipitation of the grafted polystyrene chains or the swelling of the polypropylene and thus its role cannot be satisfactorily explained in terms of the Odian[24] or Silverman[25] models. Acid can however lead to an increase in G(H) by a radiation chemistry process (Equations 4 and 5), the higher H atom yields leading to

$$CH_3OH \ + \ H^+ \longrightarrow CH_3OH_2^+ \qquad\qquad (4)$$

$$CH_3OH_2^+ \ + \ e^- \longrightarrow CH_3OH \ + \ H \qquad (5)$$

enhancement in grafting by abstraction reactions (Equation 1) with the trunk polymer. A similar mechanism for the acid enhancement in radiation grafting to cellulose has also been proposed[27]. Confirmatory evidence for this hydrogen atom mechanism is provided by detailed molecular weight distribution studies on the grafting solutions[28] where it is found that inclusion of acid leads to larger numbers of oligomer chains of lower molecular weight. Because there is a dynamic equilibrium between bulk monomer solution and monomer solution absorbed by trunk polymer, any change in properties of the oligomer in bulk solution must affect the properties and nature of the graft. Thus, although the oligomer chains are shorter in acid grafting solutions, the numbers of these chains are higher and, because of the smaller size of these chains, the radical intermediates associated with them can diffuse more readily into the swollen trunk polymer to give enhanced yields owing to higher concentration and higher mobilities associated with their smaller size. The lower $\overline{M}_n$ values and larger numbers of shorter chains observed in the presence of acid in the grafting solutions are also consistent with the acid enhancement in the Trommsdorff peak (Figures 1 and 2) in the present grafting work, since the lower molecular weights lead to a high solubility of oligomer radicals in solution, resulting in an increase in viscosity at the Trommsdorff peak and a reduction in chain termination, leading to a marked acceleration in grafting. This interpretation would also explain why the gel effect in radiation grafting is strongly dependent on structure of solvent, those solvents that are strong hydrogen atom donors in a radiation chemistry context being the most efficient in accelerated grafting. The molecular weight distribution data also clarify the controversy[18,19] concerning the mechanism of the original acid effect which some authors believed to be associated with the possible presence of water during sample preparation. The roles of water and acid are now

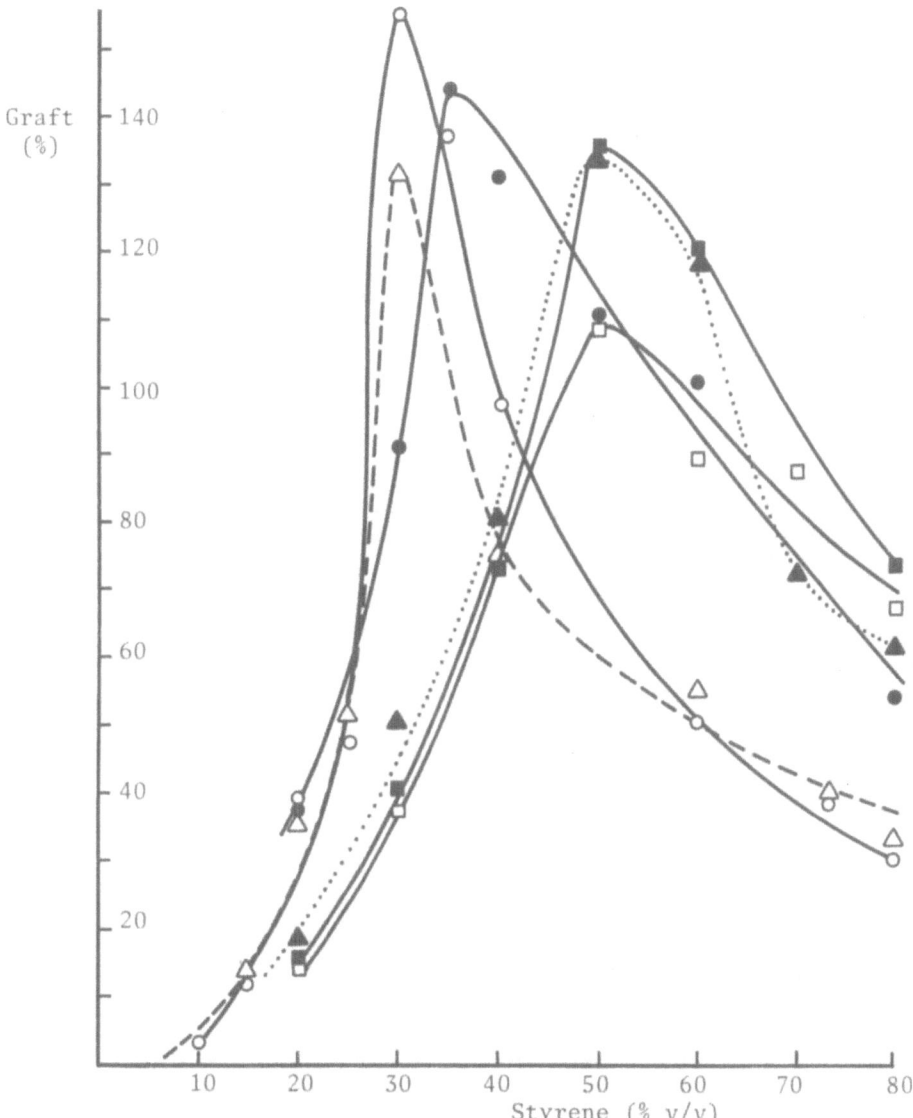

Fig 7. Effect of divinylbenzene (1% v/v) on the grafting of
        styrene in methanol to polypropylene (PPE) and poly-
        ethylene (PE) film at a dose rate of 4.0 x 10^4 rad/hr
        (4.1 x 10^4 for PE) to a total dose of 2.3 x 10^5 rad.
        —O—styrene, methanol, H_2SO_4 (0.2M), PPE;—●—styrene,
        methanol, DVB, PPE;--△-- styrene, methanol, PPE;
        —■—styrene, methanol, DVB, PE;····▲···· styrene, methanol,
        H_2SO_4 (0.2M),PE;—□—styrene, methanol, PE.

known to be quite different in these reactions[28], a conclusion
substantiated by the work of Chappas and Silverman[19]. The
possible role of hydrogen atoms in this grafting work is also
consistent with the observed acid effects in the dose-rate studies
(Figure 4) and also the grafting efficiency results (Figures
5 and 6).

Although these recent molecular weight distribution
experiments[28] indicate that H atom yields are the predominant
mechanistic pathway for the acid enhancement in radiation
grafting, recent work[18] on the radiolysis of binary mixtures
indicates that acid can also affect the yields and stability of
the styrene-methanol grafting intermediates (MR'). An increase
in yield of MR' could also lead to an increase in grafting sites
by H atom abstraction. Such a mechanism may also make a finite
contribution to the overall grafting process.

Grafting with Divinylbenzene as Additive

The inclusion of divinylbenzene (DVB) enhances the
radiation grafting of styrene in methanol to polypropylene
(Figure 7) especially at concentrations above 30% monomer where
there is generally a 100% increase in yield. Compared with acid
and neutral solutions, the Trommsdorff peak in the DVB solution
has moved slightly to higher monomer concentration for polypropylene grafting. The acid enhancement is greater than the
DVB effect at monomer concentrations lower than 30%, but the
reverse holds above this styrene level with polypropylene.

When polyethylene is used as trunk polymer (Figure 7),
the inclusion of DVB in the styrene-methanol grafting solution
leads to a significant enhancement in copolymerisation, especially
at the Trommsdorff peak. The effect of acid and DVB are similar
with polyethylene grafting except above 60% monomer where the
DVB enhancement is significantly higher.

DVB has previously been used extensively to cross-link linear
polymer chains. In the present experiments, the DVB has a dual
function, namely to enhance the copolymerisation and to cross-
link the grafted polystyrene chains. In the presence of a
polyfunctional monomer such as DVB, branching of grafted chains
can occur. Branching of the growing grafted polystyrene chain
takes place when one end of the DVB, immobilised during grafting,
is bonded to the growing chain. The other end is unsaturated
and free to initiate new chain growth via scavenging reactions.
The new branched polystyrene chain may eventually terminate,
cross-linked by reacting with another polystyrene chain or an
immobilised divinylbenzene radical. Grafting is thus enhanced
mainly through branching of the grafted polystyrene chain. This

conclusion is important especially for renewable resources applications since the structure of the graft copolymer from the DVB work should be significantly different to the analogous material obtained from acid enhanced grafting. The product from the latter process should be essentially linear in structure and not extensively cross-linked as is to be expected for the DVB material. Thus even though both acid and DVB increase grafting yields, the properties of the resulting copolymers from preliminary experiments appear to be significantly different[21] and hence the two type additives act in a complementary manner to produce enhanced yields of grafted products. Brief studies[21] with other polyfunctional monomers, particularly the acrylates, indicate that the enhancement in grafting observed with DVB is a general phenomenon of polyfunctional monomers when used as additives under specific radiation conditions in grafting.

Comparison of Polypropylene with Cellulose and Leather for the Radiation Grafting of Styrene

The radiation grafting properties of polypropylene using styrene in methanol as monomer system have been compared with those of the natural materials, cellulose and leather. Of the three, polypropylene is the most reactive trunk polymer (Table 2). Cellulose and leather exhibit comparable styrene grafting properties and all three trunk polymers show acid enhancement in copolymerisation especially in the 20-40% monomer concentration region. The polypropylene and cellulose data illustrate the

Table 2.　Comparison of Polypropylene with Cellulose and Leather for the Radiation Grafting of Styrene in Methanol in Presence of Acid[a]

Styrene (%v/v)	Polypropylene	Graft (%) Cellulose	Leather, % Graft		
20	29 (218)	8 (24)	5.7[b]	-[c]	32[d]
30	94 (150)	17 (33)	4.6	21.5 (27)	56
40	50 (85)	22 (31)	4.0	57	58
50	37 (55)	21 (27)	-	-	-
60	36 (45)	28 (25)	7.5	-	-

a.　Radiation conditions for (i) polypropylene were 0.3Mrad with 0.2M H_2SO_4 (ii) cellulose were 0.2Mrad with 0.4M H_2SO_4 and (iii) leather were 0.1, 0.3 and 0.5 Mrad successively with 0.1M H_2SO_4. Dose rate in all systems was 0.04 Mrad/hr. Figure in brackets graft with acid.

differences between synthetic and natural polymers in radiation grafting in a unique manner. The radiation conditions used in Table 2 for polypropylene lead to a Trommsdorff effect at 30%

monomer concentration. Inclusion of acid not only enhances the intensity of this peak, it also shifts the maximum to a lower monomer concentration. By contrast, with cellulose, no Trommsdorff peak is evident under the radiation conditions used with neutral solutions, however addition of acid to the solution not only enhances the original grafting yields it also induces the formation of a gel effect at 30% monomer concentration. Thus the polypropylene and cellulose data indicate two significant features of the acid effect in grafting. Firstly acid can enhance the magnitude of a Trommsdorff effect if it is already present in the system,or secondly,inclusion of acid can induce the gel effect if no such peak was previously observed.

Mechanistically the role of acids in enhancing grafting should be similar for all three trunk polymers. The polar nature of cellulose and leather could lead to additional pathways for grafting sites to be formed in these particular trunk polymers. Thus hydrogen ions from the added acid can equilibrate with the hydroxyl group of the cellulose (CeOH) to form ionic species which can capture thermalised electrons from the primary radiation act to give excited cellulose molecules and H atoms (Equations 6 and 7).

$$CeOH \;+\; H^+ \longrightarrow CeOH_2{}^+ \qquad\qquad (6)$$
$$CeOH_2^+ \;+\; e^- \longrightarrow CeOH^2 \;+\; H\cdot \qquad (7)$$
$$CeOH^2 \longrightarrow CeO\cdot \;+\; H\cdot \qquad\qquad (8)$$

Decomposition of the excited cellulose molecules yields radical sites for grafting and additional H atoms (Equation 8). In like manner leather (LeNH$_2$) can participate in hydrogen ion reactions (Equation 9), both the NH$_2$ and COOH functional groups in the

$$LeNH_2 \;+\; H^+ \longrightarrow LeNH_3{}^+ \qquad\qquad (9)$$

leather being capable of activation.

As a preparative technique, the present radiation grafting process is useful for polypropylene and cellulose, however with leather care is needed, since the solvents tend to dry out this trunk polymer and embrittle it with time. Thus for many leather treatments, the radiation rapid cure processes mentioned at the end of this paper are to be preferred.

Photosensitised Grafting to Polypropylene and Cellulose

Instead of using ionising radiation to initiate these grafting reactions to polypropylene and cellulose, UV can also be utilised[6,15,29]. Inclusion of sensitisers enhances the rates of such UV reactions as the data in Table 3 show for the grafting of styrene in methanol to polypropylene.

Table 3.   Photosensitised Grafting of Styrene to Polypropylene
in Alcohol Solvents[a]

Styrene (% v/v)				Graft (%)		
				Methanol	Ethanol	n-Butanol
20	0[b]	0[c]	2[d]	51	41	62
30	0	0	8	170	151	263
35	0	-	-	176	163	250
40	1	0	14	163	72	119
60	6	0	9	44	42	36
80	6	2	13	26	23	27

a. Benzoin ethyl ether (BEE, 1% w/v) as sensitiser; Irradiated 24 hrs at 24 cm from 90 W high pressure UV lamp.
b. This column, no sensitiser in methanol.
c. This column, benzophenone (BPO) in methanol (1% w/v).
d. This column, uranyl nitrate (1% w/v) in methanol.

Benzoin ethyl ether (BEE) is better than benzophenone (BPO) or uranyl nitrate as sensitiser in this system. The results are analogous to the ionising radiation work in that gel effects are observed (30/35% monomer concentration) and the lower molecular weight alcohols are very efficient for the process.

If the acrylates and their esters are used instead of styrene, homopolymerisation can become a serious competing problem. Novel sensitisers have been developed for these monomers in order to overcome this problem. The charge transfer complex between dimethyl aminoethanol and benzophenone (DMAE:BPO) has been used successfully to graft methyl methacrylate in butanol to poly-propylene (Table 4), neither of the two components in the complex being active when used alone. Butanol was used as solvent in Table 4, because it was the most reactive of those studied for grafting styrene in Table 3. BEE gives higher grafting yields than DMAE:BPO (Table 4), however the slower grafting rate with the latter sensitiser leads to lower homopolymer yields and an improvement in the uniformity of graft in the copolymer films.

When cellulose is used as trunk polymer, a wide range of monomers can be grafted by photosensitisation processes (Table 5). Uranyl nitrate is a better sensitiser for cellulose grafting than either of the other two discussed in Table 3, only relatively short exposure times of three hours being needed with the 90W high pressure lamp to achieve large grafting yields with most monomers studied. Thus UV can be used a complementary initiation technique to gamma rays to modify cellulose by grafting reactions. (Table 6).

Table 4.   Sensitiser Effect in UV Grafting of Monomers such as [a]
Methyl Methacrylate (MMA) in Butanol to Polypropylene[a]

MMA	Graft (%) in Sensitiser			BEE
	DMAE: BPO Ratio			
(%v/v)	3:2	2:1	3:1	
20	6.9	5.9	11	0
30	7.3	15	22	0
40	23	25	63	22
45	-	-	-	80
50	33	39	78	174
55	-	-	-	320
60	35	35	78	502
70	-	-	-	-
80	6.4	31	77	538

a. DMAE = dimethyl aminoethanol ; BPO = benzophenone ; BEE =
benzoin ethyl ether; sensitiser concentration 1% (w/v).
No graft in DMAE, BPO alone and without sensitiser; Irradiated
6 hrs at 24 cm from 90W high pressure UV lamp.

Table 5.   UV Grafting of Miscellaneous Monomers to Cellulose[a]

Monomer	Irradiation Time (hrs)	Monomer(%)	Graft(%)
Styrene	24	80	53
a-Methyl styrene	24	79	5
o-Methyl styrene	24	79	63
Iso-Octyl vinyl ether	3	30	4
Dodecyl vinyl ether	3	30	2
Vinylidene chloride	3	30	19
4-Vinylpyridine	24	20	24
Trimethylol propane triacrylate	3	30	25
Allyl methacrylate	3	30	10
Methyl methacrylate	3	30	305

a. All runs used uranyl nitrate (1% w/v) except last monomer
(3% w/v); all solutions 24 cm from 90 W high pressure UV
lamp.

The mechanism of the sensitised UV grafting process is
similar in many respects to the radical pathway involved in the
analogous ionising radiation reaction. Thus using cellulose as
representative trunk polymer and UO_2^{2+} as sensitiser, grafting
sites can be formed in cellulose by the reactions depicted in

$$UO_2^{2+} + h\nu \longrightarrow (UO_2^{2+})* \qquad (10)$$
$$(UO_2^{2+}) + CeOH \longrightarrow UO_2^+ + H^+ + CeO^\bullet \quad (11)$$

Table 6. Comparison of UV with Gamma Radiation for Initiating
Grafting of Styrene in Methanol to Cellulose

Monomer (% v/v)	UV[a]	Graft %	γ
20	13	8[b]	10[c]
40	28	22	29
60	34	28	21
80	53	34	18
90	64	-	-

a. For the UV studies, solutions contained 1% w/v of uranyl
nitrate and were irradiated for 24 hr at 24 cm from a 90 W
high pressure lamp.
b. Dose rate of 0.04 Mrad/hr to total dose of 0.2 Mrad.
c. Dose rate of 0.075 Mrad/hr to total dose of 0.25 Mrad.

Equations 10 and 11 where intermolecular hydrogen abstraction of
the cellulose with excited sensitiser molecules or solvent
radicals occurs. With the organic sensitisers such as BEE,
radical formation with UV occurs by α cleavage and grafting sites
can then be formed by hydrogen abstraction reactions.

Comparison of Grafting with Radiation Rapid Cure Processing for Upgrading Materials

In terms of modifying substrates by copolymerisation
reactions with monomers as a means for renewing resource materials,
radiation grafting can be considered as a complementary technique
to the so-called radiation induced, rapid cure, essentially solvent free, polymerisation reaction. In the grafting systems already discussed, relatively long irradiation times (minutes or
even hours) are usually involved with the inclusion of solvents
to optimise the copolymerisation reaction by wetting and swelling
processes. In all such systems, the predominant purpose is to
form a true carbon - carbon bond between monomer and/or polymer
being grafted and backbone polymer. Because of the advantages of
penetration of radiation, cobalt-60 has been the preferred
irradiation source for grafting with some work being performed by
electron beam (EB) or UV. In Table 7 are shown data for the UV
copolymerisation of neat styrene to cellulose and polypropylene
utilising sensitisers previously discussed for grafting in the
presence of solvent (Tables 3 and 4). The important feature of
the results in Table 7 is that significant photosensitiser
grafting can be achieved in the absence of solvent. This
observation is of value for analogous rapid cure processes since
these involve radiation curing on substrates, mixtures of prepolymers or oligomers in reactive monomers, usually polyfunctional acrylates. These results (Table 7) thus suggest that

Table 7.   Effect of Removal of Solvent on UV Grafting
             of Styrene to Polypropylene and Cellulose[a]

Irradiation Time	Photosensitiser	Cellulose Graft (%)	Homopolymer (%)	Polypropylene Graft (%)
3	Benzoin ethyl ether	2	10	2
3	Biacetyl	1	-	2
24	Benzoin ethyl ether	38	-	11
24	Biacetyl	10	-	20

a. Sensitisers (1% w/v) with high pressure UV lamp situated at
   24 cm from irradiation vessel.

some grafting is to be expected in a large number of rapid cure
processes, many of which involve essentially homopolymerisation
of a thin film of prepolymer/monomer mixture onto a substrate.
Obviously the properties of many materials, finished by rapid cure
could be enhanced if some grafting between oligomer mix and
trunk polymer does occur. Such grafting should improve the
properties of flexibility, adhesion and thermal stability. The
present authors have rapid cured prepolymer/monomer mixtures onto
cellulose and polypropylene using both EB and sensitised UV
techniques. A special tritium labelling procedure was developed
to show that some grafting, although low (1-2%), did occur in
both systems[6]. Other rapid cure processes using both EB and
sensitised UV have been found where the grafting yields are
extremely high[6]. Such a system has been found for the coating
of leather[30]. It involves photosensitised UV or EB curing of
special butyl acrylate prepolymers on to the surface of leather,
the resulting product being flexible and water resistant and a
significant improvement over the analogous materials prepared by
radiation grafting techniques using monomer/solvent solutions.
The products from the latter process tended to be brittle and
lacked the flexibility of the rapid cure coated materials.

These results show that there is thus a need for both
radiation grafting and radiation rapid cure techniques in the area
of renewable resources. In the rapid cure procedures the
occurrence of some grafting would certainly improve the
properties of the resulting product. Obviously the possibility
of simultaneous grafting occurring with cure will only be
relevant to organic type trunk polymers, however these are the
substrates where flexibility is most needed. Thus both
radiation grafting and rapid cure procedures will continue to be
valuable complementary processing techniques for the area of
renewable resources.

ACKNOWLEDGEMENTS

The authors thank the Australian Institute of Nuclear Science and Engineering and the Australian Atomic Energy Commission for the irradiations. They are also grateful to the Australian Research Grants Committee for continued support. One of them (N.P.D.) wishes to thank Sidney Cooke Chemicals Pty. Ltd. for the award of a Fellowship.

ABSTRACT

UV and radiation induced polymerisation techniques are shown to be useful processes for enhancing the range of available renewable resource materials. In particular, grafting reactions have been examined as a means for modifying polymer structures. Thus detailed studies of the effect of solvents in gamma ray grafting of styrene to polyolefin films are discussed. Low molecular weight alcohols are the best of the range of solvents studied. Inclusion of acid or divinylbenzene (DVB) as additives markedly enhances the grafting yield under certain radiation conditions. The mode of action of each additive is shown to be different, yielding structurally different copolymers. Poly-propylene is evaluated with cellulose and leather as trunk polymers for the radiation grafting of styrene. Sensitised UV is compared with ionising radiation as initiator for these grafting process. The effect of sensitiser structure on UV grafting of monomers to polypropylene and cellulose is examined. The use of the radiation grafting technique is compared with radiation rapid cure polymerisation as processes of use in the field of renewable resources.

REFERENCES

1. A. Chapiro, "Radiation Chemistry of Polymeric Systems", Interscience, New York, (1962).
2. A. Charlesby, "Atomic Radiation and Polymers", Pergamon, Oxford (1960).
3. R.B. Phillips, J. Quere, G. Guiroy and V.T. Stannett, Modification of pulp and paper by graft copolymerization, Tappi, 55:858 (1972).
4. R.J. Demint, J.C. Arthur, Jr., A.R. Markezich and W.F. McSherry, Radiation-induced interaction of styrene with cotton, Text,Res.J. 32:918 (1962).
5. Y. Nakamura and M. Shimada, Effects of solvents on graft copolymerisation of styrene with γ-irradiated cellulose, ACS Symp.Ser. 48:298 (1977).
6. J.L. Garnett, J.Rad.Phys.Chem. 14:79 (1979).
7. A. Hebeish and J.T. Guthrie, "The Chemistry and Technology of Cellulosic Copolymers", Springer-Verlag, Berlin (1980).

8.   N.Geacintov, V.Stannett, E.W. Abrahamson and J.J. Hermans,
     Grafting onto cellulose and cellulose derivatives by using
     ultraviolet radiation, J.Appl.Polym.Sci. 3:54 (1960).

9.   A.H.Reine and J.C. Arthur,Jr., Photoinitiated polymerisation
     of Methacrylamide with cotton cellulose, Text.Res.J. 42:155
     (1972).

10. S.Tazuke, T. Matoba, H. Kimura and T. Okado, A Novel
     modification of polymer surfaces,by photografting A.C.S. Symp.
     Ser. 121: 217 (1980).

11. N.P. Davis, J.L. Garnett and R. Urquhart, The Role of solvent
     alcohol in the photosensitized copolymerisation of styrene
     to cellulose, J.Polym.Sci.Polym.Lett.Ed. 14:537 (1976).

12. M.B. Huglin and B.L. Johnson, Role of cations in radiation
     grafting and homopolymerisation, J.Polym.Sci. A-1, 7:1379
     (1969).

13. J.L. Garnett and R.S. Kenyon, Acid Effects in the styrene
     comonomer technique for radiation grafting to wool, J.Polym.
     Sci.Polym.Lett.Ed. 15:421 (1977).

14. S. Dilli, J.L. Garnett and D.H. Phuoc, Effect of acid on the
     radiation-induced copolymerization of monomers to cellulose,
     J.Polym.Sci.Polym.Lett.Ed. 11:711 (1973).

15. N.P. Davis and J.L. Garnett, Modifications to cellulose using
     UV grafting procedures, in: "Modified Cellulosics", R.M.
     Rowell and R.A. Young, eds. Academic Press, New York, (1978).

16. J.L. Garnett and N.T. Yen, Acid effects in the radiation
     grafting of monomers to polymers, particularly polyethylene,
     ACS Symp.Ser. 121:243 (1980).

17. J.L. Garnett and J.D. Leeder, Recent developments in grafting
     of monomers to wool keratin using UV and γ-radiation, ACS
     Symp.Ser. 49:197 (1977).

18. J.L. Garnett and N.T. Yen, Acid Effects in the radiation-
     induced grafting of styrene to polypropylene, Aust.J.Chem.
     32:585 (1979).

19. W.J. Chappas and J. Silverman, The effect of acid on the
     radiation-induced grafting of styrene to polyethylene,
     J.Rad.Phys.Chem. 14:847 (1979).

20. H. Barker, J.L. Garnett, R.Levot and M.A. Long, Use of
     Additives to enhance radiation grafting of monomers to poly
     (vinyl chloride) and application of these PVC copolymers to
     immobilization of enzymes and heterogenization of homogeneous
     metal complexes, J.Macromol.Sci-Chem. A12(2):261 (1978).

21. C.H.Ang, J.L. Garnett and R. Levot, unpublished work.

22. N.P. Davis, J.L. Garnett and S.V. Jankiewicz, Radiation
     grafting-modifications to cellulose by accelerated copoly-
     merisation of monomers, Proc. 1st PRI Conference Radiation
     Processing for Plastics and Rubbe, Brighton, England, Paper
     27.1 (1980, in press.

23. J.L. Garnett and J.D. Rock, Curable pre-polymer compositions
     method of making and method of coating articles therewith,
     U.S. Patent 4, 057,657 (1977).

24. G. Odian and A. Rabie, Monomer exponent in radiation graft polymerization, J.Rad.Phys.Chem. 9:495 (1977).
25. S. Machi, I. Kamel and J. Silverman, Effect of swelling on radiation-induced grafting of styrene to polyethylene, J.Polym.Sci. A-1, 8:3329 (1970).
26. C.H. Ang, Radiation induced copolymerization studies with polyolefins, Ph.D. thesis, The University of New South Wales, Sydney, Australia (1981).
27. J.L. Garnett, D.H. Phuoc, P.L. Airey and D.F. Sangster, Effect of dose rate on the radiation graft copolymerization of styrene to cellulose in the presence of acid, Aust. J. Chem. 29:1459 (1976).
28. J.L. Garnett, S.V. Jankiewicz and D.F. Sangster, Acid effects in radiation polymerisation and grafting reactions, J.Rad. Phys.Chem. in press.
29. J.L. Garnett, R. Levot and M.A. Long, UV and radiation grafting of p-styryl diphenyl phosphine to synthetic polymers and the use of the resulting copolymers in insolubilisation processes, J.Polym.Sci.Polym.Lett.Ed. 19:23 (1981).
30. J.L. Garnett and J.D. Rock, Radiation cured coating for leather, Aust.Patent Appl. No. 50690/79.

SYNTHESES AND PROPERTIES OF CHEMICALLY MODIFIED POLY(γ-METHYL-L-GLUTAMATE) MEMBRANES

Tsutomu Nakagawa and Takanori Shibata

Department of Chemistry
Faculty of Engineering, Meiji University
Higashimita, Tama-ku, Kawasaki-shi, Japan

ABSTRACT

Poly(γ-methyl-L-glutamate),PMLG, was reacted with ethylene chlorohydrin, followed with sodium N-methyl dithiocarbamate to obtain a modified poly(γ-methyl-L-glutamate) containing N-methyl dithiocarbamate group, PMLG-MD,of varying degree of substitution. The resulting polymers were irradiated with γ-rays from a Cobalt-60 source at room temperature and the evolved gaseous products were measured and analyzed with a mass spectrometer. The apparent G values for gas evolution for PMLG-MD membranes decreased remarkably. For example, a G value of 0.49 was shown for PMLG-MD which contains 13.8 mole-% dithiocarbamate group, compared with a G value of 1.62 for unmodified PMLG membrane. The ESR specrum of the irradiated PMLG-MD showed a strong anisotropy with high g values which differed significantly from the spectrum of the irradiated PMLG membrane.

The effect of N-methyl dithiocarbamate substitution on the transport of gaseous oxygen and oxygen dissolved in water was studied. The permeability of PMLG-MD showed the maximum point in permeability coefficients and the minimum point in the activation energies for permeation at about 10 mole-% of N-methyl dithiocarbamate group. These results suggest that the modified PMLG membrane is stable against the sterilization processing, and might be used as a biomaterial for artificial lung or skin.

INTRODUCTION

Poly(γ-methyl-L-glutamate) has been prepared in the industrial scale from L-glutamic acid which is obtained by the fermentation method from the molasses, one of the renewable resources. Also, poly (α-amino acid), including poly(γ-methyl-L-glutamate),PMLG, whose

structures resemble to those of peptide is called synthetic polypeptide. Although a number of papers concerning the structure of
poly(α-amino acid) such as a helix-coil trnsition have been presented,
none of the papers deal with the radiation stability of poly(α-amino
acid) membrane and very few papers deal with the transport properties
of the membranes. Poly(α-amino acid) membranes have already been
shown to be biologically compatible with blood and tissue. For
practical use such as synthetic biomaterials, they would be required
to be stable in the sterilization processing. The sterilization by
γ-irradiation is more profitable for poly(α-amino acid) membranes
which are less thermally stable. On the other hand, the transport
of oxygen through poly(α-amino acid) membranes is of special interest
because of the importance as a biomaterial for artificial lungs, skin
and corneas.

One of the authors found that N-methyl dithiocarbamate group
had the excellent protection action to poly(vinyl chloride) against
γ-irradiation from the view point of small amount of gaseous products
evolved.[1] Therefore, the authors thought that this group might have
the protection action against γ-irradiation to poly(α-amino acid) by
the same manner as that of poly(vinyl chloride).

The purpose of the present study is to synthesize the antiradiation poly(α-amino acid) by dithiocarbamate substitution, as well
as to study the effect of dithiocarbamate substitution on the
transport property of oxygen.

EXPERIMENTAL DETAIL

Materials

Poly(γ-methyl-L-glutamate),PMLG. As a poly(α-amino acid),
PMLG was chosen, because this polymer has been synthesized in the
industrial scale in Japan. PMLG was prepared by the polymerization
of N-carboxyamino acid anhydride. The viscosity-average molecular
weight is about 60,000 which was determined by the intrinsic
viscosity of dichloroacetic acid solution at 25°C.

Synthesis of PMLG containing chloroethyl group,PMLG-CE. The
PMLG containing a chloroethyl group was prepared from PMLG and ethylene
chlorohydrin with the aid of small amount of concentrated sulfuric
acid at 60°C for 24 hrs by the method described by Tanaka and
Okawara[2].

Synthesis of PMLG containing N-methyl dithiocarbamate group,
PMLG-MD. In a 200 ml three-necked flask equipped with a stirrer,
thermometer and condenser was placed a solution of 1.5g PMLG-CE in
60 ml dimethyl formamide(DMF), which was then raised up to 60°C under
nitrogen atmosphere. To the stirred solution was added 0.7g of
sodium N-methyl dithiocarbamate(NDC) in 20 ml DMF. The mixture was
stirred at 60°C for between 20 minutes and 3 hrs under nitrogen
atmosphere. Then the reaction mixture was poured into methanol-
water(2:1) mixture. A white polymer was precipitated. The polymer
was separated by filtration and again dispersed in water for one day.
After the polymer was separated by filtration, it was dissolved in
DMF and re-precipitated in a large amount of methanol-water mixture.

This precipitated polymer was filtered and washed several times with methanol-water mixture and dried under vacuum. A membrane of PMLG-MD thus obtained was prepared by casting a dilute solution of the polymer onto a glass plate, allowing the solvent to evaporate slowly, and drying under vacuum at 80°C for one day. The thickness of the membrane was 60 u.

Measurement of Evolved Gaseous Products Consequent to Irradiation. A 40-50 mg of the membrane sample was weighed precisely and placed in a about 5ml sample cell equipped with a small mercury manometer. The membrane was degassed for several days at 10^{-5} mmHg pressure until the amount of degassing was negligible and then irradiated in a sealed sample cell by γ-rays from a cobalt-60 source at a dose rate of 1.0 Mrad/hr at 23°C. The total dose ranged between 1 and 15 Mrad. After irradiation, the cell was immersed in a constant temperature bath at 25°C and the pressure increase was measured by cathetometric observation of the mercury in the manometer. At the final stage of the experiment, the volume of the cell was determined by a mercury displacement technique.

Measurements of Infrared Absorption Spectra and ESR Spectra. A Nippon Bunko A-3 Infrared Spectrometer was used. For measurement of ESR spectra, the membrane was put in a quarts tube and irradiated under vacuum with 10 Mrad at 23°C. The irradiated tube was stored in liquid nitrogen and ESR spectra were measured at -196°C with a Varian E-9 ESR Specrometer.

Permeability Measurement. The general theory of gas transport in polymers and detailed discussion of the method of measurement and calculation of the permeability, diffusion coefficients have been published elswhere[3] In this study, two kinds of permeability measurement were made. One experimental method was an adaptation of the high vacuum gas transmission technique described by Stannett and co-workers[4]. The other experimental method for permeation of oxygen dissolved in water was an adaptation of the oxygen electrode as follows.[5]

A cross-sectional view of the oxygen electrode is given in Figure 1. The electrode has an anode(A) of silver tubing(about 150 mm in length and 10 mm in diameter) and a platinum disk(about 5 mm in diameter) as cathode(F). The electrode is sheathed by an outer tube of stainless steel(B). An electrolyte solution(0.5N KCl aqueous solution) partially fills the annular space(C). The electrode was designed for easy change of membranes, to ensure favorable contact between the cathode and the membrane which is at the end of the outer tube, and to permit measurements on water-swallen membranes. Prior to each run, the cathode was polished to minimize contamination. The electrode was inserted vertically into distilled water. Distilled water was saturated with nitrogen gas to displace oxygen and other dissolved gases and then saturated with oxygen gas. The electrode was operated at -0.7v, and the reduction current was measured by use of a Hokuto Denko HM-101 anmeter. From the steady state current i_∞ of the permeation curve, the permeability coefficient P [cm^3(S.T.P.)-cm/cm^2-sec-cmHg] can be calculated by the following equation.

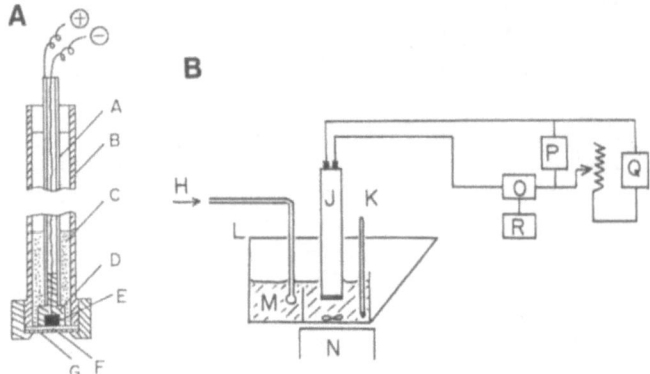

Figure 1.   (a) cross-sectional view of electrode (b) schematic dia-
gram of experiment: A, Ag anode; B, outer tube; C, 0.5N
KCl solution; D, insulator; E, Pt cathode; F, membrane;
G, support; H, N_2 or O_2 inlet; J, electrode; K, Thermometer; L, glass vessel; M, distilled water; N, magnetic
stirrer; O, microammeter; P, voltmeter; Q, dry cell;
R, recorder.

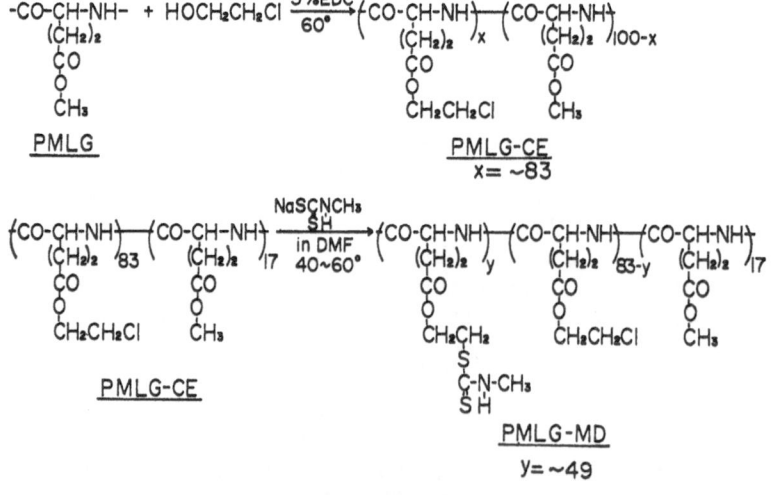

Chart 1.

$$P = i_\infty L / NFAP_s \qquad (1)$$

where L is the thickness of the membrane(cm), N is the number of electrons per molar unit of reaction (four in this work), F is Faraday's constant, A is the area of the cathode (0.190 cm^2 in this work), and P_s is the oxygen partial pressure in the distilled water (1 atm in this work). Prior to measurement, the membranes were immersed in distilled water for about one week. Sorption equilibrium with respect to water is assumed.

RESULTS AND DISCUSSION

Reaction of PMLG-CE with Sodium N-Methyl Dithiocarbamate

The reaction of chloroethyl group of PMLG-CE with sodium N-methyl dithiocarbamate was confirmed by comparison of the synthesis of poly(vinyl chloride) containing N-methyl dithiocarbamate group, covalently bonded to the polymer, which has been reported in detail elsewhere by one of us[6]. The general synthetic approach to the PMLG containing N-methyl dithiocarbamate group of interest to us is outlined in Chart 1, where X=83 and Y=7.5 to 49. Among the 100 monomer units of PMLG-MD, γ-methyl-L-glutamate component is 17 and

Table I. Synthesis of PMLG Containing N-Methyl Dithiocarbamate group

PMLG-CE (g)	NDC* (g)	DMF (ml)	Temp. (°C)	Reaction time(hr)	Sulfur content(%)	D.S.** (%)
(1) Effect of Reaction Time						
1.50	0.70	80	50	0.25	3.23	9.6
1.50	0.70	80	50	0.50	5.91	18.1
1.50	0.70	80	50	1.0	8.08	25.4
1.50	0.70	80	50	2.0	11.31	37.0
1.50	0.70	80	50	3.0	12.78	42.6
(2) Effect of Reaction Temperature						
1.50	0.70	80	40	3.0	9.84	31.6
1.50	0.70	80	50	3.0	12.78	42.6
1.50	0.70	80	60	3.0	14.42	49.1
(3) Effect of Mole-Ratio (NDC/Cl)						
1.50	0.60	80	50	3.0	10.94	35.6
1.50	0.70	80	50	3.0	12.78	42.6
1.50	0.82	80	50	3.0	14.42	49.1
(4) Effect of Polymer Concentration						
1.50	0.70	90	50	2.0	10.12	32.6
1.50	0.70	80	50	2.0	11.31	37.0
1.50	0.70	67	50	2.0	10.50	34.0

* NaSCSNHCH$_3$·2H$_2$O
**Degree of Substitution

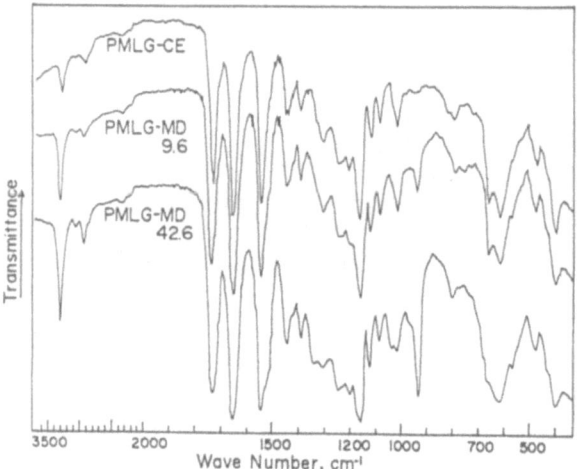

Figure 2.   Infrared Spectra of PMLG–CE and PMLG–MD.

the maximum degree of substitution of dithiocarbamate group is 49.
The degree of substitution of dithiocarbamate group, Y, was depended
on the reaction conditions and the unreacted chloroethyl–L–glutamate
component is 83–Y. The results are summarized in Table I.

Figure 2 shows the infrared spectra of PMLG–CE and PMLG–MD.
The charcteristic strong absorption bands at 3300 cm^{-1}, 1500 cm^{-1},
1340 cm^{-1} and 930 cm^{-1} attributable to νN–H, νC–N, νC=S and νC–S,
respectively, seen in PMLG–MD were not present in PMLG–CE. On the
contrary, an absorption band at 660 cm^{-1} decreased considerably.
This suggests that the chlorine of the chloroethyl group was substituted to dithiocarbamate group. In this reaction for 5 hrs at 60°C,
the polymer in the reaction mixture did not show the gel–like
substance. At this stage it is uncertain that the intramolecular
hetrocyclic intermediate may partially be formed by the neighboring
group participation of the N–methyl dithiocarbamate group.

Antiradiation Property of Modified PMLG

The Mass spectra of the evolved gas from the irradiated
PMLG–CE and PMLG–MD whose degree of substitution of MD is 36 mole–%
are shown in Figure 3. The mass spectra of the evolved gas from the
irradiated PMLG–CE shows each strong peak of $18(H_2O)$, $28(C_2H_4)$, 29
(C_2H_5), $50(CH_3Cl)$, $64(CH_3CH_2Cl)$. On the other hand, the mass spectra
of that from the irradiated PMLG–MD shows each peak of 32(sulfur),

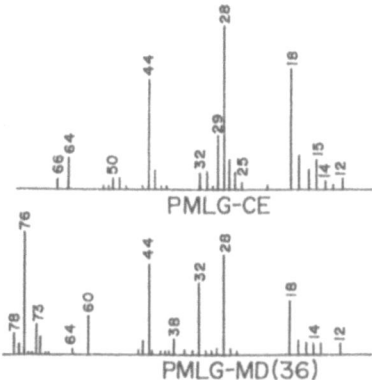

Fig. 3. Mass spectra of gaseous products evolved from
PMLG-CE and PMLG-MD with irradiation. Dosage:10 Mrad.

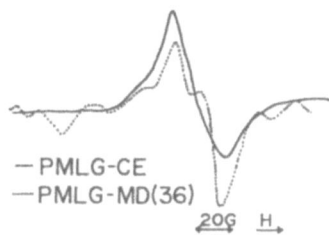

Fig. 4. ESR spectra of γ-irradiated PMLG-CE and PMLG-MD.
Dosage:10 Mrad.

$60(CH_2CH_2S)$, $76(CS2)$, except peak of 50. These data suggest that
the side chaines of PMLG-CE and PMLG-MD are decomposed by γ-irradiation.

This discussion is also supported by the ESR study. ESR
specra of the irradiated PMLG-CE and PMLG-MD are shown in Figure 4.
Both samples were kept at -196°C immediately after irradiating at
room temperature under vacuum, and measurements were made at -196°C.
The irradiated PMLG-CE shows a single spectrum of carbon radical.
The ESR signal of the irradiated PMLG-MD is different, the strong
anisotropy and the high g values of 2.075, 2.045 might indicate
a sulfur radical[1].

The amount of gaseous products evolved from PMLG-CE and PMLG-MD

with various amounts of dithiocarbaamte group,$-CH_2CH_2-SCSNHCH_3$ (MD)
is plotted versus total dose in Figure 5. In order to compare the
stability of PMLG-MD with that of another poly(α-amino acid) membrane
which contains sulfur, poly(methionine),prepared in our laboratory
was used. From the data in Figure 5, G(gas) of PMLG-CE was found
to be 1.65 and the gas evolution from the irradiated polymer was
linearly proportional to radiation dose. The gas evolution decreased
with increasing N-methyl dithiocarbamate content. The amount of the
gaseous products evolved from poly(methionine) whose sulfur content
is 30.8 wt-%, is comparable to that from PMLG-MD whose sulfur content
is only 2.8 wt-%. This means the N-methyl dithiocarbamate group has
the radiation protection against the decomposition of PMLG by γ-ray
irradiation. This is clearly demonstrated in Figure 6, in which the
G values for the gas evolution from PMLG-MD, G(gas) were plotted
against The MD content. The dotted line in Figure 6 shows the calculated G values in the case that individual component in PMLG-MD,
that is methyl glutamate component, chloroethyl glutamate component
and S-ethyl N-methyl dithiocarbamoyl glutamate component, are decomposed independently and the solid line shows the observed G values.

Therefore, the stabilization coefficient of dithiocarbamate
group to the neighboring side chains is considered. The stabili-

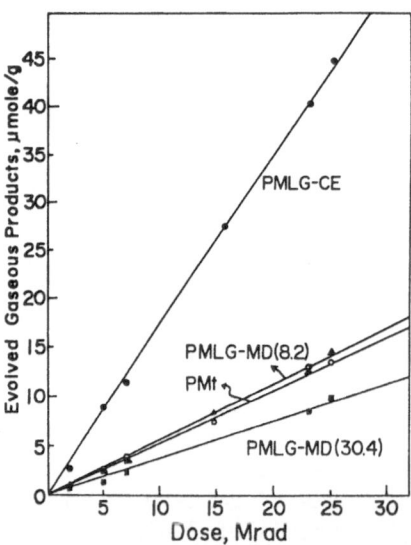

Fig. 5. Gas evolution from irradiated PMLG-CE, PMLG-MD

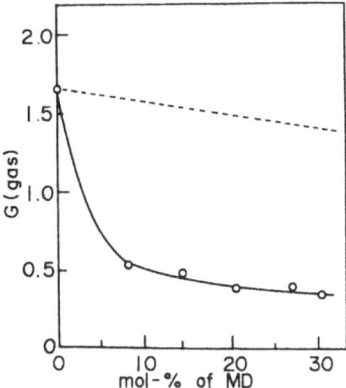

Fig. 6. Calculated G value in the case that individual residues are decomposed independently(-----), and the observed G values(—). Dosage:10 Mrad

Table II. Stability of Poly(α-amino acid) Membrane against γ-irradiation of 10 Mrad under vacuum

Polymer	MD*, mole%	q, u mole/g	G(gas)	G(gas)$_{p2}$	S.C.**
PMLG	0	16.8	(1.62)	(1.62)	–
PMLG-CE	0	17.2	(1.65)	(1.65)	–
PMLG-MD(8.2)	8.2	5.5	0.53	0.51	3.2
PMLG-MD(13.8)	13.8	5.1	0.49	0.44	3.7
PMLG-MD(20.3)	20.3	4.2	0.40	0.29	5.6
PMLG-MD(26.6)	26.6	4.3	0.41	0.26	6.3
PMLG-MD(30.4)	30.4	3.7	0.35	0.14	11.7

* N-methyl Dithiocarbamate Group
** Stabilization Coefficient

zation coefficient is as follows[1].

$$\text{Stabilization Coefficient} = G(gas)p1 / G(gas)_{p2} \qquad (2)$$

where $G(gas)_{p1}$ is the G value for the gas evolution from PMLG and PMLG-CE which was used for the preparation of PMLG-MD, and $G(gas)_{p2}$ is the G value for the gas evolution from PMLG plus chloroethyl components in PMLG-MD. It is considered that S-ethyl N-methyl dithiocarbamoyl L-glutamate itself is partially decomposed by γ-ray irradiation and evolved gaseous products. However, it is not possible to the amount of gaseous products evolved only from S-ethyl N-methyl dithiocarbamoyl glutamate component experimentally because

of impossibility of preparing PMLG-MD whose degree of substitution
is 100 %.   Therefore, the authors have estimated the amount of
gaseous products evolved from S-ethyl N-methyl dithiocarbamate group
by using the G value of 1.35 in poly(vinyl chloride) containing this
group.

$$G(gas)_{p2} = (q-W_{MD} \times 14.0 \times 10^{-8}) \times (9.6 \times 10^6/ W_{p2}) \qquad (3)$$

where q is the moles of evolved gas from PMLG-MD at 10 Mrad, W_{MD} is
a weight percent of $-CH_2CH_2-SCSNHCH_3$ in PMLG-MD and W_{p2} is a weight
percent of methyl L-glutamate plus chloroethyl L-glutamate component
in PMLG-MD.   The stabilization coefficients at 10 Mrad are summarized
in Table II.   Comparing the stabilization coefficient of MD in
PMLG-MD with that of MD in poly(vinyl chloride) containing N-methyl
dithiocarbamate group, the former is smaller than the latter, because
in the case of poly(vinyl chloride) stabilization coefficients are
more than 10.   The reason might be considered as follows.   The dithiocarbamate group in PMLG-MD protects the degradation of methyl
ester and/or chloroethyl ester groups by the external protection
action which is caused by the physical contact of side chains

Permeability of Modified PMLG Membranes to Gases
especially Oxygen

Typically determined values of permeability coefficient, P, and
diffusion coefficient, D, are shown in Table III.   Logarithms of
permeability coefficient of PMLG-MD(11.1) to various kinds of gases
are plotted against the inverse temperature in Figure 7.   These data
were obtained with a high vacuum technique, which means the membrane
was in perfectly dry atmosphere.   Permeability coefficients in the
humid atmosphere were measured by the oxygen electrode.   Logarithms
of the permeability coefficient of PMLG-MD membranes to oxygen dissolved in water are also plotted in Figure 8.   A comparison of the
permeabilities of oxygen dissolved in water with those of gaseous
oxygen obtained by the high vacuum technique shows the influence of

Table III.   Permeability and Diffusivity of
Poly(α-amino acid) Membranes

Polymer	$P \times 10^{10}$			$D \times 10^7$		
	20°C	30°C	40°C	20°C	30°C	40°C
PMLG (dry)	0.867	1.38	2.10	4.46	6.68	10.0
PMLG (wet)	2.28					
PMLG-CE(dry)	0.188	0.400	0.76	2.60	4.85	7.03
PMLG-CE(wet)	0.312	0.708	1.23			
PMLG-MD*(dry)	0.092	0.197	0.400	1.59	3.44	6.67
PMLG-MD*(wet)	0.433	0.800	1.35			

$P: cm^3(S.T.P.)-cm/cm^2-sec-cmHg$
$D: cm^2/sec$
* mole-% of MD:11.1

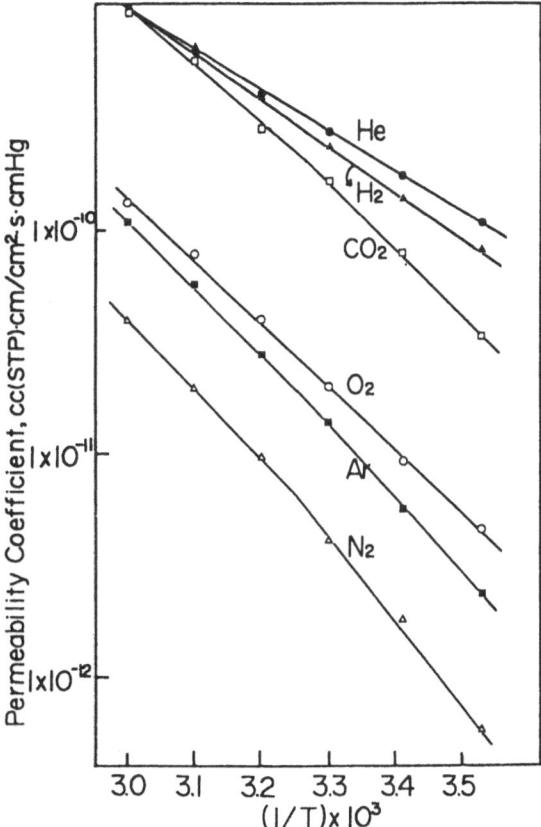

Fig. 7. Temperature dependence of permeability of PMLG-MD
to helium, hydrogen, carbon dioxide, oxygen, argon, and
nitrogen. MD content:11.1 mole-%.

water on oxygen permeation. The influence of water on the permeability of the PMLG-MD membrane is larger than that of the PMLG and
PMLG-CE membranes. One of the authors has suggested that the
sorption and the diffusion of small molecules in poly(α-amino acid)
membranes with α-helical structure take place in the side chain
regions between helices[7,8]. Accordingly, the hydration to N-methyl
dithiocarbamate groupmay be easier than that to chloroethyl group
methyl group, and the water in the membrane may play role of
plasticizer.

Figures 7,8 show that the arrhenius plots are linear for all
polymers in measurements, and the activation energies of permeation,
E_p, were calculated from the data of these Figures ,using the following equation.

$$P = P_0 \exp(-E_p/RT) \qquad (4)$$

From these data, the effect of mole-% of MD in PMLG-MD on the oxygen
permeability and the activation energy of the oxygen permeation is
shown in Figure 9. The interesting phenomena of the permeability

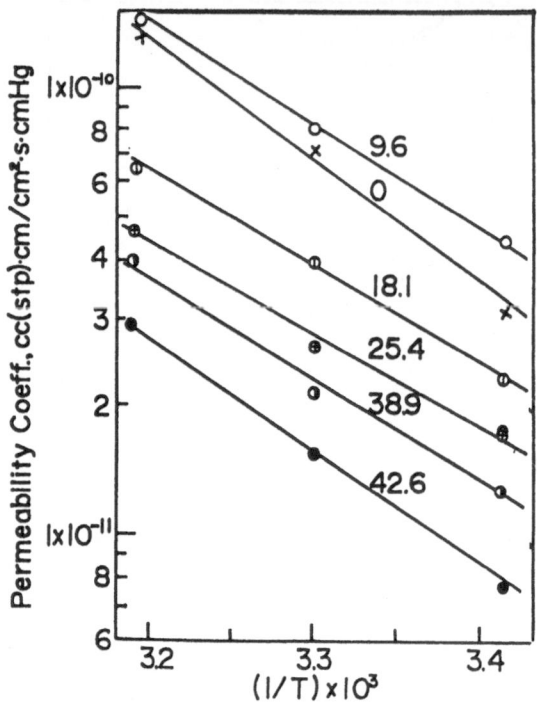

Fig. 8.   Temperature dependence of permeability of
          PMLG-MD to oxygen dissolved in water.

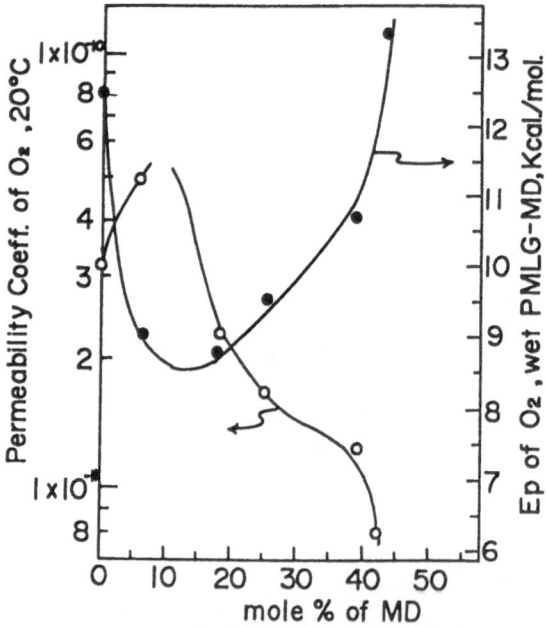

Fig. 9.   Effect of mole-% of MD in PMLG-MD on P and E_p

of PMLG-MD are indication of the maximum point in the permeability coefficient and another minimum point in the activation energies at about 10 mole-% of MD content. The relatively large group substituted in the PMLG-MD membrane extends the distance between helices and makes it easy for oxygen to diffuse, and with the progress of substitution of N-methyl dithiocarbamate group, compactness or packing of the space in the α-helical structure with this group decreases the diffusion coefficient and also increases the activation energy of diffusion, which increase, in other words, permeation.

We conclude that by 10 mole-% of the N-methyl dithiocarbamate substitution, the modified poly(γ-methyl-L-glutamate) membrane is better both in antiradiation property and the oxygen permeation.

REFERENCES

1. T. Nakagawa, H. B. Hopfenberg, and V. Stannett, J. Appl. Polym. Sci.,15, 747 (1971).
2. H. Tanaka, T. Endo, and M. Okawara, Kogyo Kagaku Zasshi, 79, 1779 (1973).
3. V. Stannett, in "Diffusion in Polymers" (J. Crank and G. S. Park, Eds.), Academic Press, London, 1968, Chpt. 2.
4. B. P. Tikhomirov, J. L. Williams, H. B. Hopfenberg, and V. Stannett, Makromol. Chem., 118, 177 (1968).
5. N. Minoura, Y. Fujiwara, and T. Nakagawa, J. Appl. Polym. Sci., 24, 965 (1979).
6. T. Nakagawa, Y. Taniguchi, and M. Okawara, Kogyo Kagaku Zasshi, 70, 2382 (1967).
7. N. Minoura, Y. Fujiwara, and T. Nakagawa, J. Appl. Polym. Sci., 22, 1593 (1978).
8. N. Minoura and T. Nakagawa, J. Appl. Polym. Sci., 23, 815 (1979).

MODIFIED FIBERS FROM REGENERATED COLLAGEN

James M. Brennan and Ferdinand Rodriguez

School of Chemical Engineering, Olin Hall

Cornell University, Ithaca, N.Y. 14853

INTRODUCTION

In a previous study, fibers were made from acid-soluble collagen coagulated in ammonium hydroxide solutions and cross-linked by glutaraldehyde[1]. The work is now extended to collagen modified by the addition of small amounts of methyl cellulose. Such fibers might be suitable as absorbable sutures if the physical properties can be enhanced to a level competitive with silk and catgut[2,3].

Collagen is a complex arrangement of protein in fibrous form found in bone, skin and connective tissues of animals. Twenty to thirty percent of the total protein weight in mammals is collagen. This has led to a great amount of research for collagen applications, particularly in the leather industry where large amounts of collagen from tissue other than the hide are wasted. Classically, collagen has been marketed commercially in gelatin, a denatured form of collagen, for use as edible sausage casing and more recently as amino acid concentrates, fertilizers and an animal feed constituent. Its use as a biomaterial is currently the topic of a large portion of collagen research.

Collagen itself is generally comprised of a tropocollagen chain with telopeptide end groups, both of which differ in primary and secondary structure. Tropocollagen is comprised of three rod-like polypeptide chains arranged in a 3000 Å long right-handed helix[4]. Electromicrographs have shown that this helical molecule is slightly flexible[5]. The primary structure of each strand consists of the repeating triplet glycine-X-Y, where X and Y are often proline and hydroxyproline, respectively[6]. It is possible for each of the three strands to differ in the exact amino acid sequence. Overall the molecule has a molecular weight of 300,000.

The end region is non-helical in structure and is the primary site of intermolecular cross-links. As the collagen ages, cross-links also form between tropocollagen molecules. It is because of these cross-linked regions that the greater part of collagen is insoluble[7]. Chemically the end region differs from tropocollagen since it is relatively rich in tyrosine and relatively lean in hydroxyproline. The telopeptide region has also been found to be the major immunologic site in native collagen[8].

As a biomaterial, collagen has an advantage over synthetic polymers since it is derived from living organisms. It is presently believed that only the telopeptide groups differentiate·collagen types from various tissues and organisms. Once these are removed, an organism can no longer distinguish implanted collagen from that internally produced[4]. Because of this, cell growth on the surface of collagen is uninhibited and actually aided. As a wound covering, collagen will not only protect but it will promote healing of living tissue. Collagen can also be absorbed by the organism, resulting in no permanent residue once recovery is complete. This is of particular importance if collagen prosthetic devices are to be used as temporary implants. These factors seem to make collagen an ideal biomedical material.

Biomedical studies have been focused on telopeptide-poor col-lagen because it is a poor antigen. Unfortunately, only insignif-icant amounts of this tropocollagen occur naturally. Therefore, it must be produced chemically. Some proteases, particularly proctase and pepsin, specifically attack and remove the telopeptide region, unlike collagenase which attacks proteins on the main helical structure. This material can now be solubilized in an acid solution. Concentrated solutions of this enzyme-treated material can be extruded in fiber form and subsequently cross-linked chemically, or with a form of irradiation. The primary use of such a fiber could be as a surgical suture or in a non-woven fabric for wound coverings. It would also be possible to produce hollow fibers for use in kidney research with similar technology.

To understand the direction of this research it is necessary to examine the current applications of sutures in surgery. The purpose of a suture is to bind and prevent the reopening of a wound during the period of healing. Ideally, once artifical support is no longer necessary, the sutures will be assimilated by the organism. Until the introduction of high strength poly-glycolic acid sutures, catgut, derived from bovine intestine, was the only truly absorbable suture. Unfortunately both catgut and polyglycolic acid have been shown to lose strength rapidly once implanted into the body[9]. The absorption of collagen on the other hand, can be controlled by the degree and type of cross-links introduced. This allows the absorption rate to be tailored to the healing rate of various tissues. Finally, while a suture does not have to promote healing, it must not cause an inflammatory reaction. Therefore, cross-linking of collagen must be limited to those agents which will not alter its biocompatibility.

Cross-linking by high energy radiation is ideal for medical applications. Unfortunately, ultra-violet light and gamma irradiation can cross-link collagen only in an oxygen free atmosphere[10,11,12]. The presence of oxygen promotes a chain scission reaction rather than the desired cross-linking. Industrially, an oxygen-free atmosphere would not only be difficult to attain but also very expensive. The cross-links are also of lower stability than those formed chemically[13,14].

Chemical cross-links can be introduced by reagents such as aldehydes, acrolein, and chromium salts[1,13,15]. Of these reagents, aldehydes have been shown to be superior in many respects. In particular, glutaraldehyde exhibits a better blood compatibility, introduces more cross-links, and yields a cross-linked collagen which is more resistant to bio-attack than other reagents[13,16]. Glutaraldehyde therefore appears to be a promising cross-linking agent for collagen sutures.

EXPERIMENTAL PROCEDURE

Acid-soluble collagen was prepared from dehaired calfhide by extraction with aqueous HCl and pepsin for four days followed by coagulation in concentrated NH_4OH[12]. The material was redissolved in 0.001 N HCl, filtered, neutralized, and defatted with acetone. All operations took place below 21°C.

Fibers were formed from a 5% solution (pH = 3) of collagen (or collagen plus methyl cellulose) extruded from a plastic syringe and stainless steel needle mounted in an Instron Testing Machine. The needle extended into a coagulating bath with 1 N NH_4OH where fibers remained for one minute. The fibers then were transferred to a cross-linking bath for a specified time and then air-dried on Teflon-coated trays. All films were conditioned for 24 hours in water before tensile testing. The standard conditions for extrusion are listed in Table 1. The tensile testing used fibers wrapped around brass spindles with a gauge length of 25 mm, a total fiber test length of about 200 mm, and a separation rate of 25 mm/minute.

Table 1. Standard Conditions

Extrusion
    Cross-arm speed   25 mm/minute
    Volumetric rate   4.4 cm^3/minute

Die
    Inside diameter   0.58 mm (0.023")
    Length   18 mm (0.71")

Coagulation   1 min. in 1 M NH$_4$OH

Cross-linking   0.2% glutaraldehyde
                             in 90/10 EtOH/H$_2$O

Methylcellulose   Methocel A, USP, 25 cps
                             (Dow Chemical Co.)

RESULTS AND DISCUSSION

Cross-linking Time

    The variation of strength with cross-linking time is shown in
Figures 2 and 3, while that for collagen is shown in Figure 1.
Schimpf found that for collagen cross-linked in one percent and
one-half percent glutaraldehyde baths, maximum UTS (ultimate tensile strength) occurred at minimum time[1]. However, the UTS of
collagen at a glutaraldehyde concentration of 0.2% did not start
at nor go through a maximum, but rather, leveled off at a maximum
UTS after three minutes of cross-linking. The cross-linking-chain
scission competing mechanism approach previously used cannot explain
this behavior satisfactorily. This mechanism assumes that cross-
linking slows down due to the exhaustion of sites for a particular
glutaraldehyde concentration and that a chain scission reaction,
independent of glutaraldehyde, takes place increasing the number
of chain ends, thus lowering UTS. The new results suggest collagen
is an elastic polymer that tends to become brittle when highly
cross-linked. Stress-strain curves show that the material undergoes strain hardening suggesting that a point is reached where
cross-linking is primarily limited to the surface of the fiber.
This greatly increases the local cross-link density but not the
cross-link density of the overall fiber. This theory allows the
low cross-link density core material to slip past other molecules
and orientate while allowing a brittle surface to exist. Further
grounds for this core-sheath concept is seen in the shrinkage tests
which show shrinkage temperature leveling off between 69 and 73°C
with increasing cross-link time. This is an indication that cross-
linking has stopped. Yet, shrinkage temperatures reported by
Cater[13] are in excess of 80°C for glutaraldehyde cross-linked

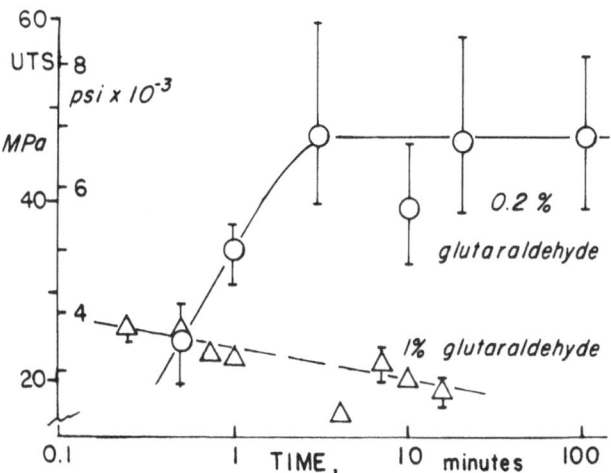

Figure 1.   Effect of cross-linking time on ultimate
tensile strength (UTS) for collagen alone.

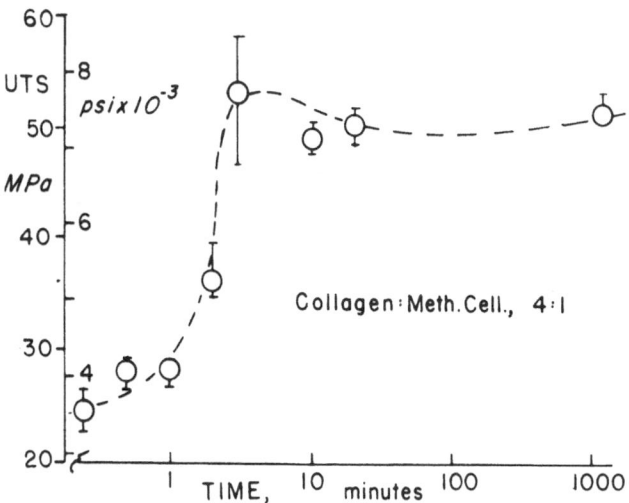

Figure 2.   Effect of cross-linking time on ultimate
tensile strength (UTS). Data for 0.2%
glutaraldehyde solution.

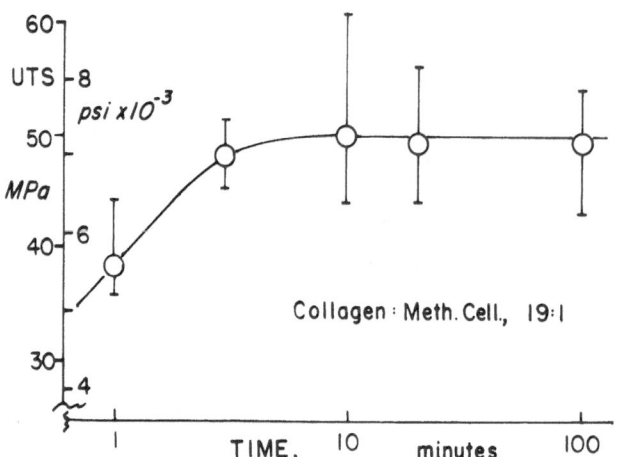

Figure 3. Effect of cross-linking time on ultimate
            tensile strength (UTS). Data for 0.2%
            glutaraldehyde solution.

fibers, evidence that possibly all available cross-link sites
throughout the entire fiber have not been exhausted in our
procedure.

     At 0.2% glutaraldehyde concentration there is a relatively
low probability of cross-linking and hence a relatively low number
of active cross-link sites. Therefore, a plateau of UTS is reached
rather than a peak since the active sites in the surface region
are used up and cross-linking ceases. Cross-linking has therefore
been stopped before the surface imperfections have become amplified
as stress concentrations. This allows fibers to be cross-linked
for the "optimum" time which heretofore was impossible due to the
manual nature of the fiber production procedure.

     Methylcellulose, a moderately substituted water soluble
cellulose ether, has been reported to have been cross-linked by
dialdehydes such as glutaraldehyde[17,18]. This result was checked
experimentally by cross-linking an air dried methylcellulose film
in 0.2% glutaraldehyde for 24 hours. The cross-linked film no
longer dissolved in water, but rather, swelled and split into
smaller fragments which remained visibly stable upon further
standing in water. This suggests that methylcellulose will undergo a degree of cross-linking, but the cross-links will be of little
aid to tensile strength.

Rerunning the cross-linking time-tensile strength experiment, now with the addition of methylcellulose to the collagen solution, UTS peaks appear for a 0.2% glutaraldehyde cross-linking bath. Figures 2 and 3 show that these peaks, although small, exist for both methylcellulose concentrations. This suggests that methylcellulose allows more cross-linking in the surface region before sites are depleted, enough so that some brittleness is incurred and UTS is decreased. Twenty percent methylcellulose remains weaker than collagen at all cross-link times, although it approaches the collagen value at times from three to ten minutes. If, as suggested earlier, the methylcellulose area is uncross-linked and as such can be deducted from the UTS calculation, the strength becomes greater than collagen alone. This means methylcellulose has some positive effect on the fiber. Five percent methylcellulose-collagen fibers have tensile strength greater than collagen at all cross-link times.

The performance of the twenty and five percent methylcellulose-collagen fibers is a combination of cross-linking and extrusion effects. The methylcellulose molecule is smaller than the collagen molecule and is able to fit in a number of voids in the overall fiber structure. After a certain quantity of methylcellulose, the voids are full, causing additional amounts to expand the distance between collagen molecules. This additional distance lowers the probability that collagen molecules are near enough to cross-link, or in effect, it decreases the number of cross-links.

Methylcellulose Concentration

The results for the methylcellulose-collagen fibers in the previous section suggest that a maximum should be reached in the UTS vs. methylcellulose concentration curve. Figure 4 shows that this maximum is at five percent methylcellulose for all glutaraldehyde cross-linking bath concentrations. Examination of stress-strain curves reveals that the modulus also peaks at five percent methylcellulose.

These peaks are a result of two separate mechanisms. At first a polymer filler interaction, in which methylcellulose fills pores in the polymer matrix, increases the tensile strength. Once the pores methylcellulose will fit into are filled, it begins to act as a diluent. At this point collagen molecules are pushed further apart, causing some cross-link sites to be separated by a distance too great for cross-linking to take place. This results in fewer cross-links and subsequently, a lowering of UTS.

Cross-link Bath Concentration

For a cross-linking time of three minutes the optimum glutaraldehyde concentration is 0.5% for both collagen and five percent

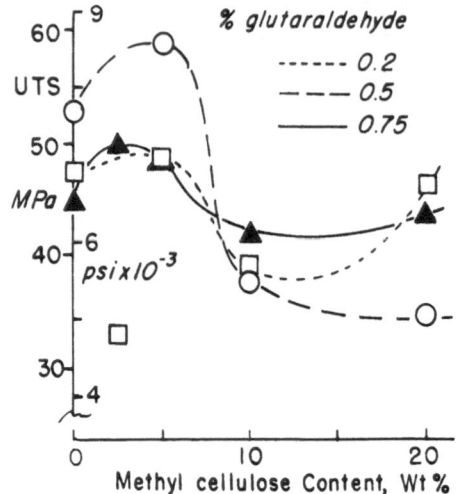

Figure 4. Effect of synthetic polymer (methyl
cellulose) addition on strength of
collagen fibers. Cross-linking time
is 3 minutes.

methylcellulose-collagen fibers (Figure 5). The primary effect of
glutaraldehyde concentration is on the number of cross-linking
sites and hence on the probability of cross-linking. As glutaraldehyde concentration increases the number of cross-link sites per
gram of collagen increases[13]. This affects the proposed model by
increasing the number of available cross-link sites in the surface
region and hence the degree of brittleness of the cross-linked
fiber. It should also be noted that the degree of strain hardening
observed falls with increasing glutaraldehyde concentration. This
means the high cross-link density region goes deeper into the
fiber. The optimum glutaraldehyde concentration is reached when
amplification of imperfections due to brittleness offsets any
increase in strength gained by more cross-links.

Extrusion

Perhaps the most dramatic effect the methylcellulose has on
the collagen fiber is in that it permits fibers to be extruded at
more favorable processing conditions, greatly increasing tensile
strength. The laboratory preparation can now be carried out at
die diameters below the previous limit and also at cross-arm

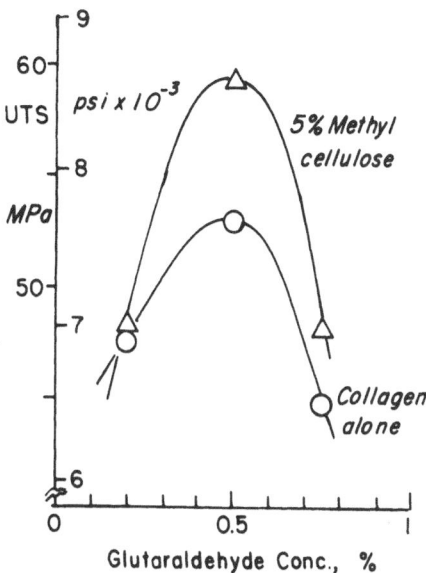

Figure 5. Cross-plot of Figure 4 indicating optimum
          strength for collagen fibers containing
          5% methyl cellulose and cross-linked 3
          minutes in 0.5% glutaraldehyde.

speeds greater than previously practical[1]. In short, it enables
an increase in orientation along the axis of extrusion to be realized. This is particularly important in the current laboratory
procedure in which no means has been devised to draw the fiber
during extrusion and coagulation, an operation classically used to
orient materials[19,20].

    Figure 6 shows how the tensile strength of 1:4 methylcellulose-
collagen fibers varies with cross-arm speed for cross-link times
of one and ten minutes. As expected, the tensile strength rises
with increasing cross-arm speed. The upper limit in cross-arm
speed of twenty inches per minute is due to the inability to handle
a higher jet velocity in the coagulation apparatus rather than to
any fiber weakening flow instability. Interestingly, it is at the
upper limit where fibers cross-linked for one minute approach the
tensile strength of those cross-linked for ten minutes. This is
due to the high degree of orientation in the precross-linked fiber.
In a highly oriented fiber cross-links are more effective and therefore fewer are needed to increase tensile strength. Also, as can

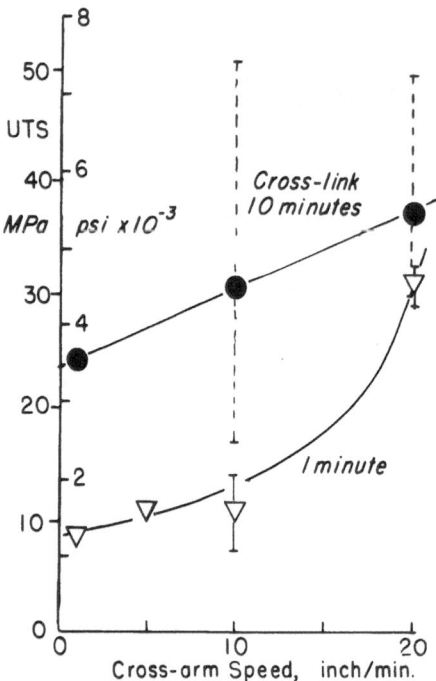

Figure 6. Strength of collagen fibers extruded through
          a No. 20 needle at various rates. Ratio of
          methyl cellulose to collagen is 1:4.

be seen from the relative size of the range bars, the lower cross-
link time minimizes fiber imperfections. This result presents an
extremely important alternative method of increasing UTS in that
bio-absorbability is a function of the number of cross-links[4,14].
It would now be possible to control tensile strength and the length
of time a fiber will remain intact in an organism.

     It was then desired to see how five percent methylcellulose
fibers behaved when extruded at different cross-arm speeds and
through different die sizes. Figure 7 and Table 2 list the
results for various extrusion conditions.

     By comparison, commercial spinneret diameters for solution
wet spinning vary between 0.002 in. to 0.006 in., or slightly
smaller than those presently used in the laboratory procedure[19].
Again, the practical limit in die size was dictated by backups in

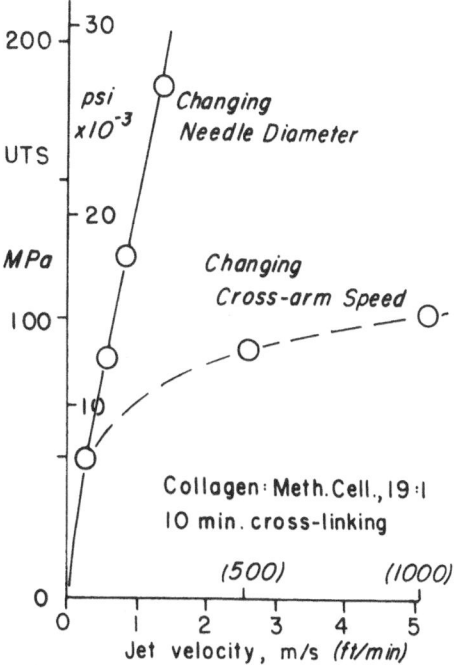

Figure 7. Strength of collagen fibers extruded at
           constant diameter (changing cross-arm speed)
           or at constant volumetric flow rate with
           decreasing diameters. Ratio of methyl
           cellulose to collagen is 1:19. Cross-linking
           time is 10 minutes.

Table 2. Extrusion Rate Tests

Needle No.	Inside diameter		Cross-arm speed	Jet velocity		Shear rate at wall
	mm.	inch	inch/min	ft/min	m/s	sec^{-1}x10^{-3}
20	0.58	0.023	1	55	0.28	3.9
20	0.58	0.023	10	550	2.8	39
20	0.58	0.023	20	1100	5.6	78
22	0.41	0.016	1	115	0.58	11
23	0.33	0.013	1	175	0.89	21
26	0.25	0.010	1	295	1.5	47

the coagulation bath rather than extrusion problems. Realistically, the lower limit on die size will be set by flow instabilities which cause catastrophic defects in a fiber[19,20]. Experimental jet velocities do extend into the range used industrially, that is, 300 to 450 ft/min[20].

In Figure 7 it is seen that changing die diameter is more effective than extruding at higher rates at constant die diameter, even at the same rate of shear. This result is to be expected since at constant extrusion rate the smaller diameter die yields a greater degree of orientation in the fiber. Decreasing fiber diameter also increases the surface area to volume ratio which in turn results in higher overall cross-link density. Qualitatively, this is seen in the stress-strain relationship by a change from upward to downward concavity. A final observation about the difference between the series of fibers extruded at changing cross-arm speed and the alternative series deals with the regularity of the fiber cross-sectional area. Fibers made at high cross-arm speeds were visibly irregular but those made from smaller diameter dies were seen to be reasonably uniform when examined under a microscope.

CONCLUSIONS

The combination of methylcellulose with collagen in the ratio of 1:19, extruded through the smallest die gave a tensile strength of 186 MPa (27,000 psi or 1.6 g/denier). This is getting into the range for chromated collagen (1.7 g/denier) and catgut (1.8 g/denier)[1]. However, it still is far short of the values reported for silk (2.5 g/denier) and poly(glycolic acid)(4.8 g/denier). The methylcellulose appears to permit high extrusion rates which could lead to even stronger fibers in commercial-scale equipment.

REFERENCES

1. W. C. Schimpf and F. Rodriguez, I&EC Pdt. Res. Dev., 16, 90 (1977).
2. A. L. Rubin and K. H. Stenzel, MIT Technol. Rev., 71 (2), 3 (Dec., 1968).
3. M. Chvapil, R. L. Kronenthal, and W. van Winkle, Jr., Int. Rev. Connect. Tissue Res., 6, 1 (1973).
4. A. L. Rubin and K. H. Stenzel, in "Biomaterials," Plenum, New York, 1969.
5. A. G. Walton and J. Blackwell, "Biopolymers," Academic Press, New York, 1973.
6. E. A. Balaz (Ed.), "Chemistry and Molecular Biology of the Intracellular Matrix, Vol. 1," Academic Press, New York, 1971.
7. A. L. Rubin and K. H. Stenzel, Biochemistry, 4, 181 (1965).
8. F. O. Schmitt, et. al., Proc. Nat'l. Acad. Sci. U.S., 51, 493 (1964).
9. J. B. Hermann, Archives of Surgery, 106, 710 (1973).
10. F. A. Bovey, "The Effects of Ionizing Radiation on Natural and Synthetic High Polymers," Interscience, New York, 1958.
11. A. Chapiro, "Radiation Chemistry of Polymer Systems," Interscience, New York, 1962.
12. G. Hamed and F. Rodriguez, J. Appl. Polym. Sci., 19, 3299 (1975).
13. C. W. Cater, J. Soc. Leather Trades' Chemists, 47, 259 (1963).
14. T. Miyata, T. Sohde, A. L. Rubin, and K. H. Stenzel, Biochem. Biophys. Acta, 229, 672 (1971).
15. F. M. Richards and J. R. Knowles, J. Mol. Biol., 37, 321 (1968).
16. A. Veis in "Treatise on Collagen, Vol. 1," (G. N. Ramachandran, Ed.), Academic Press, New York, 1967, chpt. 8.
17. A. B. Savage in "Encycl. Polym. Sci. Tech., Vol. 3," Wiley-Interscience, New York, 1967, pp. 472, 492.
18. L. Rebenfeld in "Encycl. Polym. Sci. Tech., Vol. 6," Wiley-Interscience, New York, 1967, p. 505.
19. S. M. Atlas, H. F. Mark, and E. Cernia (Eds.), "Man-Made Fibers, Vol. 1," Wiley, New York, 1967, pp. 95-130.
20. J. M. Preston, "Fiber Science," Textile Institute, Manchester, England, 1949, pp. 51-68.

RELATION BETWEEN THE STRUCTURE OF WOOL GRAFT COPOLYMERS AND THEIR

DYNAMICAL MECHANICAL PROPERTIES

Kozo Arai

Department of Chemistry, Faculty of Technology
Gunma University
Kiryu, Gunma 376, Japan

INTRODUCTION

Wool fibers contain two types of cells, viz. cuticle cells and cortical cells. The cuticle cells consist of external epicuticle, exocuticle, and endocuticle. The cortical cells are divided into two different types of cells termed as orthocortical and paracortical cells which occupy about 90% of the wool fibers. They are separated from one another by a cell membrane complex with three layer structure. The cortex structure is constituted from the crystalline microfibril of the α-helical aggregate embedded in a matrix of high sulfur content. Wool fiber is thus a composite material with a variety of function on mechanical, chemical, and physical properties.

With research on elucidating the relation between the structure and properties of wool keratin, many promising results have been derived by the X-ray diffraction[1-4] and the electron microscopy studies[2,5] of wool graft copolymers. In addition to these physical methods, another attempt has been made to evaluate the chemical structure of keratin on the determination of the end-group incorporated into the isolated polymer[6-9] and the amino acid composition of true graft copolymers[10,11]. The important suggestion has been made that (a) whole polymers formed in the fiber are truly grafted and linked by covalent bonds with protein chains via specific amino acid residues[7-9], (b) the grafting sites occur selectively on the low-sulfur protein chains[10,11] and (c) a maximum number of the grafting sites produced is approximately two moles per low-sulfur protein molecule of the number-average molecular weight of 48,000[11]. As far as grafting is concerned, the wool fiber is <u>not chemically</u> denatured, but undergoes <u>physical modification</u> to give a system possessing specific interactions between the protein and the polymer chains.

The structural studies using X-ray and electronmicrography seem to be insufficient to describe the state of aggregation of molecules in the crystalline microfibril and especially, in the amorphous matrix regions. It has been elucidated that the α-crystallites are disrupted in various degrees by the deposition of polymer and the cystine-rich matrix regions do not allow physico-chemical modifiction by grafting[1-5]. It is believed, therefore, that dispersion behavior of wool fibers modified specifically by grafting gives some information for the actual state of molecular aggregation in the complex structure of wool.

LOCATION OF GRAFTED POLYMER IN THE WOOL STRUCTURE

Mercer[12] first investigated the location of polymer in the wool graft copolymer prepared by an aqueous initiating system and proposed that, in the case of water soluble methacrylic acid, the polymer was not located in fibrillar regions, but mainly in the intercellular regions. Andrews et al.[13,14] postulated that, under limited magnification conditions, the polyacrylonitrile was located in the keratinous regions which account for approximately 90% of the fiber. Ingram et al.[15] studied the location of polystyrene in radiation grafted wool fibers by using high resolution electron micrographs of the cross-sections stained with platinum-carbon. They postulated that, with the aid of low-angle X-ray diffraction studies, most of the grafted polymer was located in the matrix regions between the microfibrils.

Present author[5] developed a staining method to differentiate the grafted polymer from the wool component in a thin section of the grafted wool fiber and established the optimum staining condition for the observation of poly(methyl methacrylate) in the complex morphological structure of wool: after sectioning the fixed wool fiber with 2 to 3% potassium permanganate in a phosphate buffer solution at pH 7.2 for 3 h at room temperature, the thin sections were followed by the post staining with either a saturated solution (7-8%) of uranyl acetate for 10 min just prior to observation or with a lead citrate solution after the treatment with uranyl acetate. A high resolution electron micrograph could not be obtained by these staining conditions for the ungrafted wool fiber which resulted in the even distribution of heavy metals over the two different structural components of the microfibril and the matrix.

Grafted Polymers in Merino Wool Fiber

As shown in Fig. 1, a marked difference in texture among the two cortex and the cuticle cells can be clearly differentiated.

Fine-granular orthocortical macrofibrils and coarser paracortical cells are distinguishable. Exocuticle, endocuticle and cell membrane in the cuticle keratin becomes also distinguishable. Cytoplasmic nuclear remnant regions in the paracortex are densely stained.

As far as no diffusion-controlled system of methyl methacrylate (MMA) is used to synthesize the wool graft copolymer, deposition of polymer occurs preferentially in the orthocortex regions rather than in the para-type cortex regions[2,5]. A marked difference in the

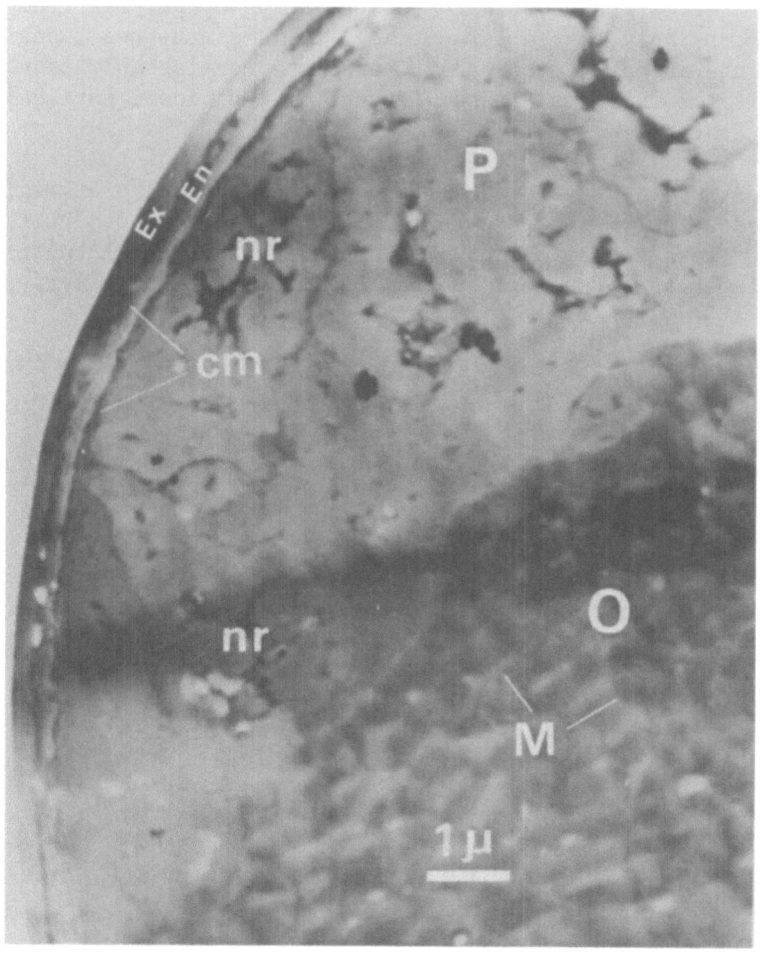

Fig. 1.   Electron micrograph of a thin cross-section of a fine Merino wool fiber, showing the cuticle, macrofibril (M) in the orthocortex (O) and densely stained nuclear remnant (nr) regions in the paracortex (P). Exocuticle (Ex), endocuticle (En) and cell membrane complex (cm) are clearly differentiated.

degree of grafting between the two cortex regions has been observed.
This also seems to be due to the difference in the cross-link density
between the two types of cortex. Fig. 2 shows the appearance of a
boundary region of the ortho- and para-cortical cells of the grafted
fiber. On the paracortical cells, variation in electron density
occurs from cell to cell. It is apparent that a higher amount of
polymer is deposited in the lightly stained cells than in the densely
stained ones, since only wool materials may be associated with the
heavy metals in the copolymer system. Diffusion of polymers into
cells mirrors the overall interaction forces in wool aggregates which
contain considerable amounts of chemical or physical cross-links.
Such a discrete difference in the distribution of polymer from cell
to cell occurs only to the paracortical cells of the two types of
cortex. This probably means that there is a considerable change in
the rigidity of the cross-linked texture of individual cortical cell
in the paracortex.

Densely stained cytoplasmic remnant and intercellular regions
are changed by polymer uptake to more light in electron density than
that of the keratinous region. The appearances in the localization
of the polymer in these regions indicate that there is a large

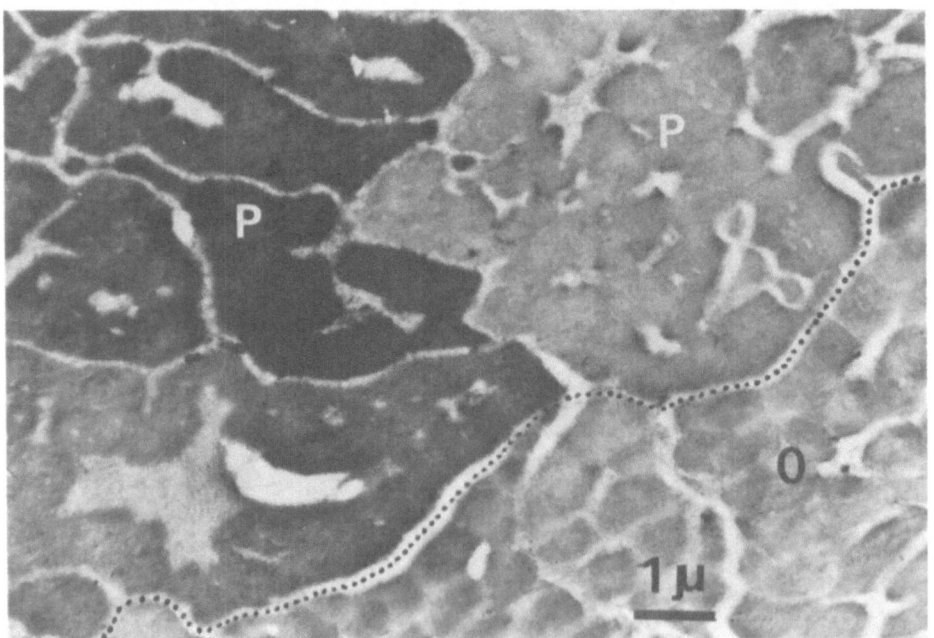

Fig. 2.   Electron micrograph of a thin cross-section of the 114%
          MMA-grafted fiber (Merino wool), showing the appearance of
          the boundary region of the orthocortex (O) and paracortex
          (P). It can be seen that the polymer occurs homogeneously
          in various degrees in the para-type cells.

enough space for growth of polymer without bringing about any distortion in geometrical sense on the cortical cells. Present author reported that, by the grafting with 250% of ethyl acrylate, the size of the whorly macrofibrils in the orthocortex was increased about 3 times in lateral dimension, but geometric form was substantially retained[2]. It should be emphasized that, with increase of polymer uptake, no perceptible change in longitudinal dimension of wool occurs, but only diameter change proceeds.

The cross-sectional view of the paracortical cells near the cuticlar region is shown in Fig. 3. Assuming that the variation of the density on the photograph reflects the amount of the deposited

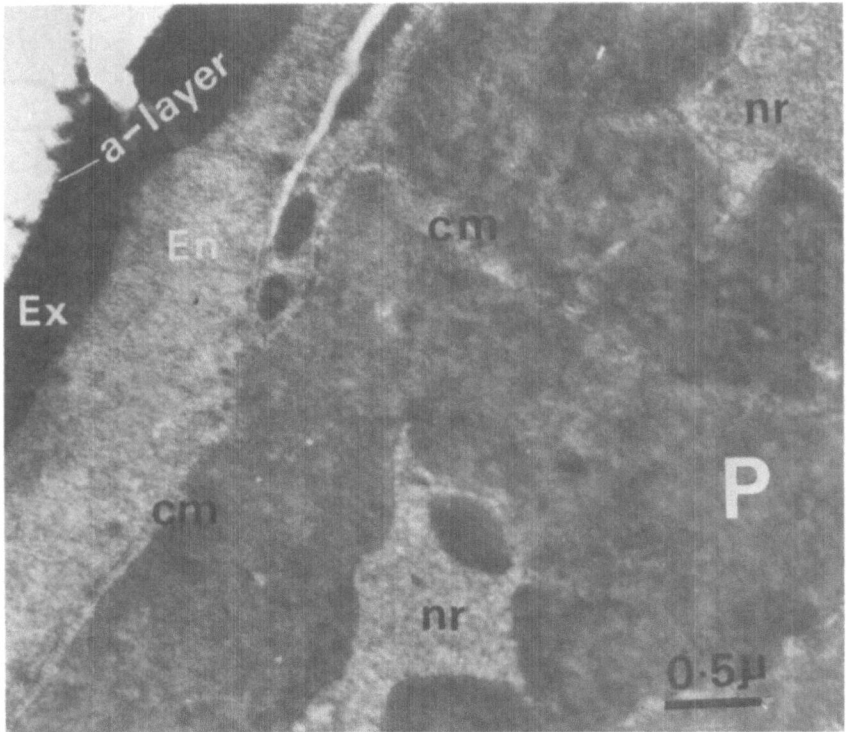

Fig. 3. Electron micrograph of a thin cross-section of the 54.0% MMA-grafted fiber (Merino wool), showing a polymer deposition pattern. It can be seen that a large amount of polymer occurs in the endocuticle (En), cytoplasmic nuclear remnant (nr), and cell membrane complex (cm) between the adjacent paracortical cells, and between the cuticle and cortex. The densely stained a-layer and the exocuticle (Ex) occurring relatively small amounts of polymer are also recognized.

polymer, visual estimate on the rigidity of many histologically
different components in the wool may be possible. The ease of the
deposition of the polymer is decreased in the order endocuticle ≃
cytoplasmic remnant ≃ intercellular membranes in cortex > cortex >
exocuticle > a-layer, which corresponds well to the inverse order
in sulfur content[16-19]. It is interesting to note that the endocuticle is presumably the same as that of cytoplasmic components
such as cytoplasmic nuclear remnant. The appearances of the two
copolymer materials are very similar to each other in the histological level. Exocuticle contains a higher amount of cystine
as comparison with the endocuticle which is more susceptible to
attack by enzymes[19,20]. The polymer penetrates into the exocuticle.
External layer of the exocuticle, termed a-layer has much high
sulfur content. However, it is not certain whether the polymer can
be located in this region. The cell membrane complex around the
cortex adjoining the endocuticle involves a clearly delineated
densely stained layer which resists to diffusion and polymerization
of monomers. This type of layer which seems to be a high sulfur
material is not present in the cell membrane complex between the

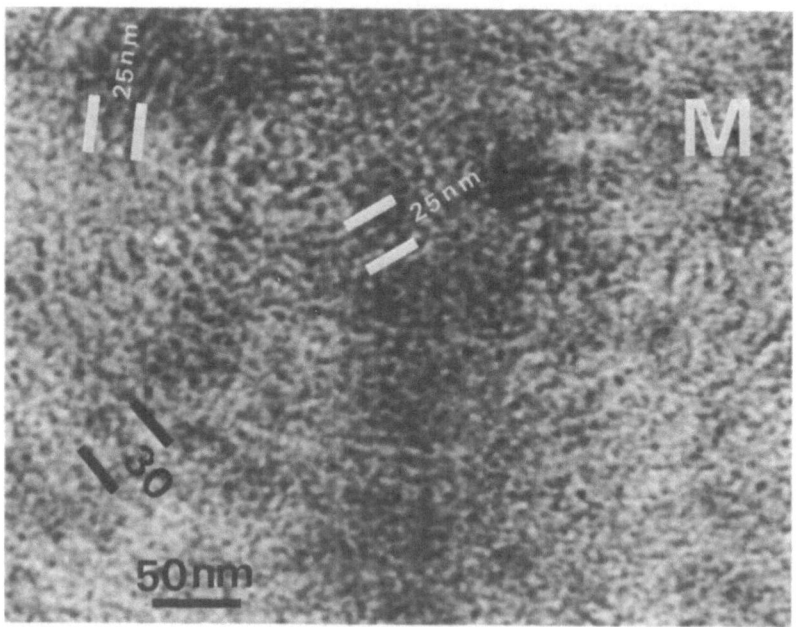

Fig. 4. A high resolution electron micrograph of a thin cross-
         section of the 114% MMA-grafted fiber (Merino wool),
         showing the appearance of the concentric layer structure
         of a orthocortical macrofibril (Mac). Polymer can be
         clearly differentiated from densely stained wool material.

endocuticle and the exocuticle. The ordered structure of the cell
membrane complex between cortical cells tends to be easily deformed
by the polymer uptake. This suggests that the components consist
of low-sulfur proteins. The polymer penetrates somewhat randomly
into the adjoining cortex cells. It is inferred, therefore, that
the plasma membrane adheres the contiguous cortex cells, but has
not resistance more than the cortex to the external forces such as
a longitudinal extension. Studies on the relation between the
structure and properties of wool in terms of fiber components is
lacking. In this respect, electron microscopy studies of the
structure of the graft copolymer promises considerable advances in
the understanding the function of many other fiber components.
The plasma membrane located between the endocuticle and exocuticle
tends to crack along the exocuticlar cell boundary owing to the
shear stress during the thin sectioning (see Fig. 3). This clearly
suggests a weak adhesion arising between the two components.

Fig. 4 shows a high magnification view of the cross-section of
a macrofibril expanded by the deposition of polymer. A maximum
width of the macrofibril is increased by the polymer uptake from

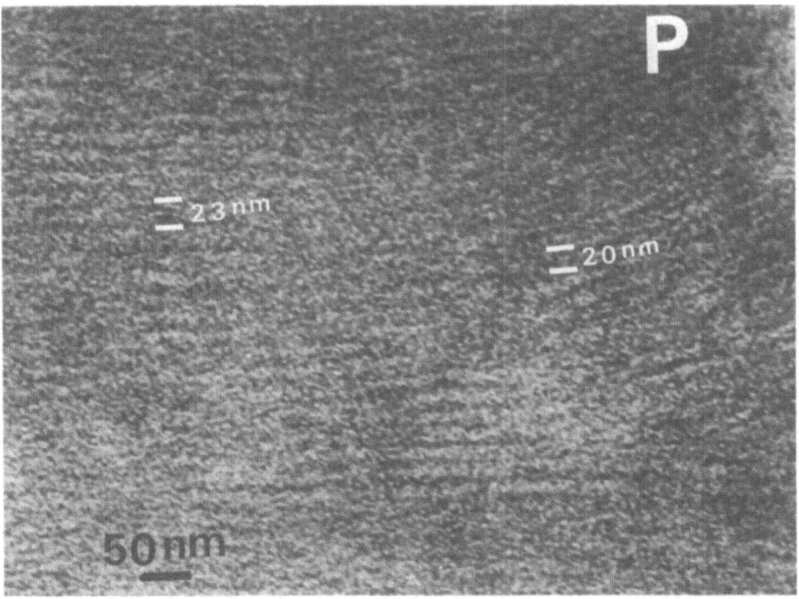

Fig. 5. A high resolution electron micrograph of a thin cross-
         section of the 114% MMA-grafted fiber (Merino wool),
         showing the appearance of lamella structure of the para-
         cortex (P). The layer structures are delineated by the
         electron light polymers.

around 0.3 μm to around 0.6, 0.8, and 0.9 μm by the grafting with
54.0%, 114% of methyl methacrylate, and 250% of ethyl acrylate,
respectively. Concentric layer lines can be clearly seen. The
deposited polymer differentiated from the electron-dense wool
material in whorl pattern can be recognized. The spacing of the
layer structure is around 25 to 30 nm, which corresponds to 3 to 4
times the spacing observed as around 8 to 10 nm for the ungrafted
wool stained by osmium tetraoxide[21]. It should be emphasized that
the ratio of the increment of the spacing by grafting is nearly
equal to that of the dimension of the macrofibril. This indicates
that a homogeneous distribution of polymer is resulted not only in
the aggregates of histologically different components, but also in
fine structural components such as microfibrils in the concentric
layer structure of macrofibril.

    A high resolution electron micrograph of the cross-section of
the paracortex region is also shown in Fig. 5. Regular striation
of lamella structure can be seen with the delineation of electron
light polymer. The spacing between the densely stained keratinous
layers or lightly stained polymer layers is around 20 to 23 nm,
which is considerably less than the dimension observed for the
orthocortical macrofibril. This is evidence that the disulfide
cross-link density in the paracortex is higher than the orthocortex.
Although the orthocortical macrofibril of concentric layer
structure expands in an isotropic manner through the grafting, it
is here uncertain of the paracortical cell how difference exists
between the degrees of swelling in the direction perpendicular and
parallel to the layer lines. However, somewhat greater swellings
would be expected in the direction perpendicular to the layer lines,
since the existence of the densely stained material continuing for
a considerable distance, suggesting the localization of polymers in
the lamella structure. This further provides important evidence
for the distribution of the disulfide cross-links at the ultrastructural level in the cortex, in that much higher amount of the
disulfide occurs in a direction parallel to the keratinous layer
which consists of sheet-like aggregates of globular components of
the high-sulfur matrix.

    Figs. 6 and 7 show the appearances of the cross-sections in
the paracortex region of partially reduced fiber grafted with
methyl methacrylate. The layer-like structure as seen in Fig. 5
is absent. A finely dispersed polymer is located around the small
subdivisions with diameter around 2 nm. There is X-ray evidence
that the intensity at the 0.98 nm equatorial reflection is considerably decreased by grafting onto reduced fibers, but the intensity
at the 0.51 nm meridional reflection is almost retained without
being affected by the deposition of polymer. This appears to be
due to a dissociation of the α-helical components constituting
microfibril, by which the initial conformation of the protein chain

has remainded almost intact[3].

Grafted Polymers in Lincoln Wool Fiber

A longitudinal section of the grafted Lincoln wool fiber with 73.0% of MMA is shown in Fig. 8. It can be clearly seen that the polymer also deposits in the aggregates of the nuclear remnant and the cell membrane complex. At the histological level it is unlikely that the grafted polymer gives any additional non-uniformity along the direction of fiber axis. It is further noted that much polymer deposits in nuclear remnant regions without any distortion on the geometric form.

Fig. 9 shows a high magnification view of the longitudinal section of cuticlar cells. The cell membrane complex in the flat overlapping cuticle cells consists of three layers, i.e. the β-layer adjoining endocuticle, the middle layer termed as δ-layer, and the β-layer adjacent to the surface of the inner exocuticle.

Fig. 6. Electron micrograph of a thin cross-section of the 93.8% MMA-grafted fiber (Merino wool) after treatment with 0.277 N thioglycolate, showing the appearance of paracortex region. It can be seen that the lamella structure disappears in homogeneously dispersed polymer particles.

Fig. 7. A high resolution electron micrograph similar to that in Fig. 6. The polymer is located around the small subdivisions with diameter around 2 nm.

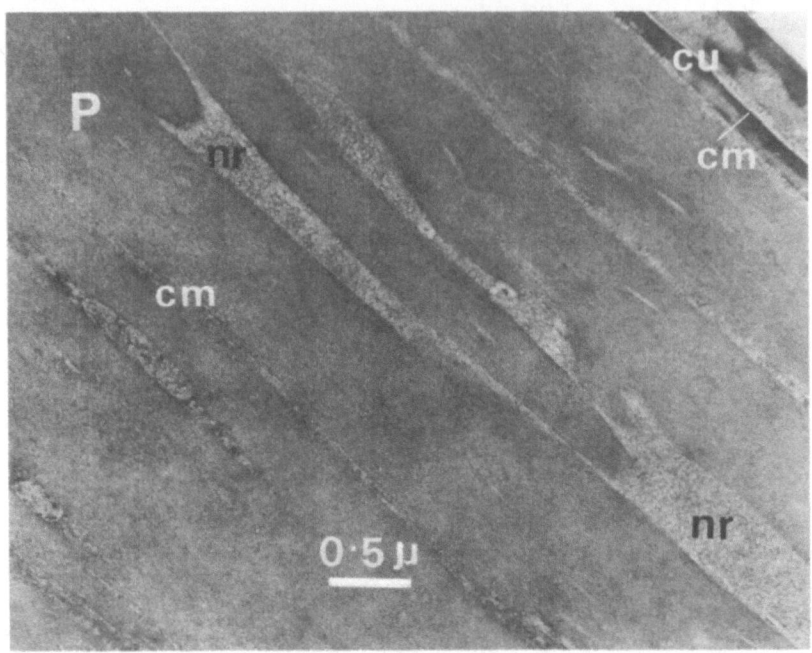

Fig. 8. Electron micrograph of a thin longitudinal section of the
73.0% MMA-grafted fiber (Lincoln wool), showing polymer
occurring parallel to the direction of the fiber axis in
the paracortex (P), and cuticle (cu), nuclear remnant
(nr), and cell membrane complex (cm).

It is clearly seen that much polymer is deposited in the latter
layer rather than the former. This seems likely to be due to the
difference in the cross-link density, probably in the disulfide
cross-link between the two β-layers. The endocuticle adheres to
the δ-layer through the disulfide bonds. However, little or no
such bonding occurs between the δ-layer and the exocuticle over
the whole distance in the longitudinal direction. It has been
reported by Nakamura et al.[22] that the δ-layers between the cuticle
and cortical cells are not identical. Similar authors asserted
that the δ-layer between the cuticle cells is similar in composition to the endocuticle and the β-layer adjoining the exocuticle
tends to be disintegrated by twisting the fiber[22]. A high concentration of polymer in the cell membrane complex between the endocuticle and cortex can be observed for the regions of cortical
cell side. The β- and δ-layers in the cell membrane complex between
the cortical cells can not be differentiated as already indicated
in Fig. 3.

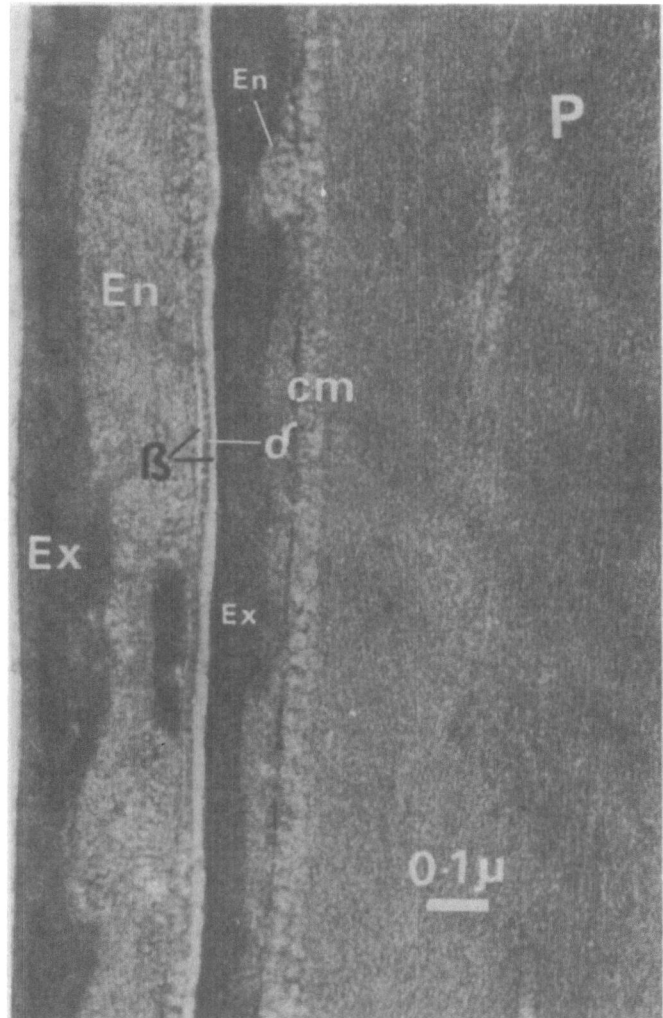

Fig. 9. A high magnification view of the cuticle similar to that
in Fig. 8. It can be clearly seen that polymer occurs in
the β- and δ-layers in the cell membrane complex between
endocuticle (En) and exocuticle (Ex), and between endocuticle and paracortex (P).

The effect of the β- and δ-layers on the mechanical properties of
wool has not been entirely clear. It is inferred that, from their
ultrastructural appearance in the grafted fiber, intramacrofibrillar
bonding is considerably stronger than the intercellular or inter-
macrofibrillar bonding as has been suggested by Orwin and
Thompson[23].

As indicated previously, the diameter of a microfibril expands
to a greater extent by grafting, but the size and shape of matrix
between sheet-like arrangement of microfibrils remain intact[5].
The polymer is located not only around and between the microfibrils
and matrices, but also at the interprotofibrillar and interstitial
regions of microfibrils. In terms of either histological or
supramolecular structure, the distribution of polymers is characteristic of the structure of wool itself as a function of cross-link
density in many type of aggregates in wool.

ALPHA AND BETA CRYSTAL FORMS IN GRAFTED WOOL FIBERS

The structure and properties of grafted wool was investigated
systematically by the present author[1-4]. The α- and β-X-ray
intensities diffracted from the grafted fibers have been analysed.
By grafting with hydrophobic monomers, such as methyl methacrylate,
methyl acrylate and ethyl acrylate, the amount of the α-crystallites
markedly decreases but the decrease in α-content is considerably
less with hydrophilic monomers[3]. On the other hand, in the grafted
wool containing the latter monomers, the production of the β-crystallites is more or less constrained throughout the fiber extension.
It has been pointed out that the location of polymer or the interaction between polymer and wool chains is considerably different
between the types of monomer and between the reaction conditions
used for grafting.

It has been demonstrated that the α-crystallites consist of
two different components: one is relatively unstable to the deposition of polymer and the other is highly stable to a large amount
of grafted polymer, which gives rise to lateral swelling without
any change of the longitudinal dimension. The experimental evidence
showed that in the so-called α — β transformation of keratin system,
about 50% of α-material disrupted by deposition of polymer is not
probably concerned with the production of the β-material, while
both stable materals, about 50%, which remain intact when a large
amount of polymer occurs in fiber and matricular components may
be regarded as the materials, originating the β-pleated sheet
structure.

VISCOELASTIC BEHAVIOR OF WOOL GRAFT COPOLYMERS

There is evidence that little or no perceptible change in
chemical nature, viz. disulfide group content, amino acid compositions, and oxidative main chain scission have occurred during
graft copolymerization[24]. Wool fibers can be specifically modified
physically by grafting to break down hydrogen bonds in the α-helix.
In such circumstances, a specific interaction may arise between

Table 1. Relative Intensities at α-Reflection of Wool in Grafted Fibers.

Treatment	Lincoln wool sample used	Disulfide and thiol contents, mole/g·wool [SS]	[SH]	Time of reaction,[e] h	Graft-on, %	I_{rel}, %[h] α-reflections 0.98 nm	0.51 nm
Untreated	A	428	32	—	—	100	100
	B	388	21	—	—	100	100
Controlled	B	390	10	5[f]	—	103	95
Reduced and S-β-cyano-ethylated[a]	B	87	35	—	—	90.2	—
Grafted with MMA	A	428	32	1.5	33.8	46.2	42.5
	B	388	21	2	42.8 ·	80.4	74.1
	B	388	21	2	43.4	72.0	—
	A	428	32	2.5	52.8	50.8	60.4
	A	428	32	7	83.1	42.7	42.5
	A	428	32	12	109	13.9	22.0
S-β-carboxy-methylated[b]	B	390 337[d]	8 88[d]	24	41.0	21.0	25.0
Reduced and methylated[c]	B	331	28	3[g]	83.7	72.3	98.0
Grafted with MA	B	388	21	2	41.4	78.0	79.0
	B	388	21	3	60.0	65.0	68.0
Grafted with EA	B	388	21	3	46.5	66.5	76.0
	B	388	21	5	82.7	96.0	100

[a] Reduction:0.1 N thioglycolate(pH=6.0);20°C;10 h;liquor ratio 100:1. Cyanoethylation:0.5 g wool,0.5 g acrylonitrile,25 ml n-PrOH,25 ml borate-phosphate buffer(pH=8.0);20°C;24 h.

[b] Carboxymethylation:0.5 g wool,0.2 g ICH_2COOH,25 ml n-PrOH,25 ml borate-phosphate buffer(pH=8.0);20°C;5 h.

[c] Reduction:1 g wool,0.0484 mmole tri-n-butyl phosphine,50 ml borate-phosphate buffer(pH=8.0);20°C;24 h. Methylation:1 g wool,0.5 g CH_3I, 50 ml n-PrOH,50 ml the same buffer;20°C;24 h.

[d] In graft copolymers.

[e] Grafting system:1 g wool,27.5 g LiBr,0.2 g $K_2S_2O_8$,22.5 g diethylene glycol monobutyl ether(BC),44.8 g H_2O,and 5 g monomer;30°C for MMA;40°C for MA & EA.

[f] Treated with the same reaction liquor as the grafting system without monomer for 5 h at 40°C.

[g] Grafting system:1 g wool,0.2 g $K_2S_2O_8$,22.5 g BC,72.3 g H_2O,5.0 g MMA;30°C.

[h] Index of the x-ray intensity diffracted from the crystallites in grafted wool, i.e. $I_{rel} = 100(I_{cr}/I_{cr}^o)$; where I_{cr}^o is the intensity in the 0.98 nm α-reflection for the ungrafted and unextended wool[4].

grafted polymer chains and wool chains constituting many types of aggregates, especially the aggregate components with low-sulfur proteins or with a lower cross-link density.

With the aid of the X-ray and electron microscopy, a dynamical study on viscoelasticity for the grafted wool fibers may give many suggestions for the interpretation of molecular mechanism associated with the molecular motion of wool itself, and further the role of the α-crystallites for thermal stability of wool structure may be clarified through the measurements of dynamical viscoelastic properties as a function of temperature.

The relative X-ray intensities diffracted from the grafted and chemically modified wool used for the measurements of dynamical properties are summarized in Table 1. No substantial difference has been observed between the values of I_{rel} at 0.98 nm and at 0.51 nm except the case for the wool grafted after reduction and

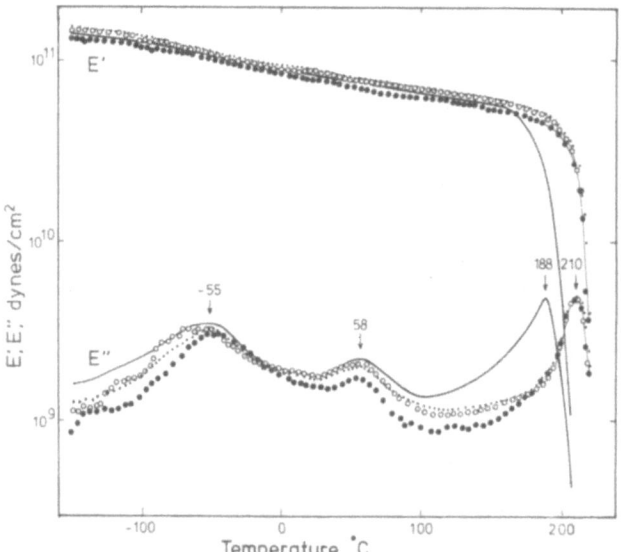

Fig. 10.   Tensile storage modulus, E' and loss modulus, E" versus temperature for the untreated and chemically modified wool: o, first run for untreated wool; •, second run for untreated wool after raising temperature upto 202°C; ···, controlled wool; and ——, reduced and S-β-cyano-ethylated wool. Dynamical properties were measured at 110 Hz.

methylation[3]. As a simple measure of the amount of the α-crystallites, value of the I_{rel} at 0.51 nm reflection should be used[3,4].

Ungrafted Wool

    Dispersion curves of the untreated and controlled wool are
shown in Fig. 10. It is important that a controlled blank is used
in this type of work so that the significance of any change in
morphology arising from exposure to reagents other than monomer can
be assessed. The form of isochronal dispersion curve of the control
sample is similar to the untreated wool. It is inferred, therefore,
that no substantial structural change has occurred during the treatment by the system in the absence of monomer. This result is the
same as the conclusion derived from previous studies on X-ray and
stress-strain properties of the control wool. Three absorption
peaks are found at about -55°C, 58°C, and 210°C. The dominant
feature of the peak near 210°C in loss modulus and fall in E' may
be mainly associated with the disordering of the α-crystallites in

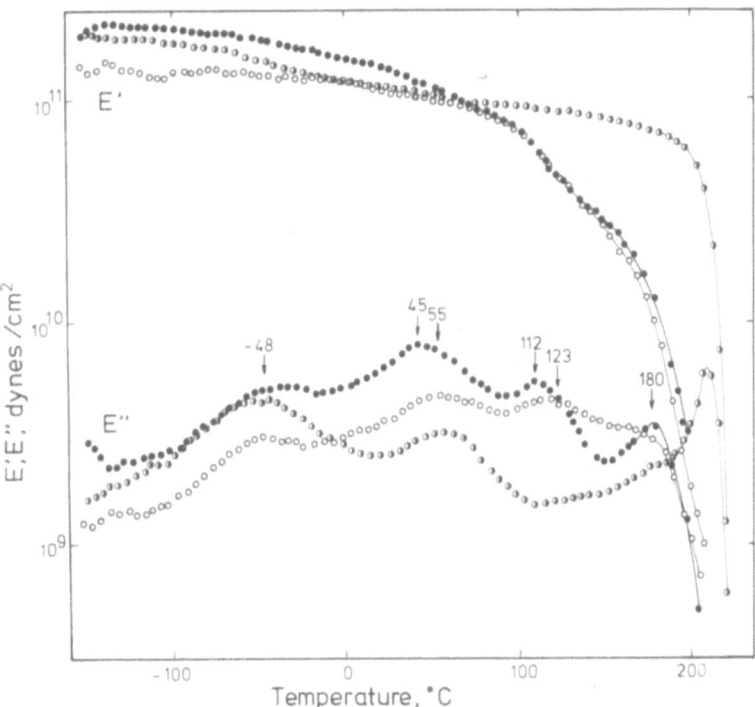

Fig. 11. Tensile storage modulus, E' and loss modulus, E" versus
        temperature for MMA-grafted wool: o, first run for the
        109% grafted wool; ●, second run for the 109% grafted
        wool after raising temperature upto 210°C; o, 83.7%
        grafted wool after reduction and methylation.

keratin. Disordering of the α-keratin commences at 210°C to 230°C
at which major cystine loss occurs[25].

The absorption curve for reduced and subsequently S-β-cyano-
ethylated wool is also shown in Fig. 10. The corresponding peak is
shifted to lower temperature at about 188°C. This probably due to
a decrease in the thermal stability of the α-helix in the low cystine
environment. The viscoelastic behavior enhanced above around 100°C
seems to suggest that molecular movements in the amorphous matrix are
unstabilized by break down of the cystine bridges.

As seen from the inspection on the curves of E' and E" for the
first and the second heating of the wool which has been heated upto
202°C, no perceptible change is observed as a whole. It is inferred,
however, that the decreasing tendencies observed for E"-values at
the temperatures from about 0°C to 180°C are due to a rearrangement
of molecules during the first heating to more stabilized confor-
mational position.

Wool Grafted with Methyl Methacrylate

Fig. 11 shows the variation of E' and E" with temperature for
the samples of the 109% methyl methacrylate (MMA)-grafted wool, and
the 83.7% MMA-grafted fibers after reduction and methylation treatments. Two samples are greatly different in the α-content in wool,
namely, 22.0% as I_{rel}-value in the former and 98.0% in the latter.
The loss peaks of the latter all coincide with those of the untreated
wool, irrespective of the presence of a large amount of graft add-
on. This supports the idea that the absorption appearing at 210°C
seems to be not dependent on the polymers occluded, but on the α-
materials present in wool.

Dispersion behavior of the 109% MMA-grafted wool is quite
different from the ungrafted wool. A softening process continuing
to above 210°C begins at around 110°C. The absorption corresponding to the temperature of 210°C is shifted to lower temperature
and diffused at around 180°C. A new broad peak centered at 123°C
appears, which is accompanied by a discrete change in E' associated
with the glass transition temperature of poly(methyl methacrylate)
(PMMA). Absorptions observed on the ungrafted wool at -55°C and
58°C are detected without much shifting and merging. Here, it is
important to note that the intensity of the absorption at around
60°C is higher than either the ungrafted and the grafted wool
containing higher amounts of the α-materials.

The second heating of the grafted wool with 109% MMA which has
been heated upto 210°C was carried out and the resulting curves are
shown in Fig. 11. A discete peak appears at 180°C and the broad
peak at 112°C is some 12°C lower than the corresponding absorption

peak in the first heating. As seen in the first heating process,
the Tg of the PMMA in the wool is some 15°C higher than the corresponding bulk materials. This is due to hindrance of the PMMA-α
process motion and would be interpreted as due to high interaction
between the polymer and the stable wool chains. This interpretation
is fairly supported by the results of electron microscopy studies[2,5].
The decrease in the temperature associated with Tg of the PMMA in
the wool observed on the absorption curve of the second heating is
probably due to the formation of some amounts of polymer phase or to
a separation into regions rich in PMMA during the first heating
process. Lower temperature regions, especially below 100°C, on the
storage modulus curve, the magnitude in E' observed in the second
heating process is considerably higher than that of the first
heating. This is ascribed to the formation of continuous polymer
phase in the wool. The E"-peak markedly increases in the intensity
and shifts to some 10°C lower temperature than the corresponding
peak for the first heating. This seems to be due to the decrease
in the content of the α-materials during the heating and to the
effect of the PMMA-β process motion. With respect to the PMMA-α
process in the reduced and methylated wool, there is no discrete
absorption on the E"-curve. This could be explained by the marked

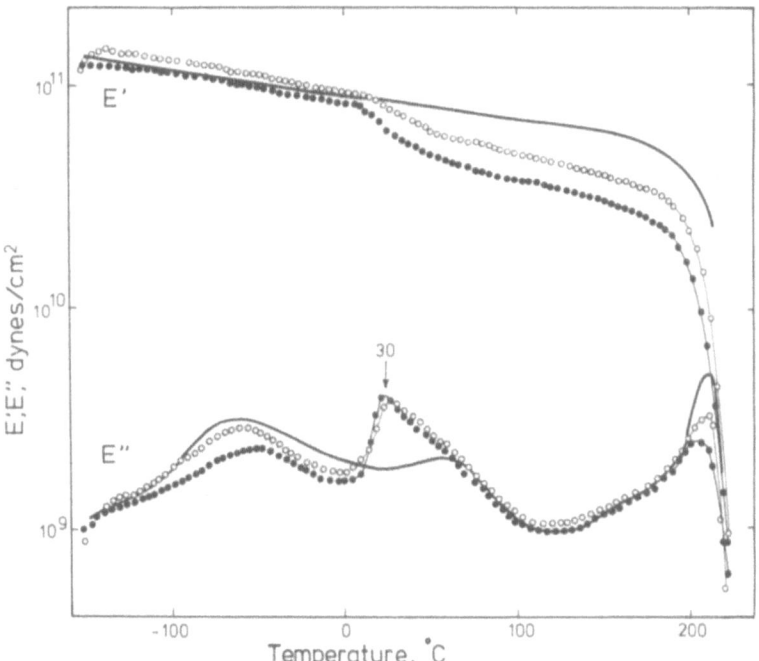

Fig. 12. Tensile storage modulus, E' and loss modulus, E" versus
          temperature for MA-grafted wool: o , 41.4%;● , 60.0%; and
          ——, untreated wool.

hindrance of the segmental motion of the grafted polymer in the
wool. The broad absorption peak at the lowest temperature on E"-
curve for any samples can be found in a temperature range around
-50°C to -55°C, which seems to be independent of the amounts of
polymer and the α-crystallites in the fibers.

Wool Grafted with Methyl and Ethyl Acrylates

The variation of E' and E" with temperature for the wool grafted
with methyl acrylate (MA) is shown in Fig. 12. With increasing
add-on of the polymer, the amount of α-crystallites is decreased.
The intensity of the absorption at around 210°C tends to decrease
and the peak position moves to lower temperature side with decrease
of the amount of the α-crystallites. As compared the form of the
absorption curve of the untreated wool with that of the grafted
samples, no discrete change is observed over the broad temperature
range of 70°C to 180°C. Absorption in this range of temperature
is associated with a motion of the amorphous matrix[26]. It is
inferred, therefore, that the matrices do not modified by the
polymer enough to change the absorption behavior. With respect to
the location of polymer in wool, it has been proved that the grafted

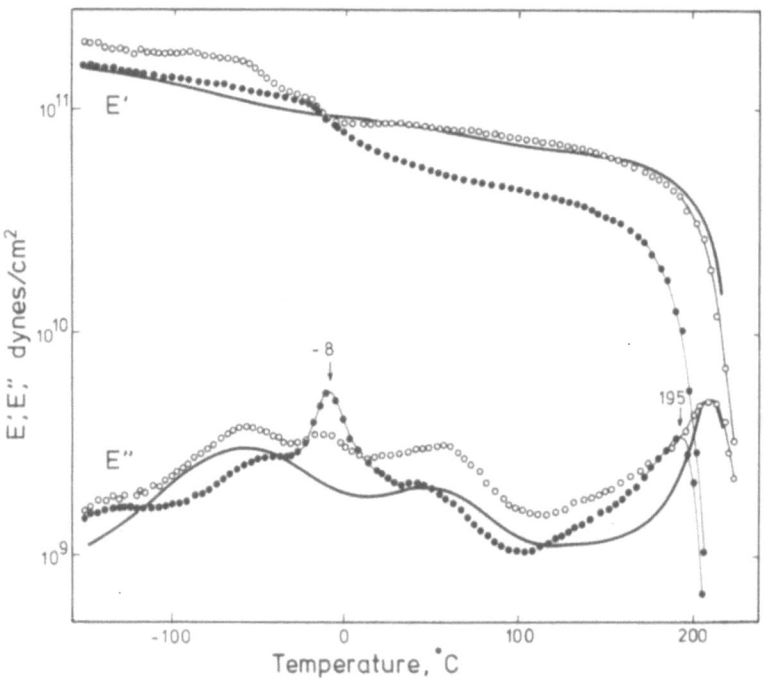

Fig. 13.   Tensile storage modulus, E' and loss modulus, E" versus
           temperature for EA-grafted wool: o, 46.5%; •, 82.7%; and
           ——, untreated wool.

polymer occurs within the microfibrillar aggregates rather than the highly cross-linked matrices[2]. Softening in E' and a new loss peak centered at about 30°C appears for the grafted wool at which the grafted polymer chains are thought to aquire considerable mobility in the PMA-α process. The T_g of the PMA in the wool is some 25°C higher than the corresponding bulk homopolymer. This indicates that the grafted polymer is interacted with ordered microfibrillar materials through a finely dispersed state of the molecules.

As shown in Fig. 13, a similar behavior can be also observed for the dispersion curves of the ethyl acrylate (EA)-grafted wool. The storage modulus is remarkably decreased at about -10°C. With increase in the polymer content, the decreasing tendency in modulus is pronounced. The softening process is followed by an increase in absorption at around -8°C which is associated with the T_g of the PEA in the wool. The mobility of polymer may be considerably restricted in the fibers, since the T_g of the corresponding bulk polymer is about -22°C. It is futher noted that the increase in the absorption intensities above around 100°C is remarkable, especially for the case of the wool with highest level of grafting, and the absorption maximum at 210°C for the untreated wool is shifted to a lower temperature at about 195°C, irrespective of the fact that at 20°C, the α-crystallites have been remainded intact in the grafted fibers (see Table 1). These suggest that the thermal stabilities of both matrix and microfibril are greatly reduced as a result of their specific interactions with PEA segments. On the temperature dependence of E" in the higher temperature regions, characteristic difference is observed between the fibers grafted with MA and EA, which is ascribed to the differences in the location of polymer and characteristics in the interaction of polymer with wool chains. As has been shown in a previous study of X-ray diffraction, the randomization of the α-crystallites due to grafting depends on the nature of monomer used[3].

Effect of α-Helical Content on Viscoelasticity

Temperature dependence of E' and E" for various MMA-grafted fibers are shown in Fig. 14. With increasing graft add-on, the absorption peak extends from 210°C until it merges and tends to shift to lower temperature. As seen in Table 1, the relative intensity of the α-reflection is decreased with increasing the graft add-on. Thus, it might be said, that the absorption intensity is increased with decreasing the degree of crystallinity of wool. Thermal stability is also decreased with increasing graft-on or with decreasing the crystallinity as seen in the dispersion curves at about 100°C.

The absorption peak found at about 60°C is increased with increasing the extent of grafting. This seems to be associated

with the motion of the amorphous materials in which the disrupted
sections of the α-crystallites is involved. While the absorption
at around -50°C may not be directly concerned with both crystallinity of the wool and polymer add-on. This interpretation was
confirmed by the experiments for the grafted wool with a similar
amount of polymer and with different amount of the α-crystallites.

X-ray diffraction study has presented the evidence that the
α-crystallites consist of two different components: one is highly
stable to a large amount of grafted polymer and the other is
relatively unstable components which occupy about one-half of the
α-materials and exist in a less laterally cross-linked region in
wool[1-4]. By usual means, it is, therefore, impossible to obtain
the grafted fibers with a lower content of the α-crystallites at
a lower content of polymer. For the requirement to the present
experimental purpose, carboxymethylated wool prepared by the
treatment with iodoacetic acid (conversion of free thiol groups
into S-β-carboxymethyl groups is about 60%) was used. A lower
X-ray intensity, I_{rel} = 25.0%, was obtained for the graft copolymer

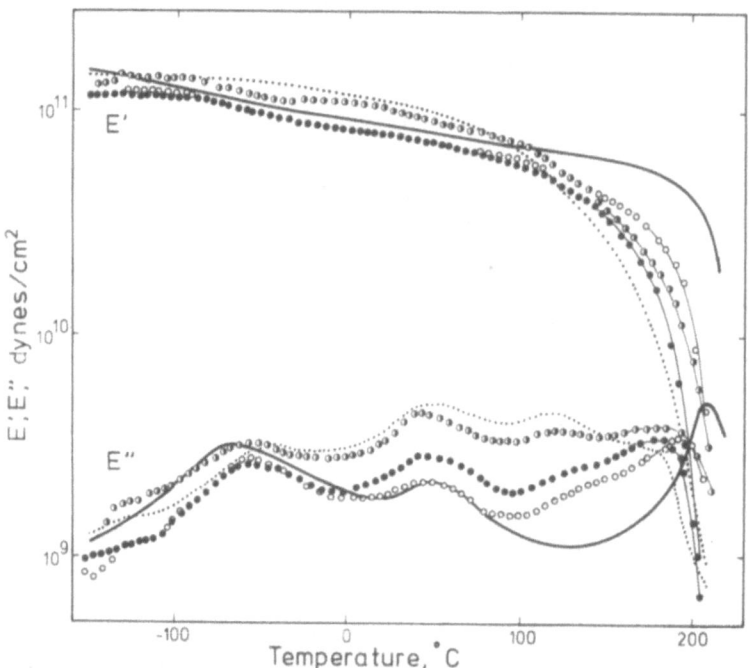

Fig. 14. Tensile storage modulus, E' and loss modulus, E" for
          MMA-grafted wool with a different amount of polymer:
          o, 42.8%; ●, 52.8%, ◐, 83.1%; ···, 109%; ———,
          untreated wool.

with 41.0% of MMA. As shown in Table 1, structural changes are observed on the thiol and disulfide contents. After grafting, the content in the former is greatly increased and in the latter is inversely decreased. Production of new thiol groups is clearly attributed to a homolytic mechanical scission of disulfide linkage. Such mechanically induced structural changes scarcely occur in the usual grafting conditions for unmodified wool, since the strained disulfide network imparted by the swelling due to the deposition of polymer can be free from the stresses through the so-called thiol and disulfide interchange reactions. On the contrary, for the S-β-carboxymethylated wool, a lower concentration of thiol groups which act as catalysts tends to retard the interchange reaction and the retardation will be further enhanced by the acidic environment of carboxyl groups introduced. When the concentration of localized stress produced by such a mechanical straining process can exceed in such situations the critical value of the bond energy of disulfide linkage, rupture may occur and for the stable components of the α-crystallites as well.

Turning now to the results shown in Fig. 15. The curves of the temperature dependence of E' and E" for the S-β-carboxymethylated and grafted fibers show that (i) the absorption maximum at 210°C disappears and appears as a shoulder at around 180°C, (ii) a discrete peak appears at around 120°C which is associated with some restricted PMMA-α process motion accompanied by a softening process begining at about 110°C and continuing above 210°C, and (iii) an intensified broad peak appears at around 60°C. These characteristic features are quite similar to those of the graft copolymers with the highest grafting yield examined and with the lowest content of the α-materials (see Fig. 14). It is important to note that the intensification of the absorption at around 60°C is independent of the extent of grafting, but depends on the content of the amorphous materials. However, it is not enough clear whether this absorption is only associated with a motion of non-hydrogen bonded segments in any different structural environment.

To elucidate this uncertainty, temperature dependence of loss modulus for a permanent set wool was investigated. Wool fibers grafted with 43.4% of MMA (I_{rel} = 72.0% at 0.98 nm reflection) were extended at 60% to the original fiber length in water and held in boiling water for 3 h and then relaxed for 1 h in the boiling water. The permanent set fibers possessed no α-crystal form on the X-ray diffraction pattern in which the extended β-crystal form was involved. The observation that in Fig. 15, the absorption maximum is little shifted and lowered at 210°C suggests high thermal stability of the β-structure. The absorption at around 60°C is somewhat intermediate level between the other two samples possessing different amounts of the α-crystallites, i.e. 25.0% and 74.1% as I_{rel}-value.

It is worthwhile to note that as pointed out in preceding
discussion, absorption at around -50°C is totally independent of
the contents of the α- and the β-materials, and also of the grafted
polymer. However, some intensity difference is observed among the
samples. This seems to be due to the difference in the <u>average
cross-sectional area</u> occupied by the wool components in each test
sample which consists of exactly 35 single fibers, since the
diameter of the wool and longitudinal uniformity is somewhat
different from fiber to fiber.

Now, we can construct the relationships between the intensity
of the absorption at around -50°C or 60°C and I_{rel}-value. The
results are shown in Fig. 16(a). The values of E''_{max} at around
-50°C are scattered, but approximately constant to be of 3.1 x 10^9

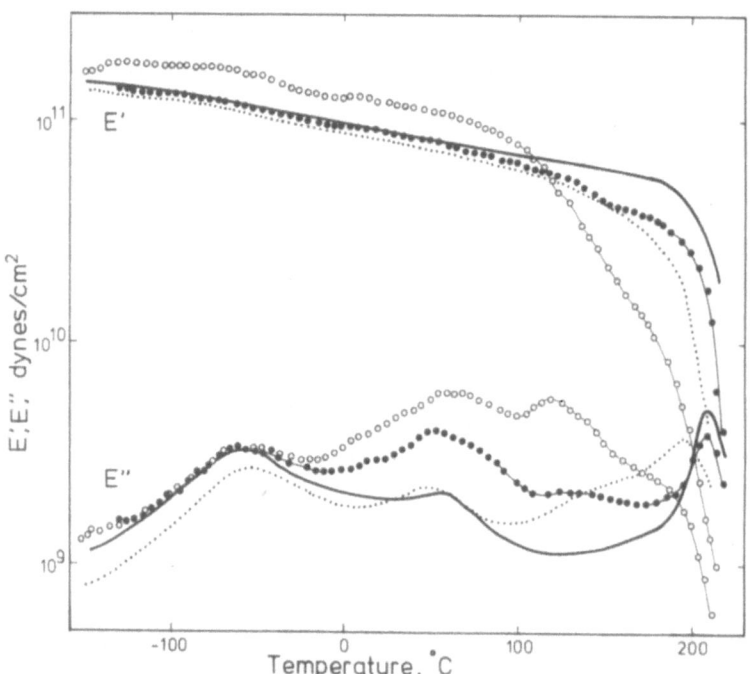

Fig. 15. Tensile storage modulus, E' and loss modulus, E" for
MMA-grafted wool with approximately similar amounts of
polymer which contains a different amount of the α-
crystallites in the wool: o, 41.0% grafts (I_{rel}=21.0%);
●, 43.4% grafted and set wool (I_{rel}=0%); ···, 42.8%
grafts (I_{rel}=80.4%); and ——, untreated wool
(I_{rel}=100%).

dyne/cm^2. Approximately linear curve decreasing against I_{rel}-value
is also obtained by plotting the E''_{max}-values at around 60°C.
In order to avoid the experimental uncertainty resulted from the
difference in the <u>cross-sectional area</u>, the maximum E''-value at
around 60°C, E''_{60}, is normalized by the maximum E''-value at around
-50°C, E''_{-50}. As shown in Fig. 16(b), a very linear relation holds
between the ratios of E''_{60}/E''_{-50} and I_{rel}-values for the ungrafted
and the grafted samples except the set wool.

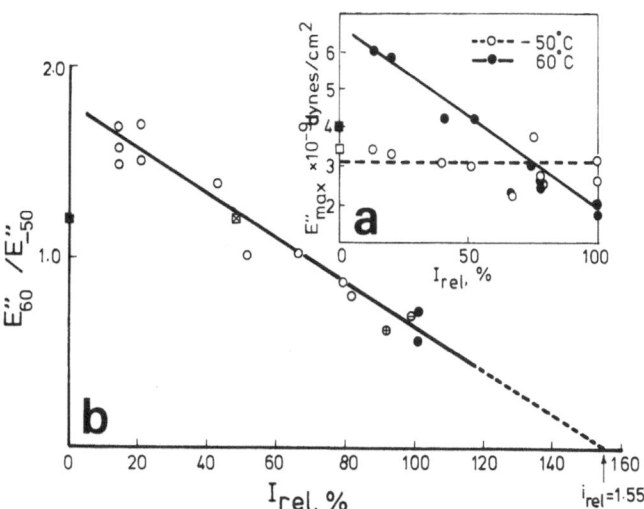

Fig. 16. (a) Relationships of maximum loss modulus, E''_{max} at 60°C
          and at -50°C versus I_{rel} for various grafted and un-
          grafted wool: ○, -50°C; ●, 60°C; □, -50°C for the
          43.4% MMA-grafted and set wool; and ■, 60°C for the
          43.4% MMA-grafted and set wool.
          (b) Relationships of the ratio of the maximum loss
          modulus at 60°C to that modulus at -50°C, E''_{60}/E''_{-50}
          versus I_{rel} for grafted and ungrafted wool: ●, untreated
          wool; ⊕, reduced and S-β-cyanoethylated wool; ○, MMA-
          grafts; ⊖, 83.7% MMA-grafted wool after reduction and
          subsequent methylation treatments; ■, 43.4% MMA-grafted
          and set wool; ⊠, plot against relative intensity
          calculated from the intensity of the β-reflection of
          43.4% MMA-grafted and set wool.

Effect of β-Crystalline Content on Viscoelasticity

It is thus important to assess quantitatively the amount of the β-crystallites in the set wool.

Total X-ray scattering intensity from a constant weight of wool does not depend on the structure of the wool itself. The weight fraction of the α-component, M_α, in the untreated and un-extended wool is given by eq.(1):

$$M_\alpha = I_\alpha^\circ / I_{tot} \qquad\qquad (1)$$

where I_α° is the diffracted intensity from the α-crystallites involved and I_{tot} is the total intensity, i.e. the sum of the intensities diffracted from the crystal and the amorphous materials.

And for the set wool in which α-component is absent, the fraction of the component, M_β is also shown by eq.(2):

$$M_\beta = I_\beta^\circ / I_{tot} \qquad\qquad (2)$$

where I_β° is the X-ray intensity from the β-crystallites.

Taking the unit weight of wool in the samples irradiated by X-rays, the ratio of M_β to M_α is represented by eq.(3):

$$M_\beta / M_\alpha = I_\beta^\circ / I_\alpha^\circ \qquad\qquad (3)$$

Here, I_α° or I_β° is the sum of the intensities diffracted from the α- or the β-crystallites over the whole of the Bragg angle, 2θ and the azimuthal angle, ψ. Assuming that I_α° could be taken to the sum of the intensities at $\psi = 0°$ (equator) and at $\psi = 90°$ (meridian), and further I_β° to the intensity at $\psi = 0°$, the ratio of the M_β/M_α can be evaluated by eq.(3).

The diffraction intensity-versus-2θ curves of the untreated wool are shown by broken lines in Fig. 17(a). The sum of the area enclosed by the intensity curve at $\psi = 0°$ with the line of 0%-level in the relative intensity and the corresponding area at $\psi = 90°$ could be taken as I_α°-value in question. The diffraction intensity-versus-2θ curve of the wool which has been held at 60% extension in boiling water for 3 h is also shown by solid line in Fig. 17(a). The corresponding enclosed area is equal to the I_β°-value. The ratio thus obtained was 0.893 or 89.3%.

Now, we can estimate the amounts of the β-crystallites in the 43.4% MMA-grafted wool which has been treated with boiling water for 3 h at 60% extension and then for 1 h at relaxing. A similar profile to the β-X-ray intensity constructed from the extended

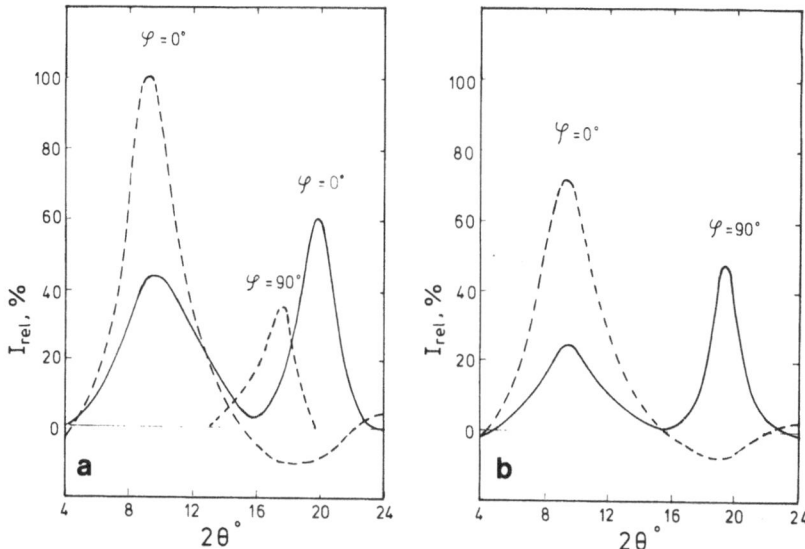

Fig. 17.   (a) Relative intensity-versus-2θ curves for unextended
           and extended wool in boiling water for 3 h: ···, un-
           extension; ——, 60% extension.
           (b) Relative intensity-versus-2θ curves for the 43.4%
           MMA-grafted and set wool: ···, unextension; ——, 60%
           extension for 3 h and relaxing in boiling water for 1 h.

wool can be recognized in Fig. 17(b). From the experiments of the
orientation of the β-crystallites, little difference was obtained
as compared to the case of the ungrafted wool shown in Fig. 17(a).
The ratio of the β-X-ray intensity, I_β to I_α° is 0.477 or 47.7% as
represented by the relative intensity. This plot in Fig. 16 holds
the linearity obtained for the grafted wool containing various
amounts of the α-crystallites. Thus important results can be
obtained that the absorption at around 60°C is certainly associated
with a motion of the non-hydrogen bonded sections of backbone chain,
and is independent of their structural environment, such as the
differences in crystal form, and the types of cross-links. The
latter is confirmed from the fact that during the boiling processes,
the cystine cross-links are changed to new types of cross-links,
i.e. intra- and inter-molecular cross-links of lysinoalanine and
lanthionine[27].

Relationships between Absorption Intensity and Degree of Crystallinity

    On the assumption that the absorption intensity at around 60°C
is proportional to the number of the non-hydrogen bonded -CONH-
groups in the system, it is possible to write eq.(4):

$$\gamma_o (1 - i_{rel}) + (1 - \gamma_o)\sigma = kD \qquad\qquad (4)$$

or

$$\sigma = [kD + (i_{rel} - 1)\gamma_o]/(1 - \gamma_o) \qquad\qquad (5)$$

where γ_o is the degree of crystallinity of the unextended and un-grafted wool, i_{rel} is the relative X-ray intensity ($=10^{-2}I_{rel}$), D is the value of E''_{60}/E''_{-50}, σ is the mole fraction of non-hydrogen-bonded -CONH- groups to the total number of the groups in the intact matrix + non-keratinous materials, and k is a constant. Here, it has been assumed that the densities of the molecular aggregates; crystalline microfibrils, matrix, and the disrupted microfibrillar materials are all equal. The first term of the left hand side in eq.(4) is the fractional amount of the amorphous materials resulted from the disrupted sections of the α-crystallites, and the second term is the fractional amount of the intact matrix + non-keratinous materials containing no hydrogen bond.

In Fig. 16(b), when the straight line is extrapolated to the relative intensity axis, a value of 1.55 is obtained as i_{rel}-value at the intersection, which would be the value for the zero intensity of the absorption. Taking i_{rel} = 1.55 and D = 0 in eq.(5), we obtain eq.(6):

$$\sigma = 0.55\gamma_o/(1 - \gamma_o) \qquad\qquad (6)$$

At present, we have not a reliable data of this γ_o-value for the Lincoln wool fiber. The α-helix content in Lincoln wool fiber was calculated by Bendit[28] and estimated to be about 40% of the "microfibril + matrix" unit, which contains 48% microfibrils. Similar data has been obtained by Bradbury[29], i.e. the amount of crystalline microfibrils is 44.5% of the fiber, matrix is 29.6%, and the other is 25.9% for small amount of cuticles and non-keratinous materials. The γ_o-value of Merino wool which contains higher amount of microfibrils than Lincoln wool has been evaluated by X-ray diffraction method to be 0.596[1]. On the assumption that γ_o-value is in the range of 0.4 to 0.5, from eq.(6), the σ-value lies from around 0.4 to 0.6. This implies that about one-half of the -CONH- groups in the wool components other than the α-crystal forms hydrogen bonds. It might be expected, therefore, that a considerably organized structure occurs in the matrix and the non-keratinous materials. However, there has not yet been work on the feasibility of the ordered structure on the X-ray diffraction patterns. As the σ-value is equal to unity when all of the -CONH- groups does not form hydrogen bonds, the magnitude of the γ_o-value should become of 0.645. This is an unexpectedly larger value for the Lincoln wool fibers.

With respect to the conformation of the non-helical parts of the low-sulfur or the high-sulfur proteins, it has been postulated that a repeating pentapeptide pattern of residues which has the form -Cys-Cys-X-Pro-Y- is present in wool[30,31]. Parry et al.[32] suggested that a high proportion of the disulfide linkages in the matrix are intramolecular rather than intermolecular and the peptide chain can adopt so-called β-bend conformation being stabilized by the hydrogen bond[33,34]. Although these are now indirect evidence, the present result obtained from dynamic mechanical properties of the grafted fibers provides an obvious evidence for the presence of a considerably organized structure in the matrix or non-keratinous materials.

THERMAL EXPANSION ALONG FIBER-AXIS DIRECTION

Fig. 18(a) shows the length-versus-temperature relationships for the grafted fibers with lower contents of the α-materials such as the S-β-carboxymethylated wool grafted with 41.0% of MMA (SGMG) and the grafted and set wool (GS). Their plots have the two temperatures T_1 and T_2 which consist of each two approximately straight lines intersecting at T_1 and at T_2 and possessing slopes of λ_{11} above T_1, λ_{12} below T_1, and of λ_{21} above T_2, λ_{22} below T_2, respectively. For SCMG wool, T_1 is around 60°C and T_2 around -48°C. Similar but relatively higher temperatures are also observed for GS wool; T_1 at around 80°C and T_2 at around -40°C.

The length-versus-temperature relationships for the grafted fibers containing relatively higher amounts of the α-materials are also observed at around 55°C to 68°C. However, the temperature corresponding to T_2 does not appear. The slopes of the linear curves, represented as λ_2 continuing from -150°C to around 0°C, tend to increase only slightly with increase of the extent of grafting, as seen in Table 2.

It should be noted that the transition temperature, T_2 at a break on the length-temperature curve appears when the stable α-helical componens, which correspond to about one-half the amount of the oriented α-materials, were disrupted and randomized, even if the oriented β-crystallites were present. This suggests that on morphological point of view, the stable α-crystallites give rise to a continuous phase throughout the fiber, which would suppress the thermal expansion along the fiber-axis direction. The transitions observed for the SGMG and GS wools, which would have lost such anisotropy, are characteristic of the high slopes of the curve above T_2.

Empirical equations for the relationships between the change in volume and temperature have been proposed by many authors[35-37]. Simha and Boyer[38] derived an equation which adequately describes

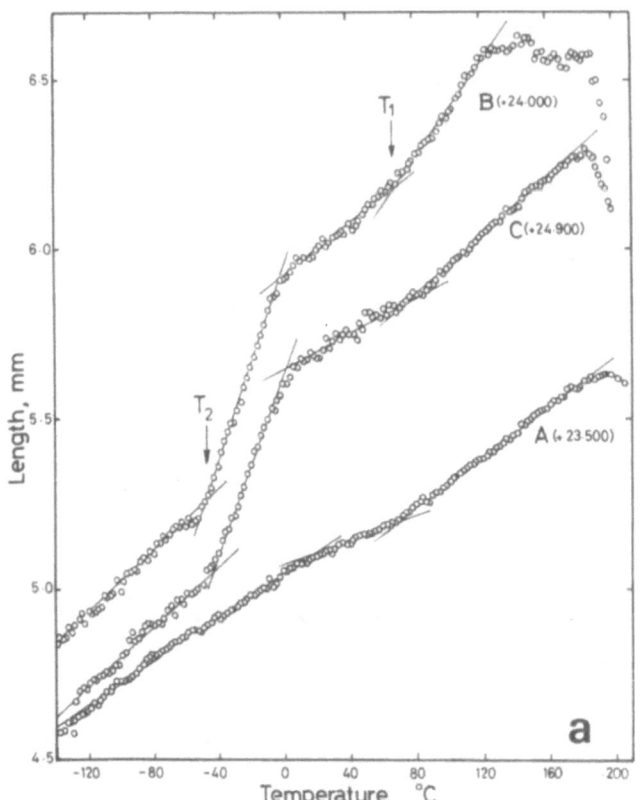

Fig. 18. Length-versus-temperature relationships of various grafted wool. (a) A, ungrafted wool which has been preheated upto 202°C; B, S-β-carboxymethylated wool grafted with 41.0% of MMA (SGMG); and C, grafted and set wool (GS).
(b) A, untreated wool; B, 42.8% MMA; C, 52.8% MMA; D, 83.1% MMA; E, 60.0% MA; and F, 82.7% EA. A figure shown by plus sign in parenthesis should be added to the scale in length-axis for the total fiber-length.

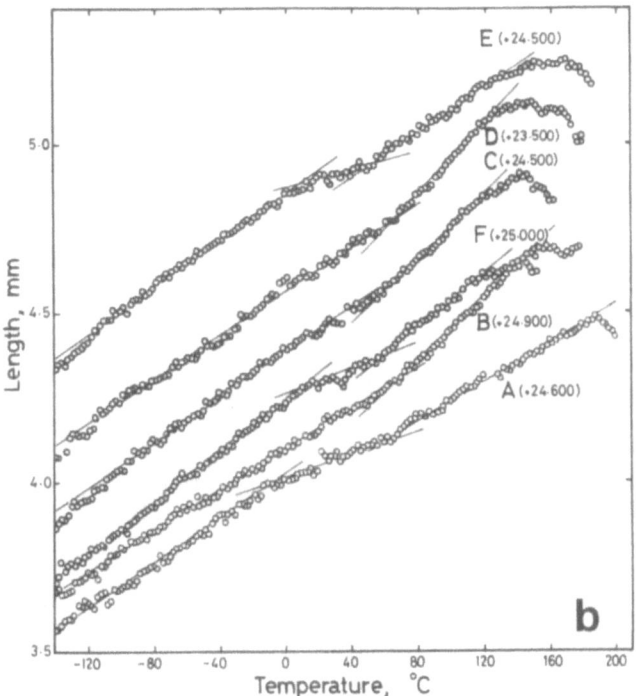

the relation between the glass transition temperature, T_g and the
difference of the thermal expansion coefficient above and below
T_g, $\Delta\alpha$, i.e. $\Delta\alpha T_g = K$, where K is a constant of 0.113. This
relation holds regardless of variations in structure such as intermolecular forces, chain flexibility, and geometry of polymer.
It is of interest to examine this relation for the transitions of
SGMG wool which would be considered to be nearly isotropic in the
thermal expansion. The differences between λ_{11} and λ_{12}, $\Delta\lambda_1$, and

Table 2. Thermal Expansion Coefficient of Grafted Wool
          along the Fiber-Axis Direction

Samples	Untreated wool	Preheated wool[a]	Grafted wool					SGMG wool[b] MMA	GS wool[c] MMA
			MMA	MMA	MMA	MA	EA		
Graft-on, %	—	—	42.8	52.8	83.1	60.0	82.7	41.0	43.4
I_{rel}, %	100	∿100	74.1	60.4	42.5	68.0	100	25.0	0[d]
T_1, °C	65	68	60	60	65	58	55	60	80
λ_{11}, $10^4/°K$	1.03	1.31	1.71	1.55	2.00	1.10	1.32	2.32	1.41
λ_{12}, $10^4/°K$	0.77	0.76	1.04	1.10	1.07	0.57	0.57	1.43	0.90
$\Delta\lambda_1$, $10^4/°K$	0.26	0.55	0.67	0.45	0.93	0.53	0.75	0.89	0.51
T_2, °C	—	—	—	—	—	—	—	-48	-40
λ_{21}, $10^4/°K$	—	—	—	—	—	—	—	4.93	4.75
λ_{22}, $10^4/°K$	—	—	—	—	—	—	—	1.60	1.56
λ_2, $10^4/°K$	1.01	1.09	1.04	1.38	1.34	1.21	1.30	—	—
$\Delta\lambda_2$, $10^4/°K$	—	—	—	—	—	—	—	3.33	3.19

[a] Ungrafted wool which has been preheated upto 202°C.
[b] S-β-carboxymethylated wool grafted with MMA.
[c] Grafted and set wool.
[d] I_{rel} of β-reflection = 47.7%.

λ_{21} and λ_{22}, $\Delta\lambda_2$ are also shown in Table 2. The calculated values
of $3\Delta\lambda_1 T_1$ is 0.09. For the transition at T_1, we have an approximate agreement with the predicted value of the glass transitions
of polymer. This transition temperature is nearly the same as the
temperature at which the absorption peak appears on the dynamic
loss modulus curve of the wool. Evidence from the dynamic and the
thermal expansion experiments suggests that the transition is
caused by the onset of the microbrownian motion in the amorphous
regions of wool.

While the calculated value of $3\Delta\lambda_2 T_2$ is 0.23. This is considerably larger than the value associated with glass transition
temperature. The absorption in dynamic loss modulus at the vicinity
of -50°C is related to this abrupt change of the thermal expansibility. It seems reasonable to assume that the absorption is
associated with the initiation of side chain motions. There is a

fact that the absorption-peak intensities are not correlated with the amounts of the α- or the β-crystallites in fibers. Deducing the molecular mechanism from the magnitude of the thermal expansibility, this relaxation process might be attributed to the occurrences of side chain motion followed by some melting phenomena of compactly packed side chains in the regions of the intact matrix which has hardly been affected by the deposition of polymer.

ACKNOWLEDGEMENT

This work was partially supported by Scientific Research Funds provided by the Education Ministry of Japan, No.440005, 1981.

REFERENCES

1. K. Arai, M. Negishi, T. Suda and S. Arai, J. Appl. Polym. Sci., 17, 483 (1973).
2. K. Arai, in "Block and Graft Copolymerization" Vol. 1 (R. J. Ceresa, Ed.), Wiley-Interscience, London-New York, 1973, Chpt.8.
3. K. Arai and K. Hagiwara, Int. J. Biol. Macromol., 2, 355 (1980).
4. K. Arai and S. Arai, Int. J. Biol. Macromol., 2, 361 (1980).
5. K. Arai and M. Negishi, J. Polym. Sci. A-1, 9, 1865 (1971).
6. K. Arai, S. Komine and M. Negishi, J. Polym. Sci. A-1, 8, 917 (1970).
7. K. Arai, M. Negishi, S. Komine and K. Takeda, Appl. Polym. Symp. No.18, Vol. I, 545 (1971).
8. K. Arai, in "Block and Graft Copolymerization" Vol. 1 (R. J. Ceresa, Ed.), Wiley-Interscience, London-New York, 1973, Chpt.7.
9. K. Arai, Polymer, 18, 211 (1977).
10. K. Arai and H. Tabei, Proc. Int. Wool Text. Res. Conf., Aachen, III, 416 (1975).
11. K. Arai and H. Tanabe, Polymer, 18, 220 (1977).
12. F. H. Mercer, J. Text. Inst., 40, T629 (1949).
13. M. W. Andrews, R. L. D'Arcy and I. C. Watt, J. Polym. Sci. B, 3, 441 (1965).
14. M. W. Andrews, J. Roy. Microscop. Sci., 84, 439 (1965).
15. P. Ingram, J. L. Williams, V. Stannett and M. W. Andrews, J. Polym. Sci. A-1, 6, 1895 (1968).
16. J. Sikorski and W. S. Simpson, Nature, 182, 1235 (1958).
17. G. E. Rogers, Ann. N. Y. Acad. Sci., 83, 408 (1959).
18. J. A. Swift, J. Roy. Microscop. Soc., 88, 449 (1968).
19. J. H. Brudbury and K. F. Ley, Aust. J. Biol. Sci., 25, 1232 (1972).
20. J. A. Swift and A. W. Holmes, Text. Res. J., 35, 1014 (1965).
21. R. D. B. Fraser, T. P. MacRae, G. R. Millward, D. A. D. Parry, E. Suzuki and P. A. Tulloch, Appl. Polym. Symp., No.18, Vol.I, 65 (1971).

22. Y. Nakamura, T. Kanoh, T. Kondo and H. Inagaki, Proc. Int. Wool
     Text. Res. Conf., Aachen, II, 23 (1975).
23. D. F. G. Orwin and R. W. Thompson, Proc. Int. Wool Text. Res.
     Conf., Aachen, II, 173 (1975).
24. K. Arai, M. Negishi, T. Suda and K. Doi, J. Polym. Sci. A-1, 9,
     1879 (1971).
25. J. S. Crighton and P. N. Hole, Proc. Int. Wool Text. Res. Conf.,
     Aachen, II, 499 (1975).
26. A. Konda, M. Tsukada and S. Kuroda, J. Polym. Sci. B, 11, 247,
     (1973).
27. A. Robson, M. J. Williams and J. M. Woodhouse, J. Text. Inst.,
     60, T140 (1969).
28. E. G. Bendit, Text. Res. J., 38, 15 (1968).
29. J. H. Bradbury, in "Advanced in Protein Chemistry" Vol.27 (C. B.
     Anfinsen, J. T. Edsall and F. M. Richards, Eds.), Academic
     Press, New York-London, 1973.
30. T. C. Elleman, H. Lindley and R. J. Rowlands, Nature, 246,
     530 (1973).
31. L. S. Swart, Nature (New Biol.) 243, 27 (1973).
32. D. A. D. Parry, R. D. B. Fraser and T. P. MacRae, Int. J. Biol.
     Macromol., 1, 17 (1979).
33. C. M. Venkatchalam, Biopolymers, 6, 1425 (1968).
34. P. N. Lewis, F. A. Momany and H. A. Scheraga, Proc. Nat. Acad.
     Sci. USA, 68, 2293 (1971).
35. T. G. Fox, Jr. and P. J. Flory, J. Appl. Phys., 21, 581 (1950).
36. S. Rogers and L. Mandelkern, J. Phys. Chem., 61, 985 (1957).
37. R. F. Boyer and R. S. Spencer, J. Appl. Phys., 15, 398 (1944).
38. R. Simha and R. F. Boyer, J. Chem. Phys., 37, 1003 (1962).

POLYMER-LEATHER COMPOSITES V. PREPARATIVE METHODS, KINETICS, MORPHOLOGY, AND MECHANICAL PROPERTIES OF SELECTED ACRYLATE POLYMER-LEATHER COMPOSITE MATERIALS

Edmund F. Jordan, Jr., Bodhan Artymyshyn, Mary V. Hannigan, Robert J. Carroll, and Stephen H. Feairheller

Eastern Regional Research Center, Agricultural Research Service, U.S. Department of Agriculture Philadelphia, Pennsylvania 19118

INTRODUCTION

This paper reviews an extensive study of largely reported work on polymer-leather composite materials (1-4). In this work two methods were utilized, for depositing selected acrylic polymers into the leather matrix by free radical polymerization (1). In one method polymer was introduced by emulsion polymerization into expanded hydrated panels. In the other, polymer was formed by bulk or solution polymerization in unexpanded acetone-dried panels. The widest feasible range of composite compositions was investigated for both methods. The kinetics of the emulsion process (2), together with the morphology (3) and mechanical properties (4) of both types of composites, were also investigated. In the present paper we attempt to combine the more important aspects of these detailed works into a comprehensive whole and to integrate this peripheral area of composite material science into the main body of work in the field.

It is well established today (5a-e) that the most industrially important composite materials containing polymer have only low concentrations (3-5%) of actual graft polymer. The core-shell model (5a) specifies that growth of the modifying polymer is restricted to discrete domains and fed by sorbed monomer; high grafting density is found only at the interfacial regions between domains, where, fortunately, it serves to insure interdomain adhesion, and thus greatly improve mechanical properties. In graft polymerization from emulsion, the polymer domains retain their initial colloidal size, but still possess core-shell

morphology. More intimate mixing can be achieved with anionic
branched or block copolymers (5b), which have high graft density,
or with free-radical initiated interpenetrating networks (5e), or
their variants, where grafting is relatively infrequent. In the
latter, domain sizes are often segmental (~100 A) but remain
colloidal when prepared by emulsion polymerization. In all types
of core-shell morphology, unextractable polymer results largely
from being entrapped in cells surrounded by a matrix skin that is
immiscible in the invading solvent. Consequently, this unretriev-
able polymer was formerly considered to be grafted.

     With the matrix consisting of natural fibers of cotton, wool,
or other types, an extensive literature (6-12) suggests that
free-radical chemical grafting of the monomer to the fiber component, proceeding from primary radical attack to substrate, was the
guiding mechanism affecting growth and morphology of the resulting
composites. High energy radiation grafting (6,7) of cotton (13-20)
and wool (9) probably induced high grafting incidence because the
sole means of initiating chains by this technique depends on
long-lived substrate radicals (post-irradiation grafting) or the
relative G values for monomer and polymer (mutual grafting). High
graft frequency is not necessarily inherent in chemical grafting,
however.

     Generally in chemical grafting of polymer to cotton and wool,
a small quantity of fiber (1 g) was used with an excess of monomer
(2 to 10 g) dispersed in a large quantity of water (or an aqueous
non-solvent for the polymer) ~25-100 parts. The residual polymer
in the fiber after solvent extraction was considered to be grafted
and the extracted polymer used to estimate grafting efficiency
(7,9,12). Saturation of fiber by polymer appeared to occur because
graft yields were generally limited to 25 to 300%, even with
excess monomer, and other forcing conditions (18-29). Thus an
upper limit of fiber expansion (6,7,9,15-30) appears to restrict
further deposition. This apparent saturation held both graft
efficiency (13,17,27,30) and apparent graft frequency (13,15,18,29)
to values much smaller than permitted by strict application of
theory (8,9,11,13,18). Grafting by redox methods (20,21), by
thermal decomposition of initiator (22), and other methods (7,9,11,
23-27), including use of oil soluble initiators (28,31) all gave
similar, but limited, yield of graft polymer. Surprisingly, this
same yield limitation appeared also to apply to the mechanistically
favorable decomposition of Ce^{IV}-cellulose complex (32) to yield
grafted composites with cotton (14,17-19). However, when the
presence of popcorn polymer was suspected (9), yields reached the
thousands (33,34) and even the millions (9). These results suggest
that maintenance of the domain integrity of the modifying polymer
within the fiber as polymerization proceeds (aided by free energy
repulsions between substrate and polymer), to develop a morphology

analogous to the core-shell model (5a,c), is a viable alternate
mechanism to grafting. Restriction of the deposition to a composition gradient extending throughout the microfibrillar and higher
aggregation levels (3), accompanied by only interfacial grafting,
would provide such a model. Polymer deposition should then be
limited in volume by restraints imposed in maintaining most
features of the original fiber, while allowing some limited lateral
expansion. Tortuosity restrictions (5d) and interchain entangle-
ments would limit the amount of homopolymer that could be extracted
(12,35). Because precipitation polymerization (36a) results from
the methods used to prepare all of these composites, the accompanying gel effect nearly always produced high molecular weight branches
(6,7,9,12,15,18,29,30). Finally, preferential location of the
available polymer within the fiber could only result from irrevers-
ible adsorption of initiator. This tendency is well documented
(2,9,14,20,21,37). However, in spite of the intensive effort
outlined above, the actual mechanism, or mechanisms, for producing
bound polymer in fibers remains elusive (9,12).

A similar mechanism of deposition of polymer formed in situ
in leather might be expected because the aggregation levels of
cotton, wool, and leather are similar (1,3). Levels in leather
proceed from nearly visible fiber bundles, 15 to 200 μm in diameter,
composed of fibers, 1.5 to 5 μm in diameter, with the main structural elements crystalline fibrils, 650-2000 A in diameter. These
are subdivided further into protofibrils 35 A, and, finally, into
triple helical tropocollagen chains, 15 A in diameter and 3000 A
long. The obvious difference between leather and other natural
fibers (12) resides in the three dimensional fabric-like structure
of leather. The matrix consists of densely packed fine fibers,
with generally parallel conformation in the grain or surface
region, yielding to coarser appearing, looser, diagonally inter-
woven aggregates of fibers (fiber bundles) in the corium region
below.

In this paper, three acrylate monomers, namely methyl methacrylate (MMA), a mixture of n-butyl acrylate and methyl methacrylate
(BA + MMA) containing 0.591 weight fraction of n-butyl acrylate
and pure n-butyl acrylate (BA) were polymerized into chrome-tanned
5 oz cattlehide by a convenient emulsion process developed at this
center (38-40), and a new method involving bulk or solution polymerization (1). In both processes the effect of composition on
deposition yields, efficiencies, and apparent graft frequencies,
as well as composite densities is emphasized in this work. The
kinetics of the emulsion process is presented to shed light on the
deposition mechanism. Photomicrographs of selected samples of the
composites and of negative replicas prepared by preferential
etching of the collagen fibers with dilute hydrochloric acid
illustrate the morphology. Water absorption is used to monitor

the shielding effect of polymer on the matrix. Finally, mechanical
properties are presented for all of the accumulated composite
materials. An additional purpose of this presentation is to
demonstrate certain similarities of the properties of these
systems with the corresponding properties of commercially important
synthetic composite materials, which have been extensively
investigated (5a-b).

EXPERIMENTAL

Detailed Description

Detailed descriptions of all experimental procedures were
presented in Parts I (1), II (2), III (3), and IV (4). Consequently, only brief outlines of experimental procedures and definitions are presented here to better enable the reader to follow
this work.

Polymerization Procedure

Panels (8.9 x 15.2 x 0.235 cm) were cut consecutively, at all
possible locations, from commercial chrome-tanned grain-split blue
stock cattlehide. Untreated control panels were included with
each treated panel. In the emulsion method, after a separate
30 min conditioning period with the potassium persulfate-sodium
bisulfite redox initiator system, the panels to be treated were
tumbled at ambient temperature under emulsion conditions with the
appropriate monomer for 24 hr. Composite composition was obtained
gravimetrically from methanol or air-dried panels. Homopolymer
was removed from apparent bound polymer by hot benzene extraction.
Standard conditions were: water 5:1 based on dry leather; $K_2S_2O_8$,
4 mole %, based on monomer; $NaHSO_3$/ $K_2S_2O_8$, 0.5; Triton X 100
(1.03%) 2 cc/g based on wet leather. For kinetic studies, the
conditioning period was omitted, and smaller panels (~0.2 X) were
used. Variable changes were made with the standard conditions
used as points of departure. In the bulk or solution process,
panels were saturated with pure monomers or monomers diluted with
benzene, and the polymer present was polymerized with bis azoiso-
butyronitrile (AIBN) as initiator in sealed systems at 60°C.

Morphological, Physical, and Mechanical Properties

Real densities were determined by use of a helium-air pycnometer (1) and apparent densities by measurement of panel volumes
(1). Neat polymer densities, in g cc^{-1}, were: MMA, 1.146; BA +
MMA, 1.103; BA, 1.072, respectively. Light microscopy of stained
sections was used to determine the thickness of layers of deposited
polymer at grain and corium surfaces (discussed below). Rough

estimates of diffusion constants were obtained by monitoring
the weight increase produced by monomer absorbed as a function of
time into films of BA + MMA copolymer under approximate standard
conditions. Light and scanning electron microscopies (SEM) were
used to obtain photomicrographs (3) of thin sections and surfaces
of the controls, composites, and their inverted replicas. The
latter (called simply replicas in this work) were obtained by hot
dilute hydrochloric acid digestion of the collageneous material
from MMA composites, leaving a replication of the fibrous detail
in the continuous polymer phase. Water uptake by total immersion
of the composites and controls in water was followed as a function
of time to 8 days at ambient temperature to obtain rate and equilibrium water absorptivities. Standard ASTM procedures were used
for the mechanical properties and a Williamson* apparatus (4) was
used for mechanical spectroscopy. Molecular weights were obtained
with a Mecrolab 501 membrane osmometer.

Definitions

In this work, mole fraction is designated m; weight fraction,
w; and volume fraction, ϕ. Subscripts are: 1, leather; 2, polymer;
d, deposited polymer; b, bound polymer; max, maximum; a, apparent;
r, real; p, synthetic polymer (or copolymer); f, float; t, total;
l, layered. Quantities W and V represent specimen weight in g and
volume in cc, respectively. Polymers deposited in leather are
identified by their monomers, MMA, BA + MMA, BA, respectively.
Other definitions are as follows:

$$\text{Fraction deposited} = W_2/W_1; \quad \text{fraction bound} = W_{2b}/W_1 \quad (1;1a)$$

Deposited, ε_d, and bound, ε_b, polymer efficiency fraction (1)

$$\varepsilon_d = W_d/W_T; \quad \varepsilon_b = W_b/W_T \quad (2;2a)$$

Bound polymer frequency fraction (1,18)

$$F_b = (\bar{M}_n \text{ collagen}/w_1)/(\bar{M}_{n2}/w_b) = w_b\bar{M}_n \text{ collagen}/w_1\bar{M}_{n2} \quad (3)$$

W_T is the total weight of polymer obtained. The molecular weight
of collagen was taken as 300,000 (1). Because % conversion-time
curves for polymer deposition in the matrix was linear to nearly
100% conversion for all systems (2), a pseudo-zero order kinetic
expression, equivalent to the fraction of monomer polymerized per
unit time, is used in this work to express the initial rates of

---

*Reference to brand or firm name does not constitute endorsement
by the U.S. Department of Agriculture over others of a similar
nature not mentioned.

monomer [M] disappearance at locations in matrix and float. The
expression is (2)

$$[M]_o - [M] = R_i t \tag{4}$$

where R_i is in moles l.$^{-1}$ sec^{-1}. An IBM 1130 computer was used
for curve fitting.

RESULTS AND DISCUSSION

A. Process Efficiencies and Composite Densities

In Table 1 are listed data for BA + MMA composite compositions,
densities, number-average molecular weights, and graft frequencies.
These were prepared under standard conditions in emulsion and by
the bulk (or solution) polymerization techniques. They can be
considered to be representative of the other two systems (MMA, BA)
(1). The effect of time for an emulsion polymerization having a
feed weight fraction of 0.5 is also included (section C). The
rate of change of density with increase in the weight fraction of
polymer, w_2, for the emulsion systems is considerably less than
the corresponding rate by the bulk-solution method, reflecting the
ability of the more expanded matrix of the former to accept polymer,
yet retain more free space. The molecular weights for the emulsion
systems, which generally decreased in the order: bound > deposited
> float for each composition increment, illustrate a strong gel
effect when compared with the molecular weight for the copolymer
prepared in the absence of leather (experiment 7). Considerable
microgel (20-50%) was also found in all of the isolated bound
polymer, which, of course, contributed to their inability to be
extracted by benzene before acid treatment. This provides clear
evidence for polymer to polymer grafting. However graft frequencies, F_b, last column, were low, in harmony with results obtained
for many other natural fibrous composites (7,9,13,15,18,29). It
is also of interest that $\bar{M}_n$ was generally constant with time, for
polymerization occurring in both matrix and float locations
(section C).

Under standard conditions for polymerizations taken to 100%
conversions (Fig. 1), a constant, designated D_e, representing the
efficiency of deposition for the monomer systems employed (because
$D_e = w_2/w_{2\ feed}$), varied as MMA > BA + MMA > BA. A similar order,
Fig. 2, insert A, was found for the rate of polymer deposited, eq.
(1). The reason for these orders can be seen in Figure 2 inserts
B, C, D. These inserts show % conversion-time curves for polymer
deposition in matrix and float, as well as the rate of bound
polymer formation under standard conditions. Generally, bound
polymer rate was about one half of that deposited in all three

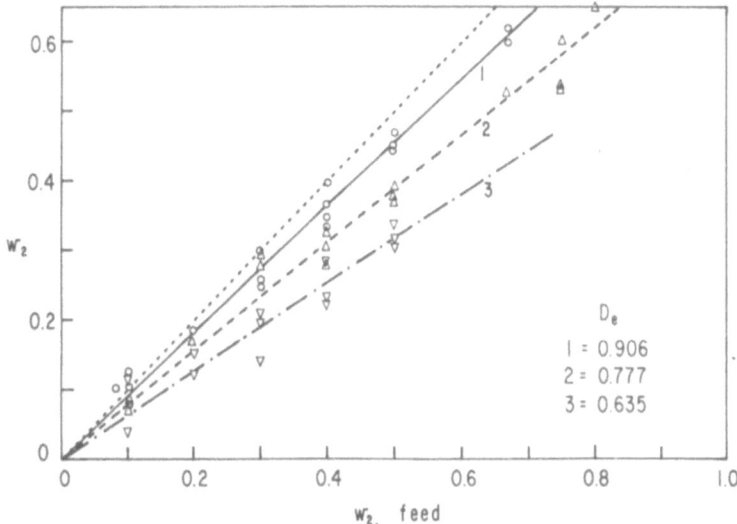

Fig. 1. The weight fraction of polymer deposited, w_2, versus the
weight fraction of monomer in the feed, $w_{2\ feed}$. Desig-
nations are: curve 1, MMA; curve 2, BA + MMA; curve 3,
BA. The slopes represent the deposition efficiency, D_e
as given in the figure. The dotted line is $D_{e\ max}$ = 1.0.

systems. In addition, the linearity of the deposition rate, which
persisted to high conversions, suggests diffusion control of
polymer growth in the matrix. For MMA, insert B, preferential
deposition in the matrix, with little polymer formation in the
float, was found at all conversions. In addition, the rate of
deposition greatly exceeded the rate in the absence of leather.
For BA + MMA, insert C, and BA, insert D, polymer initially formed
in the leather, but polymerization activity was transferred to the
float at an intermediate time before polymerization was complete.
This behavior accounts for the trends in D_e, Figure 1. In contrast
to MMA, the rate in the presence and absence of leather for BA +
MMA insert C was nearly the same, while for BA the overall polymerization rate, R_T, was retarded by leather presence till a critical
time, when a strong Tromsdorff (36a) acceleration was noted in the
float. At long times, Fig. 3, insert A, and under forcing conditions (multiple exposure to fresh monomer under standard conditions)
(1) the rate of deposition was slowed (dashed line) compared to
the initial rates (solid lines) that persisted through several
days of reaction. However, the efficiency of deposition was
greatly reduced because most polymer formed in the float. In

Table 1. Compositions, Densities, and Molecular Weights of a Selection of BA + MMA – Leather Composites Prepared by Both Methods[a]

Expt. no.	w_2 feed	Composite composition w_2 deposited	Composite composition w_2 bound	Bound pol. frn.[b] W_b/W_d	Density g cc^{-1} deposited	Density g cc^{-1} bound	Molecular weights $\bar{M}_n^c$ bound	Molecular weights $\bar{M}_n^d$ deposited	Molecular weights $\bar{M}_n$ float	F_b
A. Emulsion Polymerization										
1	0.103	0.0742	0.0338	0.437	0.622	0.618				
2	0.200	0.182	0.140	0.735	0.665	0.609	492,400		312,400	0.0993
3	0.301	0.294	0.260	0.838	0.601	0.572	605,300		606,000	0.174
4	0.401	0.327	0.279	0.795	0.614	0.573	592,900	891,200	242,400	0.195
5	0.500	0.372	0.306	0.746	0.712	0.616	1,263,200	918,200	505,000	0.105
6	0.750	0.528	0.353	0.488	0.812	0.605	1,162,000	365,100	140,950	0.141
7	1.0	1.0							168,350	
B. Bulk or Solution Polymerization										
8	0.172	0.084			0.683					
9	0.250	0.195			0.807					
10	0.315	0.263			0.845					
11	0.306	0.310			0.962			190,200	88,300	
12	0.455	0.442			1.011					

C. Emulsion Polymerization, Rate Study[e]

No.	time, min									
13		0.479	0.467		1.107				132,050	0.0339
14		0.521	0.667		1.107		744,300			0.0445
15	30	0.0618	0.0383	0.605	0.609	0.624				
16	60	0.104	0.0775	0.720	0.590	0.623	744,300			
17	90	0.149	0.116	0.746	0.646	0.617	883,800			
18	120	0.228	0.201	0.837		0.598	565,600	522,400		0.134
19	150	0.221	0.177	0.756	0.651	0.608	471,400	606,000		0.136
20	180	0.299	0.232	0.705	0.711	0.564	883,800	739,000		0.106
21	240	0.320	0.232	0.642	0.686	0.606	642,800	522,400	300,000	0.141
22	300	0.408	0.317	0.717	0.718	0.570	479,300	473,400	303,000	0.290

[a] Similar data for MMA and BA systems may be found in reference 1, Tables I and V.

[b] Weight of bound polymer divided by the weight of that deposited.

[c] Insoluble in benzene; isolated by HCl etching of collageneous material.

[d] Fraction soluble in benzene.

[e] Another set of data, reference 1, plotted in Figure 2, insert C.

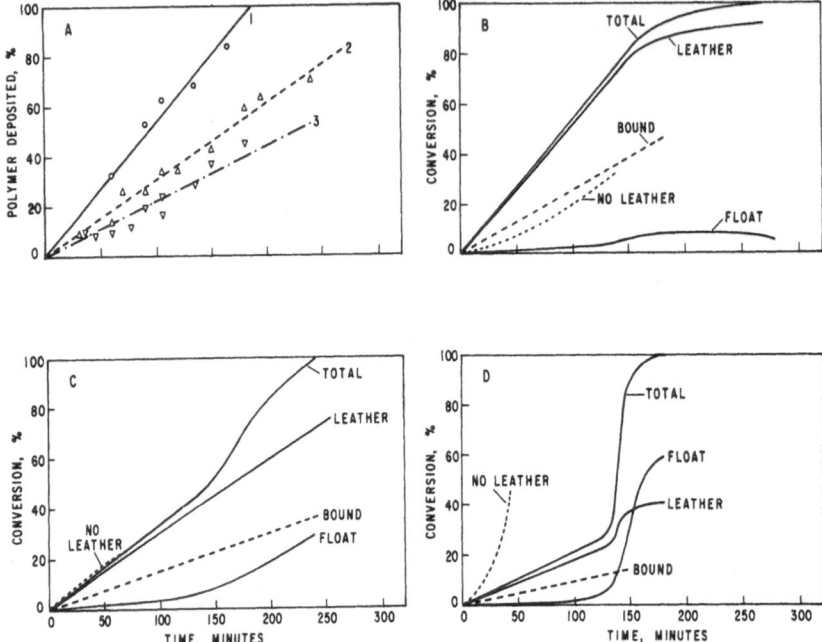

Fig. 2. Rate curves for total polymerization including composite
         formation with $w_{2\ feed}$ = 0.5. Insert A, % polymer de-
         posited in leather versus time; curve 1, MMA; curve 2,
         BA + MMA; curve 3, BA. Inserts B, C, D, percent con-
         version-time curves for: insert B, MMA; insert C, BA +
         MMA; insert D, BA. Concentrations were, in moles l^{-1}:
         MMA, $[M]_o$ = 1.84, $[I]$ = 0.0741; BA + MMA, $[M]_o$ = 1.62,
         $[I]$ = 0.0643; BA, $[M]_o$ = 1.44, $[I]$ = 0.0585; all based on
         water content.

fact, the constant for the efficiency of deposition based on the
total polymer produced ε_d (Fig. 3, insert B) eq. (2) and the
corresponding constant for bound polymer formation ε_b (insert D),
eq. (2a) declined roughly monotonically with feed weight fraction
and time, insert C. The system efficiency corresponding to D_e of
Figure 1 was still in the order MMA > BA + MMA > BA. The data in
these figures show that only a limited amount of polymer could be
efficiently deposited in the leather matrix. This saturation
amount was a function of the monomer used and declined with increase
in the molar volume of the monomer. Under forcing conditions more
polymer could be deposited, but only with greatly reduced efficiency. The effect of the polymer deposited in the residual free
space in the matrix is treated next by considering composite
densities.

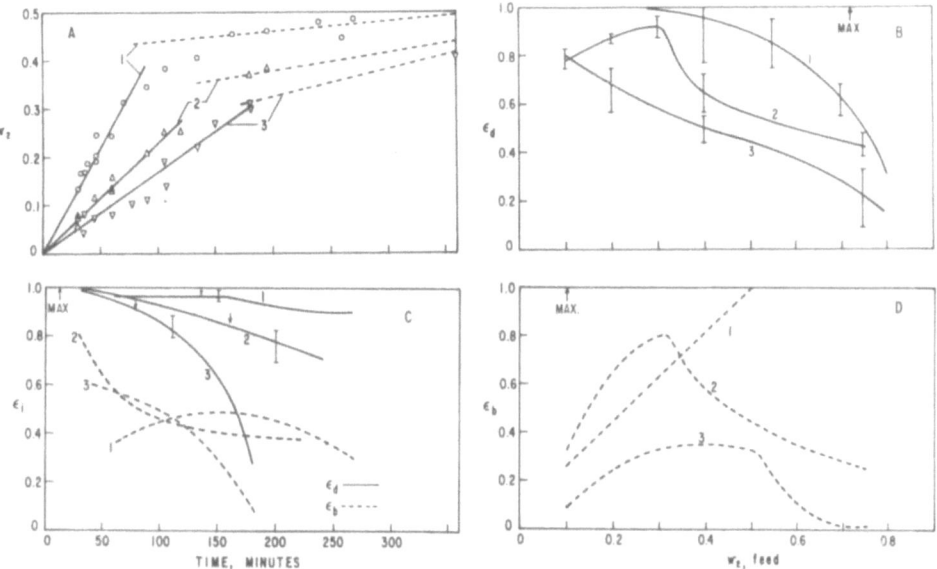

Fig. 3.   Insert A, rates of formation of w_2 for recipe conditions
with $w_{2\ feed}$ = 0.5.   Insert B, specific deposition efficiency, ε_d, as a function of $w_{2\ feed}$.   Insert D is the
same but with ε_b.   Insert C compares ε_d, solid lines, and
ε_b, dashed lines, as functions of time.   Numbers in the
figure correspond to 1, MMA; 2, BA + MMA; 3, BA.   The
maximum efficiency curve is indicated.

In the emulsion prepared composites, most of the polymer was
observed (3) to deposit in untreated leather (shown schematically
in Fig. 4, insert a) in the space around individual fibers, largely
within the confines of fiber bundles (insert f).   In this way,
experimental panels or even full sheepskins (38) increased in
thickness (insert b) without much change in area.   The coarse
polymer domains so formed (2 to 50 μm) were in marked contrast to
the appearance of polymer depositing within the individual fibers
of cotton (7,14) and wool (9), by a variety of polymerization
techniques (1,3,7,9,14).   In these systems domains as small as
20 Å (9) but usually 0.1 to 3 μm (3,14) were routinely observed by
transmission electron microscopy (TEM).   Removal of soluble polymer
by benzene extraction from the cattlehide composites (insert c)
reduced the density (Table 1) while retaining the expanded volumes
(1).   A special feature of emulsion polymer deposition in cattlehide (not found for sheepskins or other thinner, looser leathers)
(38,40) is shown in insert e, where the polymer deposited only in
layers near the outer surfaces (comprising 25 to 60% of the cross
section) (1), leaving the center section polymer free.   A thin,

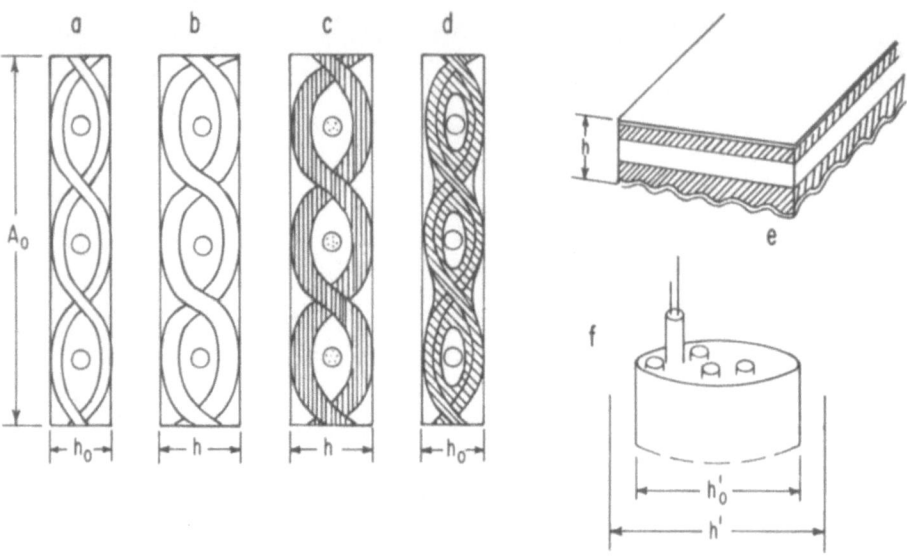

Fig. 4.   Idealized models of the composites discussed in this
           work. Insert a, untreated 5 oz cattlehide, acetone
           dried, showing two fiber bundles, eq. (6); insert b,
           polymer deposited from emulsion, eq. (5); insert c,
           insert b, but after benzene extraction; insert d, loca-
           tion of bulk or solution polymerized monomers, eq. (7);
           insert e, cattlehide panel with polymer preferentially
           deposited in layers at grain (top) and split corium
           regions (bottom) but with small clear zone at surfaces;
           insert f, schematic of fiber bundle with polymer de-
           posited around fibers (arrow). Fibrils are indicated in
           one protruding fiber; these were only occasionally
           surrounded.

clear zone was also observed at both panel surfaces by light
microscopy observation of stained sections. In contrast, polymer
deposited by bulk or solution polymerization, initiated by AIBN
(insert d) was homogeneously distributed throughout the panel,
which usually retained its original dimensions. Although benzene
extractions removed most of the PMMA from these systems, considerable amounts (BA + MMA, 17.2% and BA, 19.2%) remained bound
to the matrix, even though AIBN is a very poor initiator for
grafting (36b). In these systems, incremental space filling by
polymer is considered to be the only possible mode of deposition.
In accordance with the models illustrated in Figure 4, the variation of composite density with composition is treated first
analytically and then experimentally in the sections below.

It was shown previously (1) that the apparent density, ρ_a, of 1 g of the composite $(W_1 + W_2)$ (Fig. 4, insert b) follows the relation

$$\rho_a = 1/\{W_1/\rho_r + W_2/\rho_p + [\phi_{fo}(W_1/\rho_{ao})]\rho_{ao}/\rho_i\} \qquad (5)$$

where $W_1 = (W_1 + W_2) w_1$ and $W_2 = (W_1 + W_2) w_2$ because ρ_r is the real density of leather (1.434 g cc^{-1}), obtained with a helium air pycnometer. W_1/ρ_r is, therefore the volume of collagenous material, W_2/ρ_p is the volume of polymer, and the last term is the volume of free space remaining because ρ_{ao} is the apparent density of un-treated leather (emulsion controls, 0.5556 g cc^{-1}) and ϕ_{fo} (0.6125) is the volume fraction of free space in untreated leather. The quantity ρ_{ao}/ρ_i (0.9238) is an empirical shift factor, discussed previously (1). By similar reasoning, densities for inserts a, c, and d in Figure 4 were derived (1); some of the analytical expressions are listed below. The effect of surface and partial layering on composite density (Fig. 4, insert e) was shown in Part I to reduce to eq. (5) when the whole composite sample was considered. Consequently, the experimental data of this paper were correlated by means of the simple equations listed below.

insert a: $\rho_{ao} = 1/[W_1/\rho_r + \phi_{fo}(1/\rho_{ao})] = 1/V_T \qquad (6)$

insert d:
$$\rho_a = 1/\{W_1/\rho_r + W_2/\rho_p + [\phi_{fo}(W_1/\rho_{ao}) - (W_2/\rho_p)(\rho_r/\rho_p)]\} \qquad (7)$$

$$h = V/A = [(1/W_1)/\rho_a]/A \qquad (8)$$

where h is specimen thickness in cm, V is the volume of 1 g of composite, and A is the area. This is based on no distortion of length or width as polymer deposits (1). Volume fraction of polymer follows:

emulsion:
$$\phi_2 = W_2/\rho_p/\{W_2/\rho_p + W_1/\rho_p + [\phi_{fo}(W_1/\rho_{ao})]\rho_{ao}/\rho_i\} \qquad (9)$$

bulk-solution: $\phi_2 = W_2/\rho_p/(W_2/\rho_p + W_1/\rho_{ao}') \qquad (10)$

where: $\rho_{ao}' = 1/\{1/\rho_r + [\phi_{fo}(1/\rho_{ao}) - (W_2/\rho_p)(\rho_r/\rho_p)]\} \qquad (11)$

Replica density, after dissolution of the collageneous material with hydrochloric acid (see Experimental), follows the general equation.

$$\rho_d = W_2/(V_F + W_2/\rho_p) \qquad (12)$$

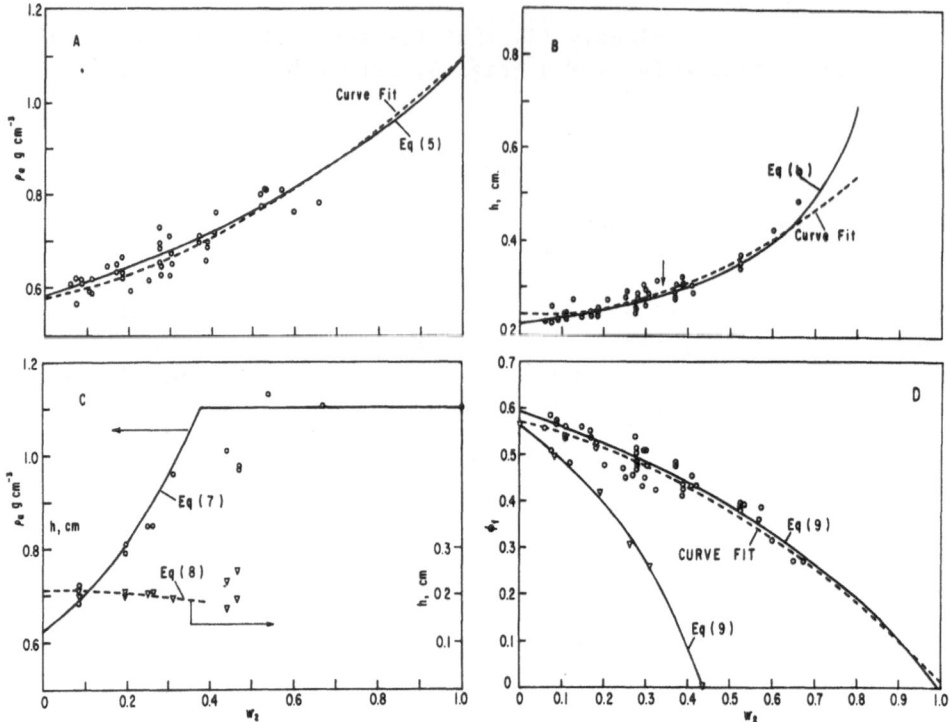

Fig. 5. Plot of apparent density, ρ_a (insert A) and specimen
thickness, h, in cm (insert B) versus the weight fraction
of polymer in the composite, w_2, for deposition of BA +
MMA from emulsion. Plot of ρ_a and h versus w_2 (insert C)
for BA + MMA composites formed by bulk or solution polymerization. Volume fraction of free space, ϕ_f, versus w_2
(insert D) for BA + MMA composites prepared by both
methods.

where V_F is the sum of the volume of free space of the matrix and
eliminated collagen. However, for the emulsion composites, the
effect of layering had to be considered (3).

Experimental data for BA + MMA composites as a function of
the polymer composition, w_2, are compared with the calculated
values in Figure 5. Similar results were found for the other two
systems (1). Curve fitted data (dashed lines) were close to the
theoretical curves (solid lines) for insert A, densities, ρ_a
(g cc^{-1}), and insert B, thickness, h (cm), versus w_2 for the
emulsion systems. The more rapid rate of change of density with
w_2 for the bulk-solution systems (insert C) as well as the influence of thickness is accurately predicted. Finally the difference
in the rate of change of the volume fraction of free space, ϕ_f,
with w_2 is illustrated in insert D for both systems. Consequently

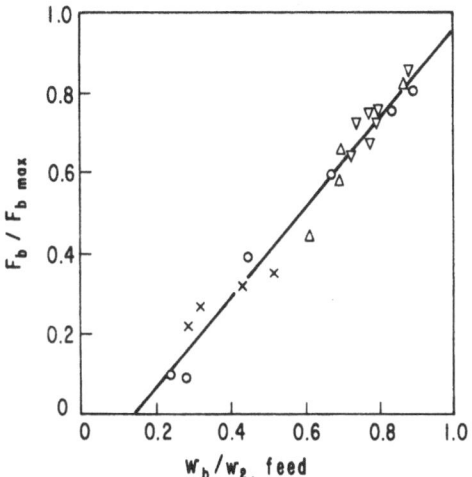

Fig. 6. The quantity $F_b/F_{b\,max}$ versus $w_b/w_{2,\,feed}$ for all three
composite systems. Symbols are: circles, MMA; triangles,
BA + MMA; ∇, BA + MMA, rate; x, BA.

the analysis of density, with due allowance for the natural variability of leather reflected in the considerable experimental
scatter (1), confirms the behavior suggested by the models in
Figure 4. Prediction of residual free space with change in composition and its proper treatment through use of volume fraction in
data comparison can now be utilized.

Graft frequencies, F_b, listed in Table 1, last column, were
estimated by use of eq. (3) from number-average molecular weights
of the bound polymer (assuming that all was grafted). All were
shown (1) to be lower than the maximum attainable, $F_{b\,max}$. The
latter was calculated by eq. (3) with the feed composition.
According to eq. (3), the number of chain branches, $F_{b\,max}$, should
increase with w_2, becoming very large as $w_2 \to 1.0$. However, the
saturation limit of polymer deposition, discussed in the Introduction, intervened. In fact, Figure 6 shows that $F_b/F_{b\,max}$ for all
three systems studied approached zero before w_b/w_2 feed. Since
the latter term is an index of grafting efficiency at 100% conversion, decreasing from right to left in the figure as monomer
feed content increased, the number of branches actually formed
declined at a faster rate, and thus demonstrated even more resist-

ance to forcing conditions. This argues strongly against a facile
grafting mechanism. In the section below, the kinetics of bound
polymer formation, in both locations, matrix and float, is treated
as a function of the reaction variables to ascertain the type of
mechanism actually operating in forming these polymer-leather
composite materials.

B. Kinetics and Mechanism of Deposition

The mechanism of grafting polymer to fibrous surfaces is
usually considered (7-12) to involve primary radical attack on
substrate, terminating initiator activity, with simultaneous
formation of an active center on the substrate that initiates a
grafted chain. A rate law, involving assumption of homogeneous
reaction kinetics, has been suggested (2)

$$R_p = k_p/k_t^{1/2} \ [M]^{3/2} \ [I]^{1/2} \ fk_{de}^{1/2} + (k_{tr}[LH])^{1/2} \qquad (13)$$

where k_p, k_t, k_{de}, and k_{tr} are specific rate constants for propagation, termination, decomposition, and transfer respectively, and
[M], [I], and [LH] are the concentrations (mole l.$^{-1}$) of monomer,
initiator and leather, respectively. This is similar in form to
that proposed for Ce^{IV} initiation on polyols (41-46).

In contrast, the rate law for the nearly steady-state rate
(interval II, conversion 5 to 50%) of polymerization in emulsion
(2,47-50) is

$$R_p = -d \ [M]/dt = (k_p/N_A)(\rho_m/\rho_p)\phi_2 NQ \qquad (14)^{\bullet}$$

where N_A is Avogadro's number, ρ, density of subscripts m (monomer)
and p (polymer), respectively; ϕ_2 the volume fraction of monomer
in the particle; N the number of particles; and Q the average
number of radicals per particle (ideally 0.5) (47). Thus, in the
ideal case, rate is proportional to only the number of particles
and the monomer concentration per particle. The number of particles, N, depends on surfactant [S] and initiator concentration
through (49)

$$N = 0.208 \ [S]^{0.6} \ R_i/K^{0.4} \qquad (15)$$

where $R_i = 2k_{de}N_A \ [I]$ and K the volume growth rate of a particle.
Equation (14) applies equally well to micellar stabilization and
self-nucleation in aqueous dispersion (51,52). If the colloid
particles coalesce, the limiting Case III condition applies (47)
and

$$R_p = k_p[M](V_p k_{de}N_A[I]/k_t)^{1/2} \qquad (16)$$

where V_p is the total volume of the polymer particles. This equation is identical in form to the normal rate law for homogeneous polymerization.

$$R_p = k_p/k_t^{1/2} [M] [I]^{1/2} (fk_{de})^{1/2} \tag{17}$$

From the foregoing, grafting (eq. (13)) is expected to show a 3/4 power dependence on [M] and is half order in initiator and leather amount, respectively. If emulsion kinetics prevail, the exponential dependence is 0.4 in [I] and 0.6 in [S], but zero in monomer disappearance from the stabilized reservoirs in the aqueous phase (2,47-50). Consequently the exponential intensity factor, a, of the general equation for rate, R,

$$R = R_o[V]^a \tag{18}$$

where [V] is the variable concentration, should permit a choice to be made of the dominant mechanism that pertains in these systems.

The effects of the initiator concentration, [I], on rate (eq. (4)) for MMA systems are shown graphically in Figure 7. These illustrate the type of information obtained and are typical of all of the kinetic curves. The curves, starting at the top, represent the total rate, R_T, the rate of deposition, R_d, the rate of bound polymer formation, R_b, and the rate of polymerization in the float, R_f. Most of the rate data were linear to about 40-50% conversion, except R_d and R_b, which were linear to 85 to 90% conversion (Fig. 2, inserts B and C). The data were maintained therefore within the interval II (3,47-50) conversion region in emulsion polymerization. After the rate data were inserted in eq. (18), the exponential intensity factors for all of the variables were obtained and are listed in Table II. The dependence of the factor, a, on monomer concentration for MMA, (Figure 8), and BA + MMA (Figure 9), was zero because rates in both matrix and float were constant as the monomer concentration in the float varied. Thus (Table II) the effect of monomer concentration is more in harmony with the theoretical requirement of emulsion polymerization than that expected of a dominant grafting mechanism. Similarly the dependence of a for R_d on [I] and [S] exhibits more typical emulsion polymerization behavior. However, the anomalously large value of a for R_d as a function of [I] is clear evidence of a gel effect (36b), which would be expected as the polymer particles coalesce on being deposited in the leather matrix (2,3), thus trapping radicals and lengthening their lifetimes. In the absence of leather, a became closer to theory. In contrast, R_b showed erratic behavior and much experimental scatter (2) in plots based on eq. (18). This suggests that the bound polymer was patterned from the deposition rate as a secondary consequence and was not the guiding mechanism as specified by eq. (13). While the initial

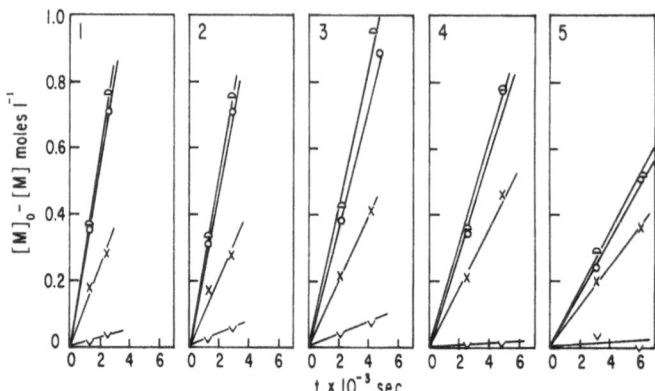

Fig. 7.   Rate of monomer consumption versus time for the MMA
          composite systems of Table 2 at varied initiator con-
          centrations.   The order of the rate curves in each insert,
          top to bottom, is: R_{T1}, R_d, R_b, R_f.   Initiator concentra-
          tions are, in mole l^{-1}: insert 1, 0.1945; insert 2,
          0.1552; insert 3, 0.0750; insert 4, 0.0582; insert 5,
          0.0402.

rates all showed an apparent dependence on leather amount (Table 2
and Fig. 2, inserts B, C, D), the magnitude of a decreased
rapidly to nearly zero (Table 3) after 1 hr reaction for BA +
MMA and BA systems.   As seen in Table 3, the total rate of
polymerization, R_T, became constant with [LH] in this interval,
when conversions were only around 17%.   Consequently the amount of
leather initially present had no permanent influence on rate for
these systems.   The apparent accelerating effect of [LH] on MMA
systems may have resulted from strong Tromsdorff retardation of k_t
in the highly viscous MMA polymer aggregate.

     From the foregoing, a mechanism of deposition that proceeds
from the observed dependence on emulsion kinetics through eq. (14)
is suggested as an alternative to grafting.   Polymer first appears
in particles in micelles or by self-nucleation (51,52) in the
float and in the aqueous phase of the leather matrix.   However,
more polymer forms initially in the matrix because of preferential
migration of persulfate ion during the reaction (2,53).   Growing
particles in the matrix and those entering from the float coalesce
and trap radicals.   The embryo deposit grows by diffusion of
monomer from the surfactant stabilized monomer reservoirs, through

Table 2. Exponential Intensity Factor, a, for All Composite Systems

Reaction variable, Moles l^{-1}	Emulsion theory eqs. (14) and (15)	Graft theory eq. (13)	Intensity constant, a					
			MMA		BA + MMA		BA	
			R_d	R_b	R_d	R_b	R_d	R_b
[I]	0.4	0.5	0.723	0.444	0.657	0.841	-	-
[M]	0	1.5	0	0	0	0	0	0
[L]	0	0.5	0.885	1.05	0.636→0[a]	0.745→0[a]	0.667→0[a]	0.702→0[a]
[S]	0.6	0	0.509	0.294	0.494	0.480	-	-
[I] no. lea.					0.500			
[M] no. lea.					0			
[S] no. lea.					0.793			

[a] Dependence of the total polymerization rate, R_T, decays to 0 by an average of 17% conversion; initial $R_d = R_T$. See Table III.

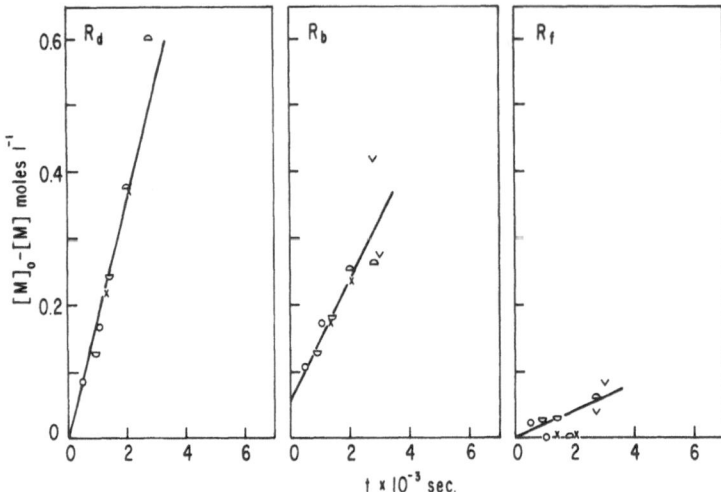

Fig. 8.   Rate of monomer consumption versus time for the MMA
          composites with monomer concentrations varied. The MMA
          concentrations used were: θ, 0.487 mole l^{-1}; ∇, 0.816
          mole l^{-1}; x, 1.28 mole l^{-1}; Δ, 1.92 mole l^{-1}; v, 0.485
          mole l^{-1}; all based on water.

the polymer layer, and is polymerized by the occluded radicals.
In the present work, 20 to 50% of the bound polymer resulted from
microgel formation. However, a considerable amount most probably
resulted from strong adsorption to collagen surfaces. It is now
known (5d) that ordered layers of up to 7000 A can form near
surfaces and become immobilized. These are difficult to remove.
This adsorption can thus contribute an enhancement to the high
tortuosity of solvent diffusing through filled polymers. Thus, in
the time scale of most extractions (35) much polymer would be
retained. Finally, actual grafting at fiber-polymer interfaces
would contribute a small amount of bound polymer.

     The data in Table 4 provide additional evidence for this
mechanism. The preferential migration of initiator, as reflected
in the quantity $[I]/[I]_0$ (2), was higher for MMA and BA + MMA
systems following a 30 min conditioning period than the statistical
ratio (0.28). However, the initial equilibrium adsorption of
BA + MMA monomer by the collagen matrix was less than the statistical amount. Consequently, insufficient monomer was initially
imbibed into the matrix to support much grafting. In contrast,
when preformed BA + MMA emulsion polymer was impregnated into the

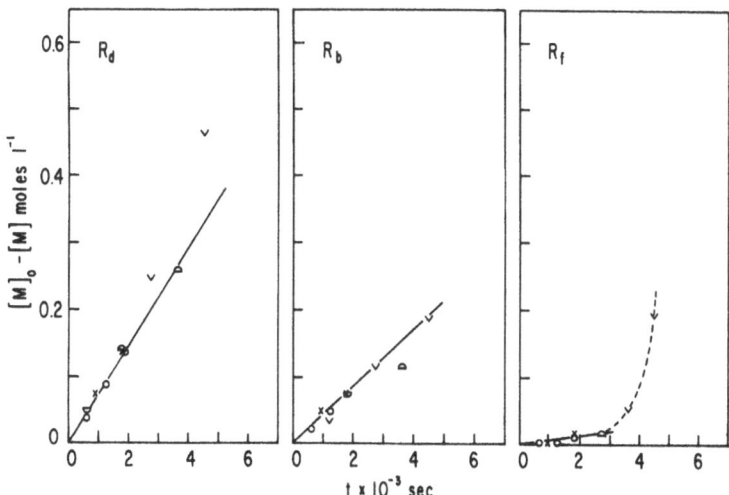

Fig. 9.   Rate of monomer comsumption versus time for the BA + MMA
           composites with monomer concentrations varied. The
           BA + MMA concentrations used were θ, 0.432 mole l^{-1}; ∇,
           0.746 mole l^{-1}; x, 1.096 mole l^{-1}; Δ, 1.70 mole l^{-1}; v
           3.34 mole l^{-1}; all based on water.

hydrated leather matrix, under standard conditions, about 50% of
this was bound. This provides strong support for polymer adsorption
because covalent bonding was impossible under the experimental
conditions used. Because rates of deposition were unaffected by
monomer amount, the ratio $R_d/[M]$ can be used to yield a propor-
tionality constant that has first order dimensions (column 2).
The effect of using ground leather on the magnitude of this constant
is shown in the last three columns. Although the deposition rate
appeared to be large in a concentrated system close to standard
conditions for intact leathers, the physical state was that of a
thick paste (2), ideally suited for polymer entrapment. At high
dilution, most activity was tranferred to the float. For both
concentrations, gel effects were greater than for initial leathers.
Because grinding should have enhanced bound polymer formation if
grafting were dominant because of the greater collagen surface
exposed, these results are also at variance with this type of
mechanism.

     Diffusion control of monomer transport through the polymer
layer gains credence from Figure 10. Insert A shows that the
system deposition constants, D_e, of Figure 1 are directly propor-

Table 3. Effect of Leather Amount [LH] on the Initial Total Polymerization Rate and the Rate after One Hour of Reaction

| | MMA | | | BA + MMA | | | BA | |
| | R_T x 10^5 moles l sec^{-1} | | | R_T x 10^5 moles l sec^{-1} | | | R_T x 10^5 moles l sec^{-1} | |
[LH] moles l^{-1}	initial[a]	1 hr[b]	[LH] moles l^{-1}	initial[a]	1 hr[b]	[LH] moles l^{-1}	initial[a]	1 hr[b]
1.96	16.5	18.2	1.96	8.40	8.96	1.98	5.00	5.56
1.37	11.0	13.6	1.34	6.50	7.61	1.26	4.00	5.67
1.03	8.50	10.4	1.02	5.45	8.78	0.960	3.00	5.08
0.690	6.17	6.95	0.711	4.08	8.94	0.625	2.40	7.26
0.319	3.20	3.62	0.341	2.75	8.31	0.302	1.45	4.81
0	2.10	2.10	0	8.25	9.75	0	14.0	27.8

[a] $R_d = R_T$ initially, because $R_f = 0$.

[b] In 1 hr the average % conversion was: MMA, 17.3; BA + MMA, 19.2; BA, 13.9.

Table 4. Steady-state Rate Constants, Initiator Distributions, Monomer and Polymer Impregnation Results, and Effect of Ground Leather on Polymerization Rate

System	$k_d \times 10^5$ sec^{-1}	$[I]/[I]_0$,[a] After 30 min	Leather impregnation, pre-formed polymer			Leather impregnation, monomer		Deposition in ground leather		
			$[P_d]^b$ moles l^{-1}	$[P_b]$ moles l^{-1}	$[P_d]/[P_T]$	$[M_d]^c$ moles l^{-1}	$[M_d]/[M_T]$	$k_d \times 10^5$ sec^{-1}	$k_f \times 10^5$ sec^{-1}	$k_T \times 10^5$ sec^{-1}
MMA	8.81	0.692								
BA + MMA	4.58	0.740	0.193	0.103	0.118	0.103	0.0644			
BA	3.56									
Powd., conc.[d]								22.1	6.18	28.0
Intact, conc.[d]								4.53	4.72	8.02
Powd., dil.[d]								4.15	32.5	40.4
Intact, dil.[d]								1.45	9.69	11.8

[a] [I] is the initiator concentration in the leather; the statistical ratio is 0.277.

[b] Initial polymer concentration in the emulsion was 1.638 Ml^{-1}. $[P_d]$ and $[P_b]$ represent the average polymer concentrations found for six impregnation experiments, of varied times between 0.5 and 5 hr.

[c] Initial monomer concentration in the emulsion was 1.638 moles l^{-1}.

[d] BA + MMA monomer. In the concentrated systems, monomer concentration was 1.13 moles l^{-1}; dilute systems, 0.277 moles l^{-1}; standard conditions, 1.67 moles l^{-1}.

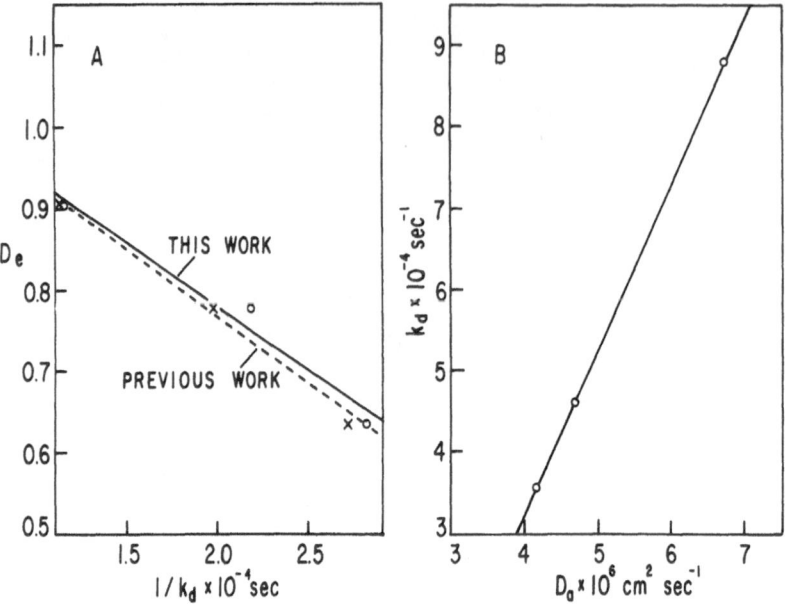

Fig. 10.   Relation (insert A) between the system efficiency constant, D_e, and $1/k_d$ of this work (from Table 4) and that calculated from the data in Table 3 of a previous work.[1a]   Insert B illustrates the relation between k_d of this work and the apparent diffusion constant, D_a, for the monomers through a BA + MMA panel. Values of D_a x 10^6 cm^2 sec^{-1} were: MMA, 6.72; BA + MMA, 4.69; BA, 4.16.

tional to the reciprocal of the deposition rate constants k_d from Table 4.   The observation that k_d was directly proportional to the diffusion constants for the three monomers through BA + MMA film (insert B) suggests that most of the monomer arrived at the active center by passage through the deposited polymer.   The preference of the polymer for the outer layers of the matrix (Fig. 4, insert e) appears to be regulated (Fig. 11) by the mean diffusive path, L in cm.   This quantity approaches a limiting value when plotted as a function of the experimental layer thickness, h_p, in cm, for all three systems, in spite of considerable differences in the rates of deposition of the monomers used.   The data for Figures 10 and 11 do not include tortuosity effects (5d). Use of a BA + MMA film as a model for the plasticized deposited polymer can, of course, lead to errors.   However, these are considered to be small if only qualitative significance is attached to the results and when compared against those of other types of possible monomer transport.

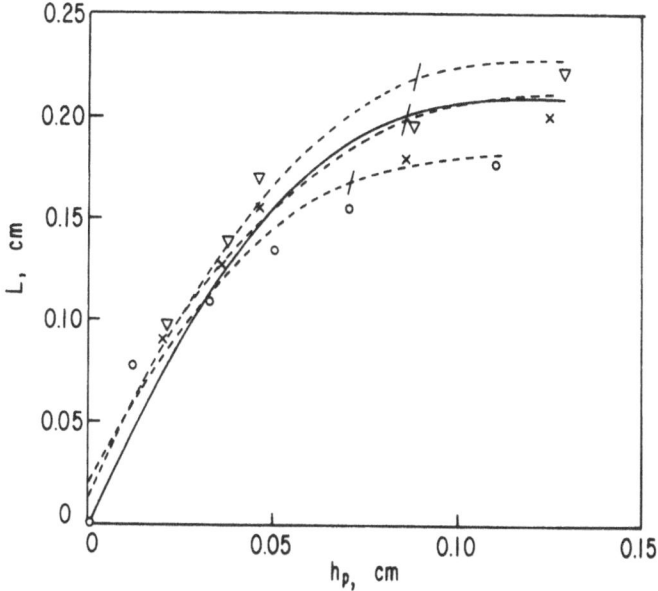

Fig. 11. Mean diffusive distance, L, versus the overall polymer
thickness in the composite, h_p, as a function of time.
Symbols are: θ MMA; x, BA + MMA; ∇, BA. Solid line is
the average for all data. Slash marks designate the
limiting h_p corresponding to the termination of significant deposition for each system.

C. Morphology of the Deposited Polymer

Extensive use has been made of the light (13,54,55) transmission (7,9,14,56-62) and scanning electron microscopes (63) in
elucidating the structure of polymer composites of cotton, rayon,
and wool. In both cotton (14) and wool (9), transmission electron
microscopy (TEM) revealed that polymer could, by proper choice of
reaction conditions, be deposited through all regions of the fiber
cross sections, residing, in some cases, even around microfibrils
in wool (64) to produce domains as small as 20 A thick (9). The
ultrafine regions of goat skins, presumably in the vicinity of
protofibrils, have been said to be permeated by depositing polymer
(65), but some doubt remains concerning this interpretation (1,3).

Replica densities as functions of composite polymer composition, w_2, are shown in Table 5, together with the extent of replica
shrinkage relative to the initial composite volume. Also included

Table 5. Replica Densities, Extent of Replica Shrinkage, and Rates and Water Absorbtivities of the Composites and Their Controls

Expt. no.	System	w_2	Replica density		Fractional extent of shrinkage[b]		Rates of water absorption[c]			Equil.[d] absorb. w_{max}/w_{so}
			ρ_d calc.[a] g cc^{-1}	ρ_d found g cc^{-1}	V_r/V_{ro}	ϕ_f/ϕ_{fo}	w_s 5 sec	w_s 10 min	w_s 8 days	
			Poly (Methyl Methacrylate) - Leather Composites and Replicas							
1*	emuls., b.[e]	0.447	0.490	0.483	0.938	0.826				
2	"	0.335	0.363	0.345	0.900	0.916				
3	"	0.203	0.236	0.326	0.783	0.865				
4	"	0.0894	0.104	0.280	0.458	0.774				
5	emuls., d.	0.470	0.543	0.453	0.995	1.02				
6	"	0.347	0.402	0.381	0.963	1.03	0.300	0.550	0.593	1.05
7	"	0.258	0.300	0.308	0.942	0.938				
8	"	0.182	0.212	0.250	0.871	0.949	0.536	0.570	0.605	0.967
9	emuls., a.	0.447	0.517	0.433	1.01	1.01				
10	"	0.366	0.424	0.410	1.04	0.955				
11	bulk, d.	0.423	0.535	0.423	1.03	0.995				
12	"	0.292	0.272	0.288	0.989	0.975				
13	"	0.234	0.199	0.219	1.01	0.995				

No.										
14	"	0.189	0.145	0.180	0.942	0.977	0.240	0.440	0.494	1.06
15	"	0.155	0.118	0.202	0.768	1.01				
16	"	0.397					0.175	0.205	0.339	1.87
n-(Butyl Acrylate-co-Methyl Methacrylate) - Leather Composites										
17	emuls., d.	0.189	0.404	0.262	0.857	0.821	0.040	0.550	0.591	1.03
18	"	0.304	0.436	0.420	1.07	1.09	0.200	0.525	0.590	1.04
19	bulk, d.	0.174					0	0.510	0.570	1.22
20	"	0.310					0	0.130	0.480	1.54
n-(Butyl Acrylate) - Leather Composites										
21	emuls., d.	0.208					0	0.300	0.527	0.902
22	"	0.303					0	0.440	0.525	0.910
23	bulk, d.	0.143					0	0.120	0.433	0.884
24	"	0.281					0	0.095	0.408	1.17
Average of All Controls										
25	–	0					0.526	0.544	0.590	0.966

[a] Eq. (12).

[b] V_r is the measured volume of the replica and V_{ro} that of the original composite, but thickness for both was calculated by eq. (8). The volume fraction of replica free space is ϕ_f, given by V_a/V_T from replica density; ϕ_{fo} is 0.6125.

(continued)

Table 5 (continued)

[c] w_s is the weight fraction of water adsorbed.

[d] w_{max} is the weight fraction of water absorbed after 8 days; w_{so} is $\phi_{fo} \, \rho_w$, where ρ_w is the density of water at 23°C.

[e] Letters designate, bound polymer, b; deposited polymer, d.; and air-dried, a.

* Denotes sample used for micrographs.

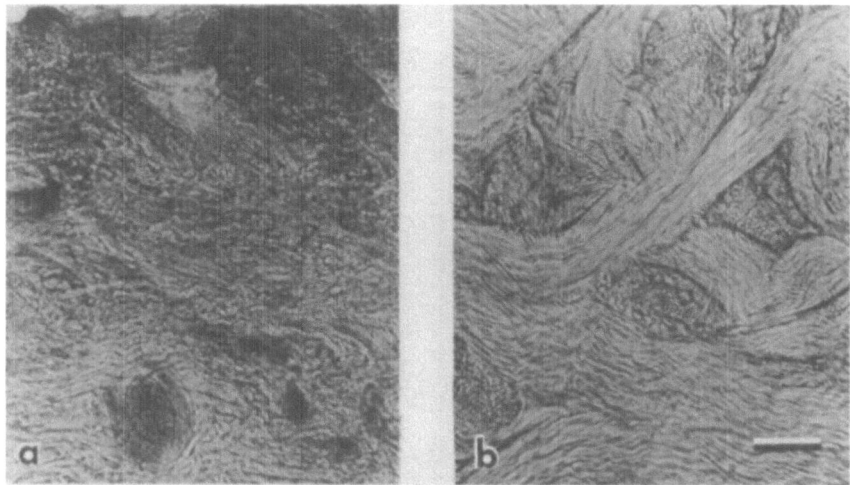

Fig. 12. Light microscope (LM) micrographs of untreated 5 oz
          chrome-tanned cattlehide (control) showing cross sections
          of portions of the grain layer (insert a) and the
          corium layer (insert b). The lower half of insert a is
          the grain-corium interface region. The dark circular
          area is the root of a hair follicle; the other irregular
          dark areas are fragments of capillaries or fatty deposits.
          The scale is 100 µm.

are significant time increments for the rates of water absorption
(columns 8, 9, 10) and the equilibrium water absorptivity (last
column) for composites compared with their controls. While replica
densities, both calculated (eq. (12)) and found, decreased as w_2
decreased, the rate of change was less for the emulsion system at
each w_2 increment than for the bulk-solution systems, reflecting
the greater exhaustion of the space brought about by concentrating
the polymer in layers. Shrinkage by both methods of composite
preparation was generally small (columns 6, 7) especially for
experiments 5 through 17. Thus the micrographs of the replicas
should reflect the actual morphology of the continuous polymer
phase present in the composites before fiber removal. Figure 12
presents light photomicrographs of untreated controls at 40X
(100 µm) for, the grain (top) and transition region below (insert
A), and the corium region below that in insert A (insert B). The
dense finely structured grain fibers and fiber bundles contrast
with the much coarser, more articulate fiber bundles. Individual

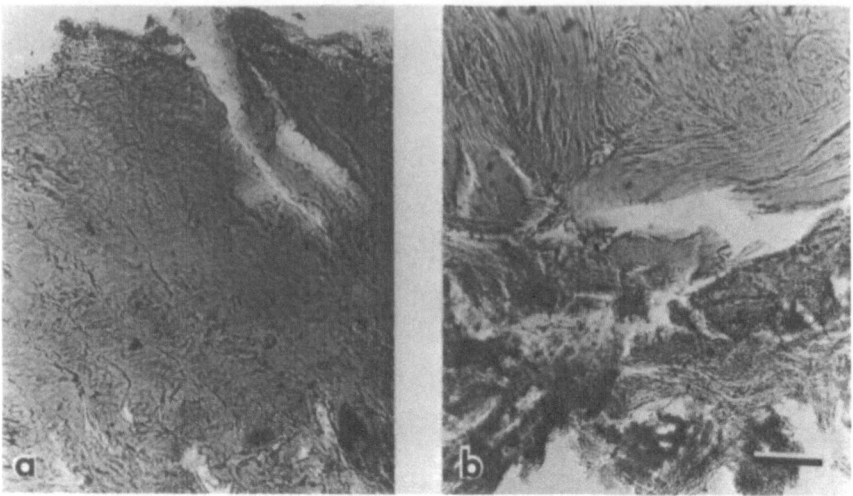

Fig. 13.   LM micrographs of the cross sections of the replicas
           prepared by hydrochloric acid etching of the composites
           made from the untreated leathers in Figure 2.   Insert a
           is the grain layer; insert b is the corium layer.   The
           scale is 100 μm.

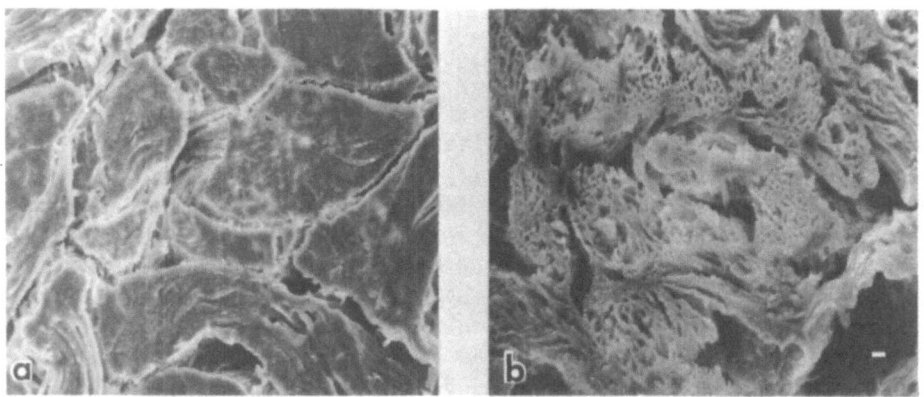

Fig. 14.   Scanning electron micrographs (SEM) of cross sections of
           bound polymer-leather composites, prepared in emulsion
           (Table V), insert a, and the corresponding replica,
           insert b, prepared by hydrochloric acid etching, showing
           the effect of removing the collageneous material, selected
           from similar regions of the corium.   The scale is 10 μm.

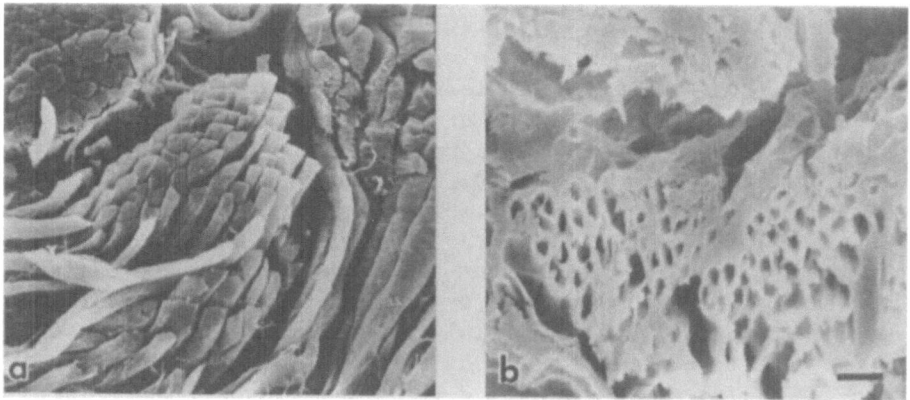

Fig. 15.   SEM of a cross section of untreated leather (insert a)
            at higher magnification (1,000X) than that of Figure 14;
            insert b (300X), compared with a typical section of the
            bound polymer replica (insert b) of Figure 14.   Fiber
            diameters correspond approximately to replica openings;
            the polymer is largely confined to the fiber bundle.
            The scale is 10 µm.

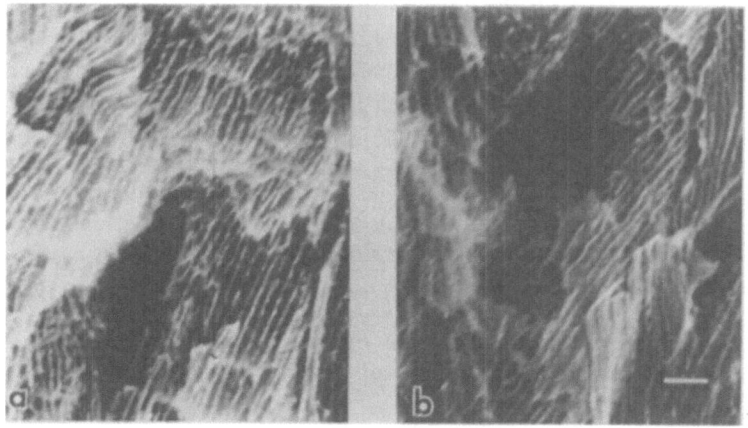

Fig. 16.   High magnification (10,000X) SEM of the cross section of
            the corium region of a replica of a composite (Table 1,
            exp. 1) inserts a and b were taken from different posi-
            tions in the replica.   The linear hollow tubes are
            approximately 500 to 2,000 Å in diameter and correspond
            to the dimensions of fibrils.   The scale is 1 µm.

fibers can be seen as fine striations in the micrographs.   Corresponding replicas of the bound polymer (experiment 1, Table 5)

from grain (insert a) and corium (insert b) in Figure 13 reflect
this fiber morphology but still reveal the plastic character of
the deposit. Scanning electron micrographs (SEM) of the cut
surfaces of the same composites (Fig. 14, insert a) showing the
expanded fiber bundles with polymer packed around individual
fibers (shown schematically in Figure 4, insert f), which, when
removed by dilute acid (insert b) reveal holes and tubes 1.5 to
4 μm in diameter in the replica, corresponding to typical fiber
diameters. This can be seen better in Figure 15 where, at higher
magnification (1000X), openings of the fiber diameter are readily
seen. Because these openings were not larger than initial fiber
diameters (insert a) and the visible polymer corresponded closely
with the composite weight fraction, most of the polymer deposited
around fibers in coarse domains, 2 to 50 μm. In this way, fiber
bundles expanded to produce the matrix expansion discussed in
terms of models in Figure 4. However, some polymer in the emulsion
systems did penetrate to lower aggregation levels. Figure 16
shows replicas at 10,000X that reveal tubular traces 300 to 800 Å
originally surrounding fibrils, which had typical diameters of 500
to 2000 Å. Other features of the morphology of these composites
illustrated by use of SEM in a more thorough study (3) revealed an
extremely complex morphology. As polymer composition in the
composites was reduced, all replicas became more porous but possessed similar morphology. Bulk and solution replicas appeared to
be rather dense, especially at high w_2, because polymer resided in
large pores. However, it was excluded from the interfibrillar
regions. Air-dried or methanol-dried replicas revealed partially
aggregated colloidal polymer spheres, 2000 to 10,000 Å in diameter,
clustered on fused polymer aggregates. These were the original
emulsion particles before coalescence. In general, the micrographs
revealed a morphology in harmony with the idealized models in
Figure 4 and with the deposition mechanism set forth in section B
above. The influence of this morphology on water absorption was
also considered.

The rate of water absorption by the untreated control in
Table 5 (average values, experiment 25) followed the simple relation (3)

$$w_s = w_{si} + klnt \tag{19}$$

where w_s is the weight fraction of water absorbed by the sample,
w_{si} is the instantaneous (1 sec) weight fraction of water imbibed,
k is a constant (0.00670 sec^{-1}), and time, t, was followed up to
8 or more days. Rates for the composites were estimated by comput-
ing a corrected absorption weight fraction, $w_{s\ corr.}$, (from w_s in
eq. (19)) to the state of pure collagen by use of $w_{s.corr.} = $
f_s/w_1, where $f_s = W_s/(W_1 + W_2)$ is the observed fractional water
absorption. Rate became vanishingly small after 8 days.

Table 6. Selected Mechanical Properties of Poly-(Methyl Methacrylate)-Leather, Composite Materials

Expt. no.	Composition		Tensile strength		Initial tensile moduli		Torsional moduli, methanol-dried		Torsional moduli, air-dried	
	w_2	ϕ_2	TS, psi	TS/TS_1	E, psi	E/E_1	E_t, psi	E_t/E_{t1}	E_{t1}, psi[a]	E_t/E_{t1}
1	0	0	1,722	1	1,290	1	1,944	1	32,000	1
2	0.0807	0.0423	1,275	0.970	3,450	2.60	2,050	1.37		
3	0.103	0.0543	1,725	1.03	5,550	4.18	1,300	1.33	31,200	0.577
4	0.182	0.0992	1,810	1.03	5,100	3.84	4,500	4.50	38,800	0.443
5	0.235	0.131	1,220	0.649	50,600	45.2	14,000	11.2		
6	0.258	0.145	1,560	0.650	8,980	5.51	7,800	5.20	26,500	0.891
7	0.347	0.202	1,215	0.631	9,840	9.93	4,800	4.00	32,000	1.10
8	0.470	0.288	1,395	0.542	41,400	20.5	29,000	20.3	39,500	2.61
9	0.504	0.314	2,170	1.28	65,400	76.0	44,000	33.8		
10	0.657	0.439	918	0.461	39,400	25.2	37,000	30.8	51,000	3.63
11	1.0	1.00	7,760	4.51	220,000	165.7	540,000	27.8		

a Approximate relative torsional stiffness for all three polymer systems at $w_2 \sim 0.25$ were: MMA, 1; BA + MMA, 0.386; BA, 0.342; controls 0.139.

b Corresponding composite w_2 and ϕ_2 were: expt. no., w_2, ϕ_2; 1, 0, 0; 3, 0.127, 0.069; 4, 0.186, 0.104; 6, 0.298, 0.170; 7, 0.342, 0.199; 8, 0.478, 0.292.

Consequently, this was the time usually taken for equilibrium absorption in Table 5. Because it was shown $(3)_1$ that wet densities remained constant after 2 sec (1.09 g cc^{-1}) the matrix expanded in proportion to the water imbibed. A limiting weight fraction of water absorbed, w_{so}, could therefore be computed from $\phi_f \rho_w$, where ρ_w is the density of water at $23°C$ and ϕ_f the volume fraction of free space in the emulsion composites, obtained by rearranging eq. (9). The ratio, $w_{s\,max}/w_{so}$, with $w_{s\,max}$ the observed equilibrium weight fraction of water absorbed in 8 days is an index of the effectiveness of deposited polymer in preventing water absorption in these composite materials. When the ratio is unity, no shielding of collagen occurs; values <1 indicate reduced absorption and values >1, enhanced absorption. For bulk or solution composites, w_{so} tended to be underestimated because of the rapid decline of ϕ_f with w_2 increase (Fig. 5, insert D). Because $w_{s\,max}$ should increase normally here since interfibrillar regions were free of polymers (3) the observed ratios (Table 5) may have increased abnormally.

Results in Table 5 show that in the emulsion systems polymer presence slowed the rate of absorption (rate data to be compared with that of experiment 25). The rates were much slower in the bulk-solution process, than in the emulsion process; some systems (experiments 14, 16, 20, 23, 24) did not reach equilibrium even after 8 days. In contrast, the equilibrium absorption index ($w_{s\,max}/w_{so}$) was nearly always unity for the emulsion systems, but was abnormally high for certain bulk systems (experiments 16, 19, 20, 24), for reasons discussed above. It may be concluded that when polymer is deposited in leather by either process, little permanent protection was observed from imbibed water under static conditions.

D. Mechanical Properties

The mechanical properties of cotton and wool and other natural fibers and fabrics containing grafted and deposited polymers have been discussed as part of reviews (4,7,9,12). In general, for treated cotton fibers, tenacity (13,57,60,61,66-67), break toughness (57,60,61,66-67), and stiffness (13,57,60,61,66-67) generally decreased, while elongation remained constant. Flex and flat abrasion resistance for fabrics was greatly improved (68-70). Similar behavior was observed for wool fibers (9,71-74) and fabrics (55,74). While many mechanical properties of leathers have been reported (75-76), including the effect of natural variability (77-79), the mechanical properties reported on polymer grafted composites have been largely limited to the determination of tensile strengths (4,80,81).

The mechanical properties selected for this work were limited to the MMA systems in Table 6 and were specifically restricted to methanol extracted and air-dried composites. Similar data were

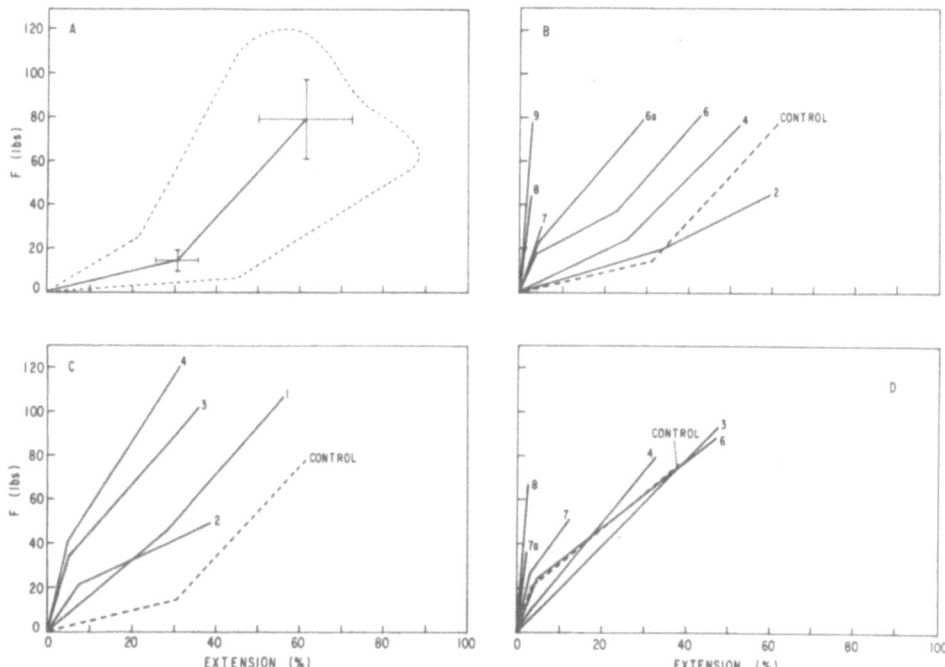

Fig. 17. Simplified force-extension curves. Insert A, averaged curve for all untreated controls. The bars designate standard deviations of both force and extension for each linear region and the dotted envelope is approximately the extreme limits observed. Inserts B (MMA) and C (BA + MMA), curves for emulsion prepared composites, methanol extracted only. Insert D, air-dried composites (solid lines) and the average curve for air-dried controls (analogous to insert A) dashed line. Numbers in inserts B and D correspond to those in Table 6. Insert C, exp. no., w_2: 1, 0.0875; 2, 0.280; 3, 0.389; 4, 0.523.

observed (4) for the BA + MMA and BA composites. Results on those systems not listed in Table 6 are treated in the figures below. Although tensile strengths did not decrease appreciably when compared to the average of the controls (experiment 1), natural variability of the controls was great. Consequently, the relative tensile strengths TS/TS_1, which compare the TS of each composite with that of its control, yielded more reliable information. Tensile strengths tended to decline with composition for the MMA systems here, but the relative values were larger for BA + MMA and BA (4). Tensile and torsional stiffness increased appreciably with increase in polymer content, both absolutely (columns 6 and 8) and relatively (columns 7 and 9). However, simply air-drying the controls (column 10) greatly increased their stiffness. This

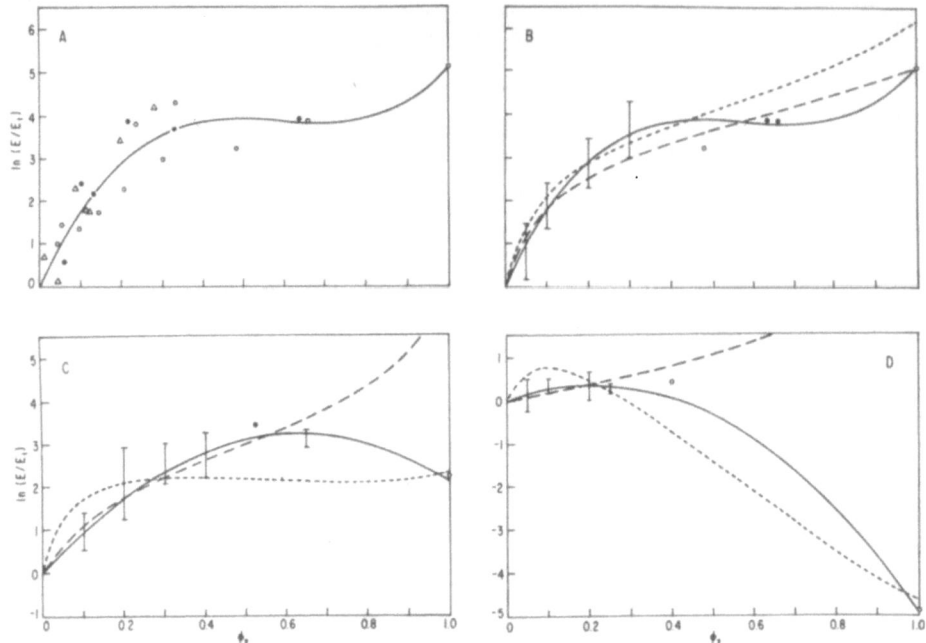

Fig. 18.   The log of the modulus ratio in tension ln (E/E$_1$) for
composites relative to their controls, (E$_1$), versus the
volume fraction of polymer in the composite, ϕ_2. Insert
A computer fitted curve (eq. (9)) drawn through all MMA
data. Symbols are: circles, methanol extracted, triangles, benzene; filled circles bulk-solution. In
inserts B (MMA), C (BA + MMA), and D (BA), curves were
fitted by computer (solid lines) and by constants of the
Halpin-Tsai equation (eq.(20)) (dashed line) and the
modified Halpin-Tsai equation (eq. (22)) (dotted line).
The bars represent the extremes of the experimental
scatter.

was thought to arise from aggregated fibers, accompanied by dis-
torted initial fiber conformations and reduction in free space.
Average densities did, indeed, increase from 0.556 to 0.649 g
cc^{-1}. Polymer presence reduced some of this aggregation because
relative moduli now generally decreased (last column). Thus, a
morphology consisting of polymer residing mostly around fibers in
fiber bundles, as was observed in section C, functions the same
way mechanically as generally aligned fibers and fiber bundles
obtained through shrinkage of the fibrous matrix.

The reason for the tensile stiffness behavior is illustrated
in Figure 17. The schematicized average force-extension curves

for all of the untreated leather controls are drawn in insert A.
The natural variability was extreme as shown by the mean deviations
(bars) and the extremes of variability (dotted envelope) but is
typical of leather (77-79). Curves for composites of MMA (insert
B) and BA + MMA (insert C) of varied w_2 are compared with this
average control tensile (dashed line). Clearly the initial modulus
increased with increase in polymer composition. Noteworthy is the
discontinuous shift in curve shape at w_2 ~0.2 for both systems.
Analogous behavior was found for the average of the air-dried
controls (MMA) dashed line in insert D, where curves for composites
and their controls were similar. Relative tensile moduli at
ambient temperature (composite/control) for all experimental data,
regardless of treatment (methanol extracted or benzene extracted)
or method of preparation (emulsion or bulk solution) are plotted
(Fig. 18) as functions of the volume fraction of polymer, ϕ_2, the
latter computed by use of eqs. (9) or (10). All data fell randomly
near a common computer fitted curve (solid line, insert A).
Consequently, both methods of preparation and all subsequent
treatments yielded similar mechanical response. However, experimental scatter, resulting largely from natural variability of the
leather (Fig. 17, insert A) was so excessive that all relationships
must be considered to be only qualitative. Curve fitted relative
moduli (solid lines) are shown for MMA composites (insert B), BA +
MMA composites (insert C), and BA composites (insert D). These
curves illustrate that, at increasing ϕ_2, a contribution of the
viscoelastic response of the specific modifying polymer became
important and finally dominated properties at ϕ_2 approaching
unity. The dashed line was drawn (4) by use of the Nielsen modification (82a,b) of the Halpin-Tsai equation.

$$E/E_1 = (1 + AB\phi_2)/(1-B\psi\phi_2) \qquad (20)$$

In this equation, the constant A accounts for the geometry of the
filler and Poisson's ratio of the matrix. The constant A increases
greatly as the filler geometry changes in shape from spheres, to
rods, to fibers (82a), and approaches an upper bound for a morphology of aligned fibers (parallel packing). The constant B, on the
other hand, is sensitive to the relative moduli for both components,
approaching zero as E_2/E_1 approaches ∞.

$$B = [(E_2/E_1)-1]/[(E_2/E_1) + A] \qquad (21)$$

Finally, the quantity, ψ, is related to the maximum packing volume,
ϕ_m (82a). Because phase inversion was never found (82a) in this
work, ψ was assigned a value of unity. For use of eq. (20), the
modulus of the untreated leather (designated to be E_1, eq. (20))
was useful only as a predictive equation for the MMA composites
(insert B). For this system, the magnitude of the constant A
(70.4) (4) clearly indicated generally parallel packing. This is

in line with a model (Fig. 4) wherein parallel packed fibers of the leather matrix are restricted in their movement by polymer encasement. The fit for the other two systems, inserts C and D, which involved an unrealistic assignment to B (4), relegated these to mere empirical correlations of limited utility.

Similar results were found for computer fitted relative torsional moduli ratios, solid lines in Figure 19, for MMA (insert B), BA + MMA (insert C), and BA (insert D). Again the Halpin-Tsai relation (eq. (20)), dotted line, was predictive only for the MMA composites. Relative moduli, similar to those for the stiff MMA composites in insert B, were obtained (insert A) for untreated air-dried control samples (Table 6) fitted by use of eq. (20) versus the fractional increase in density $(\rho_a - \rho_{a1})/(\rho_{a2} - \rho_{a1})$. Constants were: A, 268.2; B, 0.625. In view of the common mechanical behavior illustrated in inserts A and B, where fiber aggregation and polymer impregnation produced curves of similar shape, an equation was proposed (4) that linked the initial stiffness produced by permeating polymer at low ϕ_2 (as fiber movement was initially restricted, and free space depleted (Fig. 5, insert D) with the situation at higher ϕ_2 where the viscoelastic properties of the modifying polymer became manifest. For this, the constants A and B for air-dried controls were inserted in a modified Nielsen-Halpin-Tsai equation, given as (4)

$$E_t/E_{t_1} = [(1 + AB\phi_2)/(1 - A\psi\phi_2)] + C\phi_1 + \phi_2 \ln(E_h/E_c) \quad (22)$$

where $C = K\ln(E_h/E_c)$, K is a constant and E_c and E_h are the modulis of collagen and the modifying polymer, respectively (4). With small alterations of the adjustable parameters, $C\phi_1$ and ϕ_2, curves obtained by use of eq. (22) were drawn (Figure 18, dotted line) for MMA composites (insert B) BA + MMA composites (insert C) and BA composites (insert D). Similarly, curves were drawn (Figure 19, dashed lines) for MMA, BA + MMA, and BA composites in inserts B, C, D, respectively. Although the fit in Figure 19 is better than that in Figure 18, both predict only qualitatively the main features observed for ambient moduli for all of the experimental data in this paper, regardless of method of preparation or of subsequent treatment. Because the glass transition temperature (T_g) of the modifying polymers was selected to have a wide variation, eq. (22) should be useful in predicting relative moduli for many different polymers. In fact, variants of eq. (22) might have predictive merit in polymer impregnation, leather processing, and fat liquoring (83).

The effect of temperature on the mechanical response of the polymer-leather composites remains to be treated. In Figure 20 are shown the torsional modulus-temperature curves for the three

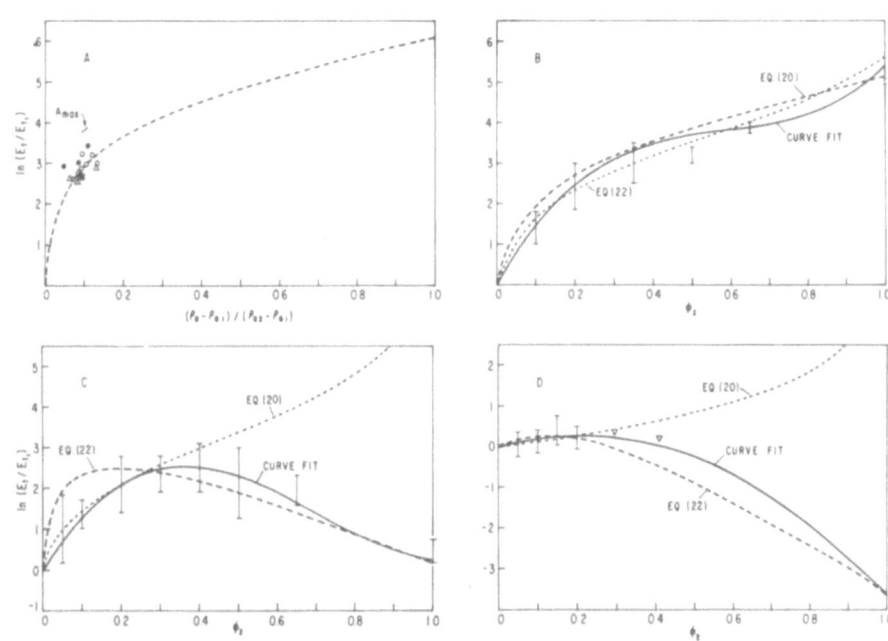

Fig. 19.   The log of the torsional modulus ratio $\ln(E_t/E_{t1})$
versus the fractional density increase $(\rho_a - \rho_{a1})/(\rho_{a2} - \rho_{a1})$ for all air-dried, untreated leather panels (Table
VI) used with methanol extracted (circles) benzene
extracted (triangles), and bulk-solution prepared composites (filled circles). The Halpin-Tsai equation (eq.
(20)) was drawn for average values. Insert B (MMA),
C (BA + MMA), and D (BA) compare curve fitted data
(solid line) with eq. (20) (dotted line) and eq. (22)
(dashed line).

neat modifying polymers of this work and the average values for
the untreated leather controls (dashed line). Natural variability
was again severe (bars). The shift in T_g, taken as the approximate
inflection temperature, $T_i - 5°C$ (4), for the modifying polymer is
again apparent. The rapid rates of change of moduli for the neat
modifying polymers contrasts with the rather featureless leather
curve. This contrast was reduced (Fig. 21) for the composites.
Curves for three polymer compositions, MMA (solid), BA + MMA,
(dashed), and BA, (broken) are now closer to that of untreated
leather (dotted curve). Gross phase separation was indicated for
these composite materials because the T_g (downward arrows) of the
parent polymer were essentially retained unchanged by the composites, and were little affected by composition (82c). This is in

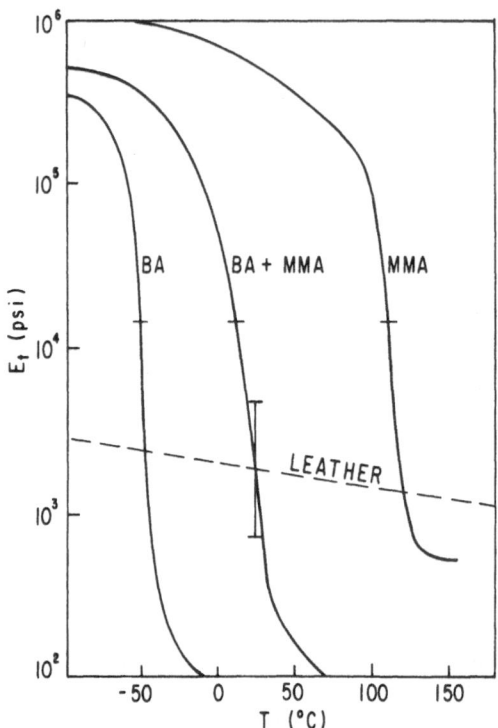

Fig. 20.   Torsional modulus (E_{t2}) versus temperature curves for
the base homo- and copolymers of this work (solid curves)
and an average curve (E_{t1}) for methanol and benzene
extracted leather controls. Slashes indicate the approximate inflection temperature, T_i at 14,500 psi (10^9
dynes cm^{-2}), with T_g estimated as T_i - 5°C. T_g was:
MMA, 105°C; BA + MMA, 14°C; BA, -55°C. Bar denotes the
extremes of variability of the leather apparent moduli
at 23°C.

harmony with the morphologies illustrated in section C. However
at low temperatures (T_i - 50°C) the torsional modulus of each
composite system became dependent on polymer composition. The low
temperature moduli increased as w_2 increased instead of reaching
a common plateau of ~10^{10} dynes cm^{-2} characteristic of most
polymer modified composite materials (82c). At high temperatures,

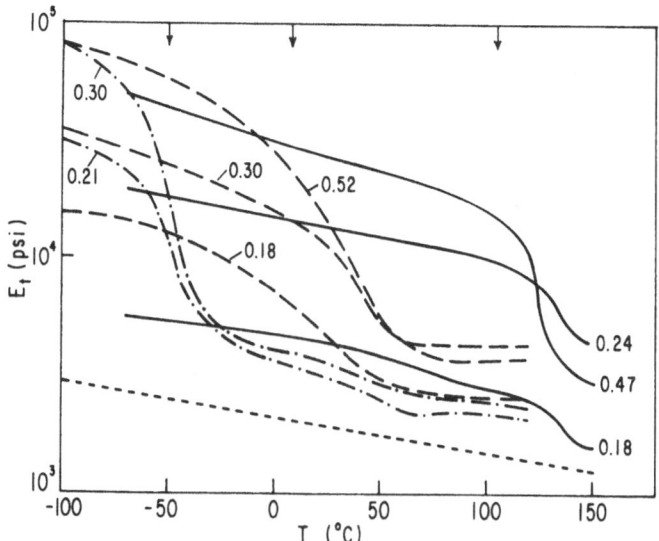

Fig. 21. Torsional modulus temperature curves for selections of
          composites for the three systems with varied composition,
          w_2. Solid lines represent data for MMA; dashed lines
          for BA + MMA; broken lines, BA. Comparative curves for
          the three systems are at similar w_2. Arrows indicate
          estimated T_g for the parent polymer from Figure 20.
          Dotted line is the average rate of change of the modulus
          for the controls, as in Figure 20.

T_i + 50°C, moduli of the starting leathers were approached. This
shifting of moduli appears to result from the free space retained
by each composition, as discussed in section A. This adds a
porosity term (5e) to the moduli of each material. The fact that
each phase was continuous, as seen in section C, and characterized
by only weak interaction between phases further contributed to
moduli residing between those of the starting materials. This
permitted the mechanical response of the polymer-leather composites
to shift from predominately plastic to predominately leather
character as temperature rose.

SUMMARY AND CONCLUSIONS

     Three acrylic monomers, MMA, BA + MMA, and BA were polymerized
in the fibrous matrix of chrome-tanned 5 oz cattlehide by emulsion

polymerization with a persulfate-bisulfite redox initiating system
and by AIBN initiated bulk or solution polymerization into acetone-
dried samples. The widest feasible range of composition of the
composites was studied. A thorough investigation was then undertaken of the parameters of the emulsion process, with emphasis on
systems and feed deposition efficiencies, the relation between
density and composition for both processes, and, finally, the
determination of graft branch frequency from molecular weights of
the bound polymer. Results showed that system efficiencies at
100% monomer conversion declined with increase in the molar volume
of the monomer and were relatively low (0.635) for n-butyl acrylate.
Likewise, increase in monomer concentration depressed deposition
and bound polymer efficiencies to small values, thus preventing
assumed graft branches from reaching their theoretical limit.
Densities were directly proportional to the volume fraction of
polymer when incremental space filling was allowed in the estima-
tion of the bulk-solution composites. Kinetic studies of the
emulsion process, involving comparison of the exponential intensity
factors for the reaction variables for polymerization in both
matrix and float with theoretical values for primary radical
grafting and normal emulsion polymerization, indicated that little
grafting was possible in these systems. Polymerization in the
leather matrix appeared to be diffusion controlled throughout the
lifetime of polymer deposition. Amplification of this conclusion
negating grafting was obtained from decreased bound polymer found
for finely divided leather dispersed in an extended float and by
similar bound polymer deposits on simple impregnation of the
leather with pre-formed polymer. Additional support was seen in
the poor graft and deposition efficiencies. The coarse morphology,
observed by light and scanning electron microscopy of composites
and corresponding replicas of the continuous polymer phase, confirmed the improbability of the existance of many attachable sites
on the collageneous fibers. Independence of the phases was confirmed by the observance of little protection of the matrix from
bulk water absorption as a function of time and at final equilibrium.
Mechanical properties revealed reduced relative tensile strengths
(compared to those of untreated leather controls). Relative
tensile and torsional moduli initially increased for composites
having a small volume fraction, ϕ_2, of polymer (~0.20) for all
systems regardless of the process and treatment used in their
preparation. At higher ϕ_2 the viscoelastic properties of the
modifying polymer dominated the composite properties. A modified
theoretical equation, based on Halpin-Tsai parameters for fiber
aggregation on air-dried controls, coupled with a term estimating
empirically polymer-collagen fiber interaction, was successful in
predicting the main trends of the relative moduli for all ϕ_2.

REFERENCES

1. E. F. Jordan, Jr., B. Artymyshyn, A. L. Everett, M. V. Hannigan, and S. H. Feairheller, J. Appl. Polym. Sci., 25, 2621 (1980).

2. E. F. Jordan, Jr. and S. H. Feairheller, J. Appl. Polym. Sci. 25, 2755 (1980).

3. E. F. Jordan, Jr., R. J. Carroll, M. V. Hannigan, B. Artymyshyn, and S. H. Feairheller, J. Appl. Polym.Sci. 26, 61 (1981).

4. E. F. Jordan, Jr., B. Artymyshyn, and S. H. Feairheller, J. Appl. Polym. Sci. 26, 463 (1981).

5. J. A. Manson and L. H. Sperling, Polymer Blends and Composites, Plenum Press, New York; 1976, a) pp. 77-119; b) pp. 121-168; c) pp. 237-270; d) 373-457; e) 363-369.

6. S. M. Atlas and H. Mark, Cellulose Chem. Technol., 3, 325 (1969).

7. J. C. Arthur, Adv. Macromol. Chem., W. M. Pasika, Ed., Academic Press, New York, 1970, Vol. 2, pp. 1-87.

8. M. S. Bains, J. Polym. Sci., C37, 125 (1972).

9. K. Arai, "Block and Graft Copolymerization," R. J. Ceresa, Ed., J. Wiley and Sons, New York, 1973, pp. 193-310.

10. W. J. Wosley, Ibid., pp. 311-333.

11. P. J. Nayak, J. Macromol. Sci., Revs. Macromol. Chem., C14, 193 (1976).

12. I. C. Watt, J. Macromol. Sci., Revs. Macromol. Chem., C5, 175 (1970).

13. R. Y. M. Huang and W. H. Rapson, J. Polym. Sci., C2, 169 (1963).

14. M. L. Rollins, A. M. Cannizzaro, F. A. Blouin, and J. C. Arthur, Jr., J. Appl. Polym. Sci., 12, 71 (1968).

15. R. Y. M. Huang, J. Appl. Polym. Sci., 10, 325 (1966).

16. V. S. Stannett, K. Araki, J. A. Gervasi, and S. W. McLeskey, J. Polym. Sci., A3, 3763 (1965).

17. A. H. Hebeish and P. C. Mehta, J. Appl. Polym. Sci., 12, 1265 (1968).

18. R. Y. M. Huang and P. Chandramouli, J. Appl. Polym. Sci., 12, 2549 (1968).

19. A. Kantouch, A. Hebeish, and A. Bendak, Europ. Polym. J., 1, 153 (1971).

20. M. Lipson and J. B. Speakman, J. Soc. Dyers & Colourists, 65, 390 (1949).

21. T. Valentine, J. Textile Inst., 46, T270 (1955).

22. B. K. Lohani, T. Valentine, and C. S. Whewell, J. Textile Inst., 49, T265 (1958).

23. M. Negishi, K. Arai, S. O. Kada, and J. Nagakura, J. Appl. Polym. Sci., 9, 3465 (1965).

24. A. Kantouch, S. Abdel-Fattah, and A. Hebeish, Polym. J. Japan, 3, 675 (1972).

25. A. Hebeish, M. H. El-Rafie, M. L. Khalel, and A. Bendak, J. Appl. Polym. Sci., 21, 1901 (1977).

26. A. Hebeish, A. Bendak, and A. Kantouch, J. Appl. Polym. Sci., 15, 2733 (1971).

27. A. Bendak, M. L. Khalil, M. H. El-Rafie, and A. Hebeish, J. Appl. Polym. Sci., 19, 335 (1975).

28. A. Bendak and A. Hebeish, J. Appl. Polym. Sci., 17, 1953 (1973).

29. R. Y. M. Huang and P. Chandramouli, J. Polym. Sci. A1, 7, 1393 (1969).

30. R. Y. M. Huang, B. Immergut, E. H. Immergut, and W. H. Rapson, J. Polym. Sci., A1, 1257 (1963).

31. K. Arai, Polymer, 18, 211 (1977).

32. G. Mino and S. Kaizerman, J. Polym. Sci., 31, 242 (1958).

33. J. L. Williams, D. K. Woods, V. S. Stannett, S. B. Sello, and C. V. Stevens, Int. J. Appl. Radiation and Isotopes, 26, 159 (1975).

34. J. L. Williams and V. Stannett, Textiles Res. J., 38, 1065 (1968).

35. Y. Ikada, Adv. Polym. Sci., 29, 47 (1978).

36. R. W. Lenz, Organic Chemistry of Synthetic High Polymers, Interscience, New York, 1967; a) pp. 364-365; b) pp. 711-713.

37. Y. Ogiwara, Y. Ogiwara, and H. Kubata, J. Polym. Sci., A16, 1489 (1968).

38. A. H. Korn, S. H. Feairheller, and E. M. Filachione, J. Am. Leather Chem. Assoc., 67, 111 (1972).

39. A. H. Korn, M. M. Taylor, and S. H. Feairheller, J. Am. Leather Chem. Assoc., 68, 224 (1973).

40. E. H. Harris, M. M. Taylor, and S. H. Feairheller, J. Am. Leather Chem. Assoc., 69, 182 (1974).

41. G. Mino, S. Kaizerman, and E. Rasmussen, J. Polym. Sci., 39, 523 (1959).

42. R. A. Wallace and D. G. Young, J. Polymer Sci. A1, 4, 1179 (1966).

43. G. Mino, S. Kaizerman, and E. Rasmussen, J. Polym. Sci., 38, 393 (1959).

44. A. A. Katai, V. K. Kulshiestha, and R. W. Marchessoult, J. Polym. Sci. C, 2, 403 (1963).

45. A. Rout, S. P. Rout, B. C. Singh, and M. Santappa, Macromol. Chem., 178, 639 (1977).

46. L. A. Gugliemelli, C. J. Swanson, and W. A. Doone, J. Polym. Sci., Polym. Chem. Ed., 11, 2451 (1973).

47. W. V. Smith and R. H. Ewart, J. Chem. Physics, 16, 592 (1948).

48. W. V. Smith, J. Am. Chem. Soc., 70, 3695 (1948).

49. J. L. Gardon, J. Polym. Sci. A-1, 6, 623 (1968).

50. J. L. Gardon, J. Polym. Sci. A-1, 6, 665 (1968).

51. R. M. Fitch, M. B. Prenosil, and K. J. Sprich, J. Polym. Sci., C27, 95 (1969).

52. R. M. Fitch and C. H. Tsai, in Polymer Colloids, R. M. Fitch, Ed., Plenum Press, New York, 1971, pp. 73-102.

53. M. M. Taylor, E. H. Harris, and S. H. Feairheller, J. Am. Leather Chem. Assoc., 72, 294 (1977).

54. T. Landells and C. S. Whewells, J. Soc. Dyers and Colourists, 71, 171 (1955).

55. T. Valentine, J. Textile Inst., 47, T1 (1956).

56. W. M. Kaeppner and R. Y. M. Huang, Textile Res. J., 35, 504 (1965).

57. F. A. Blouin and J. C. Arthur, Jr., Textile Res. J., 33, 727 (1963).

58. J. C. Arthur, Jr., and F. A. Blouin, J. Appl. Polym. Sci., 8, 2813 (1964).

59. J. L. Williams, V. S. Stannett, L. G. Roldan, S. B. Sello, and C. V. Stevens, Int. J. Appl. Radiation and Isotopes, 26, 169 (1975).

60. F. A. Blouin, N. J. Morris, and J. C. Arthur, Jr., Textile Res. J., 36, 309 (1966).

61. R. J. Demint, J. C. Arthur, Jr., A. R. Markezich, and W. F. McSherry, Textile Res. J., 32, 918 (1962).

62. M. W. Andrews, R. L. D'Arcey, and I. C. Watt, J. Polym. Sci., Polym. Letters, 3, 441 (1965).

63. W. R. Goynes and J. A. Harris, J. Polym. Sci., C37, 277 (1972).

64. J. A. Swift, Chemistry of Natural Protein Fibers, A. S. Osquith, Ed., Plenum Press, New York, 1977, pp. 81-146.

65. K. P. Rao, K. T. Joseph, and Y. Nayudamma, J. Appl. Polym. Sci., 16, 975 (1972).

66. J. C. Arthur, Jr. and R. J. Demint, Textile Res. J., 30, 505 (1960).

67. F. A. Blouin, N. J. Morris, and J. C. Arthur, Jr., Textile Res. J., 38, 710 (1968).

68. F. A. Blouin, A. M. Cannizzaro, J. C. Arthur, Jr., and M. L. Rollins, Textile Res. J., 38, 811 (1968).

69. F. L. Saunders and R. C. Sovish, J. Appl. Polym. Sci., 7, 357 (1963).

70. T. Mares and J. C. Arthur, Jr., J. Polym. Sci., C37, 349 (1972).

71. I. C. Watt, J. Macromol. Sci. Chem., A4, 1079 (1970).

72. H. L. Needles, Textile Res. J., 48, 506 (1978).

73. D. S. Varma and R. K. Sarkar, Die Angewandte Makromolekulare Chemie, 37, 177 (1974).

74. L. J. Wolfram and J. Menkert, Am. Dyestuff Reptr., 56, 110 (1967).

75. J. R. Kanagy, Chemistry and Technology of Leather, Vol. 4, F. O'Flaherty, W. T. Roddy, and R. M. Lollar, Eds., Reinhold Publishing Corp., New York, 1965, pp. 369-416.

76. I. V. Yannas, J. Macromol. Sci., Revs. Macromol. Chem., C7, 49 (1972).

77. J. R. Kanagy, E. B. Randall, T. J. Carter, R. A. Kinmouth, and C. W. Mann, J. Am. Leather Chem. Assoc., 47, 726 (1952).

78. R. M. Lollar, J. Am. Leather Chem. Assoc., 54, 306 (1959).

79. J. R. Kanagy, J. Am. Leather Chem. Assoc., 50, 112 (1955).

80. K. P. Rao, K. T. Joseph, and Y. Nayudamma, Leather Sci., 19, 27 (1972).

81. K. P. Rao, D. H. Kamat, K. T. Joseph, M. Santappa, and Y. Nayudamma, Leather Sci., 21, 111 (1974).

82. L. E. Nielsen, Mechanical Properties of Polymers and Composites, Parts I and II, Marcel Dekker Inc., New York, 1974; a) pp. 379-452; b) pp. 453-510; c) pp. 208-215.

83. T. C. Thorstensen, Practical Leather Technology, R. E. Krieger Publishing Corp., New York, 1976, pp. 190-207.

EFFECT OF CHAIN TRANSFER AGENTS DURING GRAFT POLYMERIZATION

IN LEATHERS

E.H. Harris, H.A. Gruber, P.R. Buechler,
and S.H. Feairheller

Eastern Regional Research Center, ARS and S&E
Philadelphia, Pennsylvania 19118

INTRODUCTION

In recent years intensive work has been carried out in this
and other laboratories (1-4) in attempts to modify the properties
of leather by in-situ grafting of monomers onto the leather substrate. Such work has followed the interpretations used in other
areas where grafting onto other natural materials has been employed (5-7). These interpretations classify extractable polymer
as "homopolymerized," i.e., not covalently bonded to the substrate, and unextractable polymer as "grafted," i.e., covalently
bonded to the substrate. Recent work summarized in another paper
to be presented in this book (8) casts some doubt on this
interpretation as to how much of the strongly bonded polymer is
actually covalently bonded. Nevertheless, in this paper, to
simplify discussion and to be in accord with historical perspec-
tives, the term graft polymer will be employed for the bound
unextractable polymer.

In this work the influence of the use of mercaptans as chain
transfer agents in an emulsion polymerization grafting process
was studied. Notable differences were found between water
soluble and insoluble mercaptan. These findings and their
possible significance are discussed.

Experimental

The leather employed was commercially chrome-tanned, taken wet
from the tannery drum, drained, wrapped in polyethylene and
transported to the laboratory. Moisture content was determined.
A standard redox emulsion polymerization recipe reported in an

earlier paper [9] was employed for the graft polymerization with
methyl methacrylate (MMA). In this current work the weight of
MMA offered was 1/3 the anhydrous weight of the leather. Furthermore, carefully measured amounts of alkyl mercaptans were employed as chain transfer agents in amounts from as low as 0.25 to as
high as 5.0 mole percent based on MMA monomer. The alkyl mercaptans used were n-dodecyl mercaptan (insoluble in water) and
ethyl mercaptan, n-propyl mercaptan and isopropyl mercaptan. The
last three were soluble in water at the levels used in these experiments. Solubilities were confirmed by glc analyses.

The grafted leathers were analyzed and compared with controls
(untreated leather). Ash, Kjeldahl nitrogen and ethyl acetate
extractables were determined on a dry weight basis. Total
polymer uptake was determined from increase of non-nitro organic
solids as compared with that of controls. Homopolymer was
determined from increase in extractables which were negligible
in the controls, since the leather had not been fatliquored.
Graft efficiency was calculated as the percent of the total
polymer which was bound or "grafted."

Homopolymer was isolated from the extract by low temperature
evaporation and vacuum oven drying at 48°C. Graft polymer was
obtained by removal of the protein constituents by controlled hydrolysis followed by thorough washing and by vacuum oven drying. Viscosity average molecular weights were determined for both types of
polymer by use of the Staudinger equation constants reported
previously [10].

By use of a commonly accepted molecular weight of 300,000 for
collagen, it is possible, from the total amount of graft collagen
and its average molecular weight, to calculate the average number
of grafts per mole of collagen. This also was done.

Results and Discussion

As reported previously [9] under our conditions dodecyl mercaptan had little effect on the molecular weight of grafted
polymer, although it did reduce the molecular weight of the extractable homopolymer. Table 1 illustrates this. It also shows
that the water-soluble mercaptans were much more effective in
controlling molecular weight of the homopolymer. More important-
ly they had a decided effect on controlling the molecular weight
of graft polymer, although in all cases their effect in lowering
molecular weight of the graft was less than that for the homopolymer produced in the same experiment.

Table 2 illustrates the effect of the mercaptan chain transfer
agents on the grafting process. The graft efficiency (percent of
the total polymer take-up which is bound or "grafted") when mercaptans were used was less than that of the control. However, no
consistent pattern for graft efficiency with the amount of mercaptan was found for any member of the mercaptan series.
Probably the most significant factor found in this work was that

Table 1

Mercaptan Effect on Molecular Weight of
Poly(MMA) in Leather Grafting

RSH/MMA Mole Ratio	Molecular Weight	
	Homopolymer	Graft
No Mercaptan (Control)	518,200	870,800
R=n-dodecyl		
0.05	197,600	818,700
0.025	256,000	960,200
0.0125	322,300	963,200
0.006	351,800	1,067,000
R=ethyl		
0.05	3,942	14,700
0.025	17,800	38,800
0.015	55,200	197,000
0.010	84,800	213,000
0.0025	188,500	303,000
R=n-propyl		
0.05	12,300	17,200
0.025	19,100	30,000
0.15	26,000	47,000
0.010	34,400	68,400
0.0025	104,000	214,000
R=isopropyl		
0.05	16,500	22,800
0.025	33,700	57,500
0.015	47,000	81,400
0.010	56,600	96,500
0.0025	152,000	277,400

Table 2

Effect of Chain Transfer Agents
on Grafting

RSH/MMA Mole Ratio	Average Grafts per mole Collagen*	Graft Efficiency
No Mercaptan (Control)	0.072	72.9
R=n-dodecyl		
0.050	0.062	47.0
0.025	0.058	47.2
0.0125	0.071	56.0
0.006	0.066	61.2
R=ethyl		
0.05	1.48	34.6
0.025	0.87	39.5
0.015	0.10	29.9
0.010	0.18	39.0
0.0025	0.16	55.8
R=n-propyl		
0.05	1.17	42.5
0.025	0.92	40.8
0.015	0.61	41.5
0.010	0.51	50.9
0.0025	0.15	51.5
R=isopropyl		
0.05	1.18	33.6
0.025	0.22	12.8
0.015	0.19	13.8
0.010	0.26	27.7
0.0025	0.14	41.5

---

* Based on an average molecular weight of 300,000 gms/mole.

high levels of soluble mercaptan produced a large increase in the average number of grafts per mole of collagen (to values over 1.0); without chain transfer agents values were usually below 0.1 (Table 2).

The findings in these experiments would seem to indicate that an insoluble chain transfer agent such as dodecyl mercaptan is unable to penetrate into collagen fiber bundles in its emulsified form. The water-soluble mercaptans, however, probably can and do penetrate into the fiber bundles, into the collagen fibers and perhaps, at the higher concentrations, even into the collagen fibrils. Strong adsorption of the mercaptans at sites on the protein can be expected to occur because of ion-dipole and dipole-dipole interactions. This would explain why only high levels of mercaptan give a high number of grafts per mole of collagen. As adsorption sites on the periphery of the fibers tend to become saturated, additional soluble mercaptan can progress beyond these sites into the interior of the fiber to initiate new sites on the interior fibrils. Whether the adsorbed mercaptan transfes a free radical to the protein or by its own activity initiates new growing chains in a confined space leading to increased branching and interpenetrating polymer/protein networks is debatable. There would be, in either case, less mobility to the growing chains than in the emulsion polymerized monomer forming the extractable homopolymer. Hence the possibilities for a termination step at the protein sites would be lessened and somewhat higher molecular weights for the bound polymer, rather than for the homopolymer at equal mercaptan levels, would be expected. This is in accord with the findings.

References

1) K. P. Rao, K. T. Joseph, and Y. Nayudama, J. Appl. Polymer Science, 16,975 (1972).

2) J. Kudaba, E. Ciziumaite, and D. Jonutiene Chem. Abs., 73, 4982 b (1970).

3) W. C. Prentiss, T. W. Hutton, and S. N. Lewis,JALCA, 71, 111 (1976).

4) H. A. Gruber, M. M. Taylor, E. H. Harris, and S. H. Feairheller JALCA,73,530 (1978).

5) M. S. Bains J. Polym. Sci., Part C, 37,125 (1972).

6) P. J. Nayak, J. Macromol Sci. Rev. Macromol. Chem., 14 (2), 193 (1976).

7) I. C. Watt, J. Macromol Sci. Rev. Macromol. Chem., 5 (1), 175 (1970).

8) E. F. Jordan, Jr., B. Artymyshyn, M. V. Hannigan, R. J. Carroll, and S. H. Feairheller Preceeding Chapter, this book.

9) M. M. Taylor, E. H. Harris, and S. H. Feairheller, Polymer Preprints, 19, 618 (1978).

10) G.M. Kline Ed., Analytical Chemistry of Polymers I.
    Monomers and Polymeric Materials, Vol. XII, Part I,
    Interscience Publishers, New York, NY, p. 14 (1959).

CONTRIBUTORS

Chye H. Ang, School of Chemistry, University of New South Wales,
    Kensington, N.S.W. Australia 2033

Kozo Arai, Department of Chemistry, Faculty of Technology, Gunma
    University, Kiryu, Gunma 376, Japan

Jett C. Arthur, Jr., Technical Consultant, 3013 Ridgeway Drive,
    Metairie, Louisiana 70002

Bodhan Artymyshyn, Eastern Regional Research Center, Agricultural
    Research Service, U.S. Department of Agriculture, Philadelphia,
    Pennsylvania 19118

Howard S. Blaxall, Department of Chemistry and The Brehm Laboratory
    Wright State University, Dayton, Ohio 45435

James M. Brennan, School of Chemical Engineering, Olin Hall,
    Cornell University, Ithaca, New York 14853

P.R. Buechler, Eastern Regional Research Center, Agricultural
    Research Service, U.S. Department of Agriculture, Philadelphia,
    Pennsylvania 19118

B. George Bufkin, DAP, Inc., Research & Development, Dayton, Ohio
    Ohio 45401

Robert C. Burr, Northern Regional Research Center, Agricultural
    Research Service, U.S. Department of Agriculture, Peoria,
    Illinois 61604

Charles E. Carraher, Jr., Department of Chemistry, Wright State
    University, Dayton, Ohio 45435

Robert J. Carroll, Eastern Regional Research Center, Agricultural
    Research Service, U.S. Department of Agriculture, Phila-
    delphia, Pennsylvania 19118

Chia M. Chen, School of Forest Resources, The University of
    Georgia, Athens, Georgia   30602

Shelley Coldiron, The Brehm Laboratory, Wright State University,
    Dayton, Ohio   45435

A. Conde, Ingenieria Quimica, Universidad Industrial de Santander,
    Bucaramanga, Colombia

Sajal Das, Polymer Division, Materials Science Center, Indian
    Institute of Technology, Kharagpur 721302, India

Neil P. Davis, School of Chemistry, University of New South
    Wales, Kensington, N.S.W. Australia   2033

M.T. DeCrosta, Materials Research Center #32, Lehigh University,
    Bethlehem, Pennsylvania   18015

George F. Fanta, Northern Regional Research Center, Agricultural
    Research Service, U.S. Department of Agriculture, Peoria,
    Illinois   61604

Stephen H. Feairheller, Eastern Regional Research Center,
    Agricultural Research Service, U.S. Department of Agriculture,
    Philadelphia, Pennsylvania   19118

Robert G. Fecht, Northern Regional Research Center, Agricultural
    Research Service, U.S. Department of Agriculture, Peoria,
    Illinois   61604

A.M. Fernandez, Materials Research Center #32, Lehigh University,
    Bethlehem, Pennsylvania   18015

John L. Garnett, School of Chemistry, University of New South
    Wales, Kensington, N.S.W. Australia   2033

Timothy J. Gehrke, Department of Chemistry, Wright State
    University, Dayton, Ohio   45435

Lloyd Geldard, School of Chemistry, University of New South
    Wales, Kensignton, N.S.W. Australia   2033

Jan W. Gooch, Bechtel Group, Inc., Research & Engineering, San
    Francisco, California   94101

H.A. Gruber, Eastern Regional Research Center, Agricultural
    Research Service, U.S. Department of Agriculture, Philadelphia,
    Pennsylvania   19118

Mary V. Hannigan, Eastern Regional Research Center, Agricultural Research Service, U.S. Department of Agriculture, Philadelphia, Pennsylvania 19118

E.H. Harris, Eastern Regional Research Center, Agricultural Research Service, U.S. Department of Agriculture, Philadelphia, Pennsylvania 19118

Edmund F. Jordan, Jr., Eastern Regional Research Center, Agricultural Research Serice, U.S. Department of Agriculture, Philadelphia, Pennsylvania 19118

Ju Kumanotani, Institute of Industrial Science, The University of Tokyo, 7-22-1 Roppongi, Minatoku, Tokyo, Japan 106

George R. Lightsey, Department of Chemical Engineering, Mississippi State University, Mississippi State, MS

Raymond Linville, Department of Chemistry, Wright State University, Dayton, Ohio 45435

Manoranjan Maiti, Department of Chemistry, Uluberia College, Uluberia, Howrah

Sukumar Maiti, Polymer Division, Materials Science Centre, Indian Institute of Technology, Kharagpur 721302, India

Eloisa Biasotto Mano, Instituto de Macromoleculas, Universidade Federal do Rio de Janeiro, Rio de Janeiro, RJ, Brazil

J.A. Manson, Materials Research Center #32, Lehigh University, Bethlehem, Pennsylvania 18015

Olfat Y. Mansour, National Research Centre, Dokki/Cairo, A.R. of Egypt

C.J. Murphy, Department of Chemistry, East Stroudsburg State College, East Stroudsburg, Pennsylvania 18301

A.B. Mustafa, National Research Centre, Dokki/Cairo, A.R. of Egypt

Philip D. Mykytiuk, Department of Chemistry, Wright State University, Dayton, Ohio 45435

Ahmed Nagaty, National Research Centre, Dokki/Cairo, A.R. of Egypt

Tsutomu Nakagawa, Department of Chemistry, Faculty of Engineering, Meiji University, Higashimita, Tama-ku, Kawasaki-shi, Japan.

Shahid Qureshi, Materials Research Center #32, Lehigh University,
    Bethlehem, Pennsylvania 18015

Atanu Ray, Polymer Division, Materials Science Centre, Indian
    Institute of Technology, Kharagpur 721302, India

Ferdinand Rodriguez, School of Chemical Engineering, Olin Hall,
    Cornell University, Ithaca, New York 14853

I.O. Salyer, University of Dayton Research Institute, Dayton,
    Ohio 45469

Raymond B. Seymour, Department of Polymer Science, University
    of Southern Mississippi, Hattiesburg, MS 39401

Takanori Shibata, Department of Chemistry, Faculty of Engineering,
    Meiji University, Higashimita, Tama-ku, Kawasaki-shi, Japan

K.J. Smith, Jr., Department of Chemistry, SUNY College of Environ-
    mental Science and Forestry, Syracuse, New York

L.H. Sperling, Materials Research Center #32, Lehigh University,
    Bethlehem, Pennsylvania 18015

Charles L. Swanson, Northern Regional Research Center, Agricultural
    Research Service, U.S. Department of Agriculture, Peoria,
    Illinois 61604

Thomas O. Tiernan, Department of Chemistry and The Brehm
    Laboratory, Wright State University, Dayton, Ohio 45435

A.M. Usmani, University of Dayton Research Institute, Dayton,
    Ohio 45469

Gary C. Wildman, University of Southern Mississippi, Polymer Science
    Department, Hattiesburg, Mississippi 39401